ELEMENTS OF MINERALOGY

RUTLEY'S ELEMENTS OF MINERALOGY

Twenty-Sixth Edition

H. H. READ, F.R.S.

*Professor Emeritus of Geology in the Imperial
College of Science and Technology
and the University of London*

LONDON
THOMAS MURBY & CO

NINETEENTH EDITION, THOROUGHLY REVISED,
RESET, AND NEW ILLUSTRATIONS, 1915
TWENTIETH EDITION 1918
SECOND IMPRESSION 1919
TWENTY-FIRST EDITION 1921
SECOND IMPRESSION
WITH SLIGHT CORRECTIONS 1923
THIRD IMPRESSION 1926
TWENTY-SECOND EDITION 1929
SECOND IMPRESSION 1933
TWENTY-THIRD EDITION, THOROUGHLY REVISED,
RESET, AND NEW ILLUSTRATIONS, 1936
SECOND IMPRESSION 1939
THIRD IMPRESSION 1942
FOURTH IMPRESSION 1944
FIFTH IMPRESSION 1946
TWENTY-FOURTH EDITION 1948
SECOND IMPRESSION 1949
THIRD IMPRESSION 1953
FOURTH IMPRESSION 1956
FIFTH IMPRESSION 1957
SIXTH IMPRESSION 1960
TWENTY-FIFTH EDITION 1962
TWENTY-SIXTH EDITION 1970
COMPLETELY REVISED, RESET AND MANY
NEW ILLUSTRATIONS
SECOND IMPRESSION 1971
THIRD IMPRESSION 1973
FOURTH IMPRESSION 1974

This twenty-sixth edition © George Allen & Unwin Ltd. 1970

ISBN 0 04 549005 8 (Cased Edition)

ISBN 0 04 549006 6 (Paper Edition)

*George Allen & Unwin Ltd., 40 Museum Street, London W.C.1
are the proprietors of Thomas Murby & Co.*

PRINTED IN GREAT BRITAIN
in 10 on 11 point Times
BY JOLLY AND BARBER LTD
RUGBY

PREFACE

The last thorough revision of *Rutley's Elements of Mineralogy* appeared as the 23rd Edition in 1936. In subsequent editions, an effort to keep abreast with the great progress in the science was made by small (and often awkward) modifications and, especially, by the addition of an independent chapter on the atomic structure of minerals. For this present edition, the complete re-setting of the book has made possible not only the integration of the added chapter on atomic structure into its proper place in the accounts of the chemical and physical properties of minerals, but also extensive rewriting and rearrangement of the material in the first part of the book. To this part, also, has been added a short chapter on the classification of minerals. In the second part, the Description of Minerals, numerous, if not so extensive, modifications and modernisations have been introduced. A couple of dozen new figures have been added, mostly in the early part of the book.

More specifically, the major changes in this new edition are the following. The electronic structure of atoms supplies the guide-lines for the whole account of mineral-chemistry; additional items concern the electrochemical series, of interest in the occurrence and metallurgical treatment of ores, and chemical analysis. On the physical side, the dependence of physical properties of minerals on their atomic structure is emphasized and, in addition, a brief account of radioactivity and isotopic age-determination is given. Crystal-structure, of course, enters largely into the formal presentation of the morphology of crystals; here, too, the crystallographic nomenclature is modernized and an elementary introduction to stereographic projection and its use in the description of crystal-symmetry is inserted. The chapter on the occurrence of minerals has been partly rewritten to include extended accounts of the sedimentary and meta-morphic rocks and the different types of economic mineral deposits; a summary of earth-history is given, together with some age-data. The new chapter on the classification of minerals defends that used in this book; in this, minerals are assembled into economic groups according to elements, and the elements are associated according to the Periodic Classification – this classification has stood the test of time as suitable for the wide public for whom the book is intended.

In Part II, Description of Minerals, the uses and occurrences of minerals are brought up to date, crystal structures of the rock-forming minerals are introduced and modern methods of extraction of the metals from their ores mentioned. Some metals of importance in nuclear energy and space-research are briefly noted.

My first revision of 'Rutley' was undertaken more than half a century ago – I trust that this present effort will be as useful in its limited objectives as its predecessors. During this long span I have received invaluable help from a great number of friends as recorded in previous prefaces, and to these I re-offer my grateful thanks. In this present revision I am specially indebted to Dr. F. G. H. Blyth for his critical reading of some of the manuscript and for providing certain of the figure-drawings. Dr. Janet Watson has read the whole text and contributed greatly to its clarity and, besides, has drawn many of the new figures. Finally, she has helped to see the book through the press. For such willing assistance in making it possible for me to complete this work, I offer her my sincere gratitude. My thanks are also due to Mr. J. A. Gee and Mr. R. Curtis, both of the Geology Department, Imperial College, for kindly supplying photographs for Figures 2, 13 and 14 and for Figure 17 respectively; and to James Swift & Son Ltd for the photograph for Figure 99.

H. H. READ

TABLE OF CONTENTS WITH

LIST OF MAIN TOPICS

INTRODUCTORY:

THE NATURE OF MINERALS

The Mineral Kingdom. It has long been the custom to divide nature into three great departments, the animal, vegetable and mineral kingdoms. The mineral kingdom comprises the materials that make the crust of the earth and a part of this kingdom is dealt with in the science of mineralogy. Whether or not any definite boundaries exist between the three kingdoms is a subject which remains to be investigated.

The different members of the animal and vegetable kingdoms are characterized by the development of special organs, or of certain peculiarities of structure, by means of which they pass through a series of changes known as life and growth. This latter phenomenon takes place by the absorption of various kinds of matter which then undergoes conversion by chemical processes into substances similar to those making the plant or animal. In this way the waste which accompanies life is replaced. The bones and shells of animals consists to a great extent of mineral matter. Plants are capable of deriving earthy substances from the soil in which they grow. But mineral matter which has thus been utilized by organisms passes, in the rigid interpretation of the term, beyond the pale of mineralogy, for it assumes a structure, governed by the nature and requirements of the animal or plant, that it would not possess as an ordinary portion of the earth's crust. For example, a pearl would be regarded as an organic substance and not a true mineral, although it consists of mineral matter. Again, coal, being a substance derived from the decomposition of vegetable matter, would not be rigidly classed with minerals.

Minerals. A most important characteristic of a mineral is the possession of a definite chemical composition. Some qualification of this statement is, however, necessary. Certain minerals form a closely related series in which there is a gradual replacement of one element by another, the two end-members of the series being connected by a number of transitional types of intermediate composition. In order to avoid the establishment of a great number of slightly differing mineral species, it is usual in such cases to consider the series as a whole, definite names being given to the end-members and possibly to certain intermediate types of historic or other interest. The variations of the chemical compositions of such series are not haphazard but are governed by certain rules.

The possession of a definite chemical composition does not suffice in all cases to fix the mineral species. It is found that two minerals with markedly different physical properties, such as colour, hardness, form, density and so on, may have identical chemical compositions. For example, the minerals diamond and graphite, physically very dissimilar, are both composed of carbon. In cases such as these, the two mineral species have been shown to have their constituent atoms arranged on different plans and, as a result, to have different physical properties. Under favourable conditions, the internal atomic structure of minerals finds

Fig. 1. A Group of calcite crystals.
Crown Copyright Geological Survey photograph.
Reproduced by permission of the Controller of H.M. Stationery Office.

expression in their external forms which are bounded by flat surfaces arranged in characteristic ways. Minerals with such external forms provide the beautiful objects known as *crystals*. A group of crystals is shown in Fig. 1

It follows from the requisites of a definite chemical composition and a definite atomic structure that minerals must be homogeneous, that is, each part, however small, must have the same chemical and physical properties.

Definition of a Mineral. *A mineral is a substance having a definite chemical composition and atomic structure and formed by the inorganic processes of nature.*

If we follow this definition rigidly, we are bound to consider the naturally occurring pure gases amongst the minerals. We should not include air, however, since it is a mixture of nitrogen and oxygen and is therefore not homogeneous. Again, water, snow and ice come within the definition since they are naturally occurring homogeneous inorganic substances of a definite chemical composition. The so-called mineral oils are mixtures of several hydrocarbons and therefore cannot be considered as mineral species.

What should be included within the rigid definition of a mineral is thus clear, but the term is often employed in a more extended sense, a usage which has been the cause of several celebrated law-suits. Thus, a miner considers a mineral to be anything of economic value that can be extracted from the earth. The national statistical summaries of mineral production include details of materials such as chalk, clay, coal, petroleum and granite that do not come within the definition of a mineral. In this book it is proposed to discuss not only those substances which fulfil the term, but also a few materials whose origin may not always be free from organic causes or whose chemical composition may not be constant. Coal, mineral oils, limestones and some phosphate are examples of such substances.

Bodies in no way to be distinguished from actual minerals have at various times been artificially formed, either purposely in the laboratory or by accident in industrial processes; but although identical with true minerals of like chemical composition, they are the outcome of processes controlled by human agency, and consequently are not included among minerals. They have nevertheless, a profound interest for the mineralogist inasmuch as they serve to a certain degree to elucidate the conditions under which the corresponding minerals may have been formed.

Rocks. The popular usage of the term mineral includes, as we have already seen, certain substances which are more properly called rocks. A rock is a portion of the earth's crust which has some individuality; it is the working unit of the field geologist and the distribution of the various kinds of rocks is shown upon geological maps. A rock has no distinctive shape of its own, it has no definite chemical composition and it is not homogeneous.

Examination shows that in most cases rocks consist of a mixture

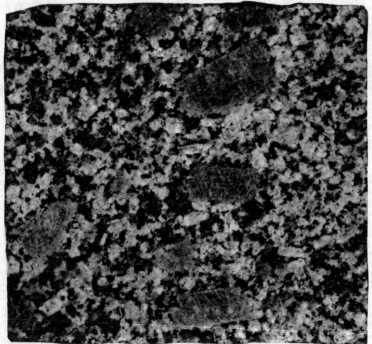

FIG. 2. The Rock Granite. A polished slab of granite composed of large crystals of feldspar in a finer base of smaller rectangular crystals of feldspar, ragged plates of black mica and shapeless grains of quartz.

of various minerals. The heterogeneous rock can be taken to pieces and the several homogeneous minerals that compose it separated out. For example, consider the well-known rock *granite* (Fig. 2). It can be seen by inspection of a hand-specimen or a polished surface of this rock that it is made up of three constituents—one white or pink and cleavable, which is the mineral *feldspar*; another, clear glassy and with no cleavage, which is the mineral *quartz;* and a third, glistening, scaly and soft, which is the mineral *mica*. Detailed chemical and physical investigation would show that the components, feldspar, quartz and mica, fulfil the requisites of minerals. They are the *mineral* units which have been aggregated together to form the *rock* granite. These three constituents occur in varying proportions in different granites and even in different parts of the same granite mass. It sometimes happens that a rock, in the geological sense of an individual portion of the earth's crust, may be composed of one mineral only. For example, a pure statuary marble consists of the single mineral calcite.

PART I
THE PROPERTIES OF MINERALS

PART 1

THE PROPERTIES OF MINERALS

CHAPTER I

THE CHEMISTRY OF MINERALS

ELEMENTS AND ATOMS

Solids, Liquids and Gases. Matter may exist in three states, the *solid*, the *liquid*, and the *gaseous*. Most minerals are solid, but some materials considered here, such as petroleum and natural gas, are fluids. Liquids and gases are 'fluids,' *i.e.* unlike solids they flow under the action of gravity: a gas entirely fills the space containing it, whereas a liquid may not, but may be bounded by an upper horizontal surface. Most pure substances can exist in all three states, and may be caused to pass from one to another by heating or cooling. At sufficiently high temperatures many minerals are melted to liquids, although some are chemically decomposed by heat before they reach their melting point. A *sublimate* is formed by the direct condensation of a gas into a solid.

Elements, Compounds and Mixtures. A *pure substance* is one that possesses characteristic and invariable properties; matter can thus be divided into mixtures and single (or pure) substances. Pure substances may be of two kinds, elements and compounds.

Elements are substances which have not so far been split up into simpler substances by any ordinary *chemical* means. About a hundred elements are at present known, but many are extremely rare and of little importance to the mineralogist. It has been estimated that the crust of the earth is composed of 46·6 per cent. oxygen, 27·7 per cent. silicon, 8·1 per cent. aluminium, 5·0 per cent. iron, 3·6 per cent. calcium, 2·6 per cent. potassium, 2·8 per cent. sodium, and 2·1 per cent. magnesium. Thus nearly 99 per cent. by weight of the earth's crust is composed of but eight elements, and most of the elements of economic value are absent from this list.

Compounds are pure substances made up of two or more elements. They are formed as a result of *chemical-change* and are different from mere *mixtures* in the following ways:
 (i) The elements constituting a compound are combined in definite proportions by weight.
 (ii) A compound cannot easily be split up, whereas the components of a mixture can usually be separated by mechanical means; these components may themselves be either elements or compounds.
 (iii) The properties of a compound are often very different from those of the elements it contains, whereas a mixture usually possesses the properties of its constituents.
 (iv) Heat is either given out or absorbed when a compound is formed; this does not in general occur when substances are merely mixed.

Minerals are compounds of their constituent elements, while rocks are mixtures of their component minerals. Thus, the mineral quartz is a compound, silica, of the elements silicon and oxygen, whereas the rock granite, as we have seen, is a mixture of several minerals, one of which is quartz.

Atoms and Molecules. The chemical and physical behaviour of substances is best explained in terms of an Atomic Theory of Matter. It is possible to break down the matter of an element into smaller and smaller particles, and at one stage of this process the particle is called an *atom*. The atoms of one element are all alike and differ from those of other elements. Chemical combination is the binding together of atoms and *an atom is the smallest part of an element that can enter into chemical combination with another element.*

The particles of a substance in the gaseous condition are widely separated from each other and in a state of rapid, random motion. These freely moving particles are called *molecules,* and they may consist of single atoms as in the gas helium or of two or more atoms of the same element, as in hydrogen or oxygen or, in the case of compounds, of two or more atoms of different elements, *e.g.* steam, carbon dioxide.

When a gas condenses to a liquid the molecules are no longer separated in space but come together and, to a certain extent, lose their identity. When the liquid is frozen to a solid, the atoms arrange themselves in a fairly rigid pattern, and it is no longer possible to segregate any one group of atoms from the rest. The term "molecule" is thus not really applicable to the solid state.

Symbols and Formulae. For convenience, an atom of each element is represented by an abbreviation called a *symbol* which is usually the first letter, or the first and second letters, of the English or Latin name of the element. The elements and their symbols are given in the Table on p. 14. The molecule of a substance is represented by a *formula*: thus, O is the symbol of an atom of oxygen, and C of an atom of carbon, and O_2 is the formula of a molecule of oxygen, and CO_2 the formula for a molecule of carbon dioxide. The proportions of the constituent elements of a solid or liquid compound are also represented by a formula; thus, calcite is $CaCO_3$. It should be clearly understood that this formula merely means that calcite is composed of calcium, carbon, and oxygen in the proportions of one atom of calcium, one atom of carbon, and three atoms of oxygen; it does not stand for a molecule.

The Structure of the Atom. According to views developed early in this century, atoms may be regarded as built up of *sub-atomic electrical particles,* named protons, neutrons and electrons. The *proton* carries a positive electric charge, which is taken as the unit charge; the *neutron*, as its name implies, has no charge. Both proton and neutron have the same mass (or weight) which is also taken as the unit of mass. Protons and neutrons together form the central part or *nucleus* of the atom, except in the lightest element, hydrogen, in which the nucleus consists of a single proton. The nucleus is surrounded by a number of *electrons,* each with a unit negative charge. The total negative charge on the electrons equals the total positive charge on the protons so that the whole atom is electrically balanced; the number of electrons must therefore equal the number of protons. The mass of the electron is about $\frac{1}{1850}$ part of that of the proton or neutron and for all practical purposes the whole mass of the atom is concentrated in the nucleus. Nevertheless, the nucleus is very small compared to the size of the atom. The essentials of the three important sub-atomic particles are thus:

		CHARGE	MASS
NUCLEUS	PROTON	1 positive unit	1 unit
	NEUTRON	none	1 unit
	ELECTRON	1 negative unit	$\frac{1}{1850}$ unit

Though other sub-atomic particles are known, those listed above are all that are dealt with here, since they are responsible for the chemical and physical properties of the different atoms.

In the Rutherford-Bohr theory of atomic structure, the electrons are pictured as revolving in fixed orbits or shells around the nucleus at the centre, rather like the planets around the sun. This picture has been found to be too simple and the fixed orbits have been replaced by a mathematical function that expresses the probability of finding electrons in certain places around the nucleus. It will be best for us to keep to the Rutherford-Bohr picture as, despite its defects, it is adequate for the majority of the elements dealt with in this elementary book.

Some Atomic Structures. The number of electrons in the atoms varies from 1 in hydrogen to 102 in the heaviest-known element; the uranium atom contains 92. This number is called the *atomic number,* denoted by Z, and is of course the same as the number of protons in the nucleus of the atom concerned. In the Table on p. 14 the elements from hydrogen to uranium are listed, with their symbols, in the order of their atomic numbers. This order is that of increasing weight and, further, each atom differs from its next lighter neighbour in having one more electron.

The shells in which the electrons can be considered to revolve around the nucleus are seven in number, designated for convenience—K L M N O P Q from the nucleus outwards. These shells can hold different numbers of electrons, K is filled by 2, L by 8, M by 18 and N by 32. Theoretically, the outer shells O P Q can contain 50, 72 and 98 respectively, but they are not fully occupied in any known element. It is obvious that the electrons of the heavier atoms occupy more shells than those of the lighter elements. In Fig. 3 there are given diagrams of the atomic structures of a number of elements which are now examined in more detail.

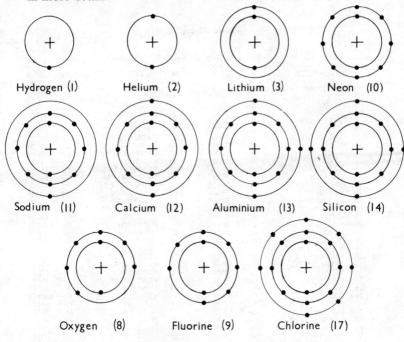

FIG. 3. Electronic Structure of various atoms.

The simplest atom is that of hydrogen ($Z = 1$) which has only a single proton and no neutron in its nucleus, the single electron occupying the K shell (Fig. 3). In the next heavier atom, helium ($Z = 2$), the nucleus is composed of 2 protons kept in order, as it were, by 2 neutrons; to attain electrical balance, 2 electrons are present in the K shell which is now full. In lithium ($Z = 3$) the nucleus has 3 protons, requiring 3 electrons, one of which must be accommodated in the L shell since K is full. In the inert gas neon ($Z = 10$) the L shell is filled so that the next heavier atom sodium ($Z = 11$) opens the M shell and calcium ($Z = 12$)

requires 2 electrons in this shell. The atomic constitution of the other atoms illustrated in Fig. 3 should be studied as the figure will be used in later paragraphs.

Table of Elements

Atomic Number	Element	Symbol	Atomic Weight	Atomic Number	Element	Symbol	Atomic Weight
1	HYDROGEN	H	1·008	47	SILVER	Ag	107·88
2	HELIUM	He	4·003	48	CADMIUM	Cd	112·41
3	LITHIUM	Li	6·940	49	INDIUM	In	114·82
4	BERYLLIUM	Be	9·013	50	TIN	Sn	118·70
5	BORON	B	10·82	51	ANTIMONY	Sb	121.76
6	CARBON	C	12·011	52	TELLURIUM	Te	127·61
7	NITROGEN	N	14·008	53	IODINE	I	126·91
8	OXYGEN	O	16·000	54	XENON	Xe	131·30
9	FLUORINE	F	19·00	55	CAESIUM	Cs	132·91
10	NEON	Ne	20·183	56	BARIUM	Ba	137·36
11	SODIUM	Na	22·991	57	LANTHANUM	La	138·92
12	MAGNESIUM	Mg	24·32	58	CERIUM	Ce	140·13
13	ALUMINIUM	Al	26·98	59	PRASEODYMIUM	Pr	140·92
14	SILICON	Si	28·09	60	NEODYMIUM	Nd	144·27
15	PHOSPHORUS	P	30·975	61	PROMETHIUM	Pm	147·00
16	SULPHUR	S	32·066	62	SAMARIUM	Sm	150·35
17	CHLORINE	Cl	35·457	63	EUROPIUM	Eu	152·00
18	ARGON	A	39·944	64	GADOLINIUM	Gd	157·26
19	POTASSIUM	K	39·096	65	TERBIUM	Tb	158·93
20	CALCIUM	Ca	40·08	66	DYSPROSIUM	Dy	162·51
21	SCANDIUM	Sc	44·96	67	HOLMIUM	Ho	164·94
22	TITANIUM	Ti	47·90	68	ERBIUM	Er	167·27
23	VANADIUM	V	50·95	69	THULIUM	Tm	168·94
24	CHROMIUM	Cr	52·01	70	YTTERBIUM	Yb	173·04
25	MANGANESE	Mn	54·94	71	LUTECIUM	Lu	174·99
26	IRON	Fe	55·85	72	HAFNIUM	Hf	178·50
27	COBALT	Co	58·94	73	TANTALUM	Ta	180·95
28	NICKEL	Ni	58·71	74	TUNGSTEN	W	183·86
29	COPPER	Cu	63·54	75	RHENIUM	Re	186·22
30	ZINC	Zn	65·38	76	OSMIUM	Os	190·20
31	GALLIUM	Ga	69·72	77	IRIDIUM	Ir	192·20
32	GERMANIUM	Ge	72·60	78	PLATINUM	Pt	195·05
33	ARSENIC	As	74·91	79	GOLD	Au	197·00
34	SELENIUM	Se	78·96	80	MERCURY	Hg	200·61
35	BROMINE	Br	79·916	81	THALLIUM	Tl	204·39
36	KRYPTON	Kr	83·80	82	LEAD	Pb	207·21
37	RUBIDIUM	Rb	85·48	83	BISMUTH	Bi	209·00
38	STRONTIUM	Sr	87·63	84	POLONIUM	Po	210·00
39	YTTRIUM	Y	88·92	85	ASTATINE	At	211·00
40	ZIRCONIUM	Zr	91·22	86	RADON	Rn	222·00
41	NIOBIUM	Nb	92·91	87	FRANCIUM	Fa	223·00
42	MOLYBDENUM	Mo	95·95	88	RADIUM	Ra	226·05
43	TECHNETIUM	Tc	99·00	89	ACTINIUM	Ac	227·00
44	RUTHENIUM	Ru	101·10	90	THORIUM	Th	232·05
45	RHODIUM	Rh	102·91	91	PROTACTINIUM	Pa	231·00
46	PALLADIUM	Pd	106·40	92	URANIUM	U	238·07

Atomic Weight. The *atomic weight* of an element is the weight of an atom of the element compared with the weight of an atom of hydrogen. It has just been seen that the weight of an atom of hydrogen will not be significantly different from that of the single proton making its nucleus. Again, the weight of an atom of helium, with its nucleus consisting of 2 protons and 2 neutrons will be 4 times that of an atom of hydrogen—the atomic weight of helium is therefore 4. It is clear that for all elements the atomic weight is the sum of the numbers of protons and neutrons that make the nucleus of the atom. The atomic weights of the elements are included in the Table on p. 14.

A glance at this table shows that few atomic weights are whole numbers as would be expected from the preceding paragraph. The reason for this is that an element, defined as a substance having a certain atomic number, may have atoms of several different weights due to different numbers of neutrons in their nucleus. Take iron ($Z = 26$) as an example: it has atoms of atomic weights 54, 56, 57 and 58—but all the atoms have 26 protons and thus all have the properties of iron. Such variants of an element are called *isotopes*. The atomic weights given in tables are the average obtained from the samples investigated. For most elements, the relative abundance of the different isotopes remains almost constant in all samples.

The *molecular weight* of a substance is the sum of the atomic weights of the atoms composing a molecule of the substance. In the case of a solid, the *formula-weight* is a convenient quantity and is the sum of the weights of the atoms making up the formula of the compound. Thus the atomic weight in round numbers of calcium is 40, of carbon is 12 and of oxygen is 16; the formula-weight of calcite ($CaCO_3$) is therefore $(40+12+3\times16)=100$.

Valency. The inert gases helium ($Z=2$), neon (10), argon (18), krypton (36), xenon (54) and radon (84) are exceedingly stable and do not combine with other atoms. Their atomic structure is characterized by 8 electrons in their outermost shell even though there may be room for more. Helium must be included for its only shell is filled with 2 electrons. This stability of the inert-gas type of structure underlies the *theory of Valency*.

In order to attain the pattern of these stable atoms, other atoms must either lose or gain electrons and these electrons are the *valency electrons*. If there are relatively few electrons in the outer shell the atoms can lose them; if there are only a few gaps in the outer shell it will be easier for it to be filled. Inspection of Fig. 3 shows that if the sodium ($Z=11$) atom were to lose one electron it would show the pattern of the inert gas neon (10); the same stable pattern is achieved by calcium (12) by the loss of two external electrons. These elements are *metals* and it can now be stated that *metals are elements in which the number of electrons is slightly greater than that in the nearest inert gas.* Now take the case of atoms that have a few electrons *less* than the stable pattern. From Fig. 3 it is seen that fluorine ($Z=9$) requires one electron and oxygen (8) requires 2 electrons to make their structural pattern like that of the nearest inert gas, neon (10); again, chlorine (17) requires one electron to become like the pattern of argon (18). These elements are examples of *non-metals, elements whose atoms are only a few electrons short of the inert-gas pattern.*

The number of electrons by which an atom differs from the nearest inert gas is the same as its *valency*. The inert gases thus have no valency. It must be understood that the nucleus of the atoms does not change with loss or gain of valency electrons, so that the elements retain their essential properties.

The hydrogen atom is unique in that it can lose its single electron and so behave like a metal or gain an electron like a non-metal. On this account the *valency of an element is measured by the number of hydrogen atoms that combine with or replace one atom of the element.* Valencies of some common elements are included in the Table on p. 24.

Ions. An atom which has lost or gained one or more electrons and is thus no longer electrically neutral is called an *ion*. With *loss* of electrons the ions become *positively* charged since the positive protons now exceed the negative electrons in number. Atoms of metals thus give positive ions, called *cations,* which carry a charge equal to the number of electrons lost. The number of charges that the cation carries is called the electrovalency and in simple circumstances can be regarded as the valency of the metal. Thus, the sodium cation has one positive charge and is denoted by Na^{1+}, the calcium cation by Ca^{2+}, the ferrous iron cation by Fe^{2+}, the ferric iron cation Fe^{3+} and so on. Conversely, the non-metals in *gaining* electrons acquire additional *negative* charges and the resulting negatively charged ions are called *anions.* For example, the fluorine anion, by gaining one valency electron, is denoted by F^{1-}, and the chlorine anion by

Cl^{1-}, the oxygen anion with 2 valency electrons is denoted by O^{2-}, and so on.

Atomic Bonding. There are four main kinds of bond that hold together the atoms in different crystal structures.

These are:

(1) The *ionic* or *polar* bond between ions of opposite electrical charge.

(2) The *covalent* or *homopolar* bond in which the atoms share electrons between them.

(3) The *metallic* bond that is responsible for the cohesion of metals and is dealt with here mainly for purposes of comparison.

(4) The *residual* or Van der Waals bond due to weak forces present in all crystals.

Though certain groups of substances are characterized by certain types of bonding, two or more types may operate between different atoms or groups of atoms in a single substance. Further, some bonds show characteristics intermediate in type.

The Ionic Bond. Atoms held together by this type of bond are in the ionised state, each atom having lost or gained one or more electrons so that they have acquired a positive or negative charge. The forces holding the ions together are those of electrical

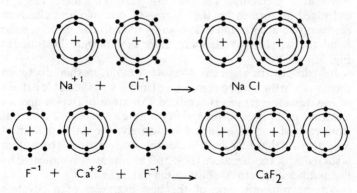

FIG. 4. Ionic Bonding.

attraction between oppositely charged bodies. Each ion is sur-rounded by ions of opposite charge and the whole structure is neutral. Crystals whose component atoms are bonded in this way may be called *ionic crystals*. They include most of the compounds of inorganic chemistry and almost all minerals.

Ionic bonding can be illustrated by reference to two common minerals, first rock-salt, sodium chloride NaCl, and second, fluorspar, calcium fluoride, CaF_2 (Fig. 4). When conditions become suitable, as for instance when metallic sodium is heated in the non-metallic chlorine gas, the Na-atom loses its one valency electron in its outer shell to the Cl-atom which has one vacant site in its outer shell. The Na^{1+} cation unites with the Cl^{1-} anion and both ions attain the stable 8-electron pattern in their outer shells. In the formation of fluorspar, the calcium atom has 2 valency electrons available, the fluorine atom has only one unoccupied site in its outer shell. Two F^{1-} anions are therefore necessary to attain the 8-electron pattern and make fluorspar, CaF_2.

The ionic bond is strong and results in crystals with a high degree of symmetry and having considerable hardness, high melting point and low coefficient of expansion.

The Covalent or Homopolar Bond. Many compounds show no evidence of consisting of oppositely charged ions of their constituent elements. The link in many of these cases is envisaged as formed by the *sharing* of one or more electrons between two atoms, and is known as the *covalent* or *homopolar* bond, the two shells overlapping. Some examples are illustrated in Fig. 5.

In hydrogen, the single electrons of two atoms are shared between them to form the hydrogen molecule; in this way the stable pattern of the nearest inert gas is acquired. An atom of oxygen has six electrons in the outer orbit and in forming a molecule 2 electrons from each of two atoms are shared to give the stable pattern of 8 electrons in the outer shell. In chlorine, one electron of each atom is shared with another atom (Fig. 5). The *covalency* is indicated by the number of electrons short in the outer shell.

Among minerals, one of the best examples of a covalent structure is provided by diamond, which is made entirely of the

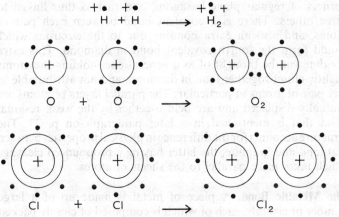

FIG. 5. The Covalent Bond.

element carbon ($Z = 6$). The carbon atom has 4 electrons in its outer shell and can form 4 covalent bonds with other carbon atoms. This is the basis of the diamond structure, in which each carbon atom is surrounded by four others arranged as shown in Fig. 6A. Atoms are located at the corners and the face-centres of the cube and also at points one-quarter or three-quarters of the way along a diagonal. This structure has great strength which is reflected in the hardness of the diamond.

It is interesting to compare the diamond structure with that of graphite, the other crystalline form of carbon. In graphite the carbon atoms lie in layers (Fig. 6B); the atoms are arranged at the

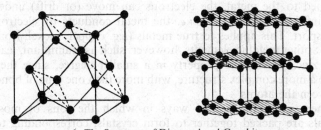

FIG. 6. The Structure of Diamond and Graphite.
A The structure of Diamond. Carbon atoms shown in black.
B The structure of Graphite. Layers of carbon atoms are shaded.

corners of regular plane hexagons, and each is thus linked to three others. There is a covalent bond between each pair of atoms, and also an extra bonding due to the electrons which would form the fourth covalent bond in diamond. This extra bonding can be thought of as a general one, holding the atoms slightly closer together than in diamond, and not attributable to any pair of atoms in particular. The parallel layers of atoms are mutually displaced and are held together by the weak residual bond that is mentioned in a later paragraph on p. 23. The structure accounts for the difference in physical properties between diamond and graphite, the latter having a pronounced cleavage which takes place parallel to the sheets of atoms.

The Metallic Bond. A piece of metal is made up of a large number of crystals, each of which is composed of closely packed atoms of the particular metallic element. Such a crystal may be regarded as an aggregate of positive ions immersed in a "gas" or "cloud" of free electrons; the atoms of the metal all contribute electrons to the common electron "gas", which serves to bind together the metallic ions. The *metallic bond* is thus an attraction between the positive ions and the "gas" of negative electrons; it differs from the ionic bond in that the attraction is exerted on *like* metal ions, not on ions of different elements. Crystals having this type of bonding between their constituent atoms are called *metallic crystals.*

Many properties which are characteristic of metals, such as their opaqueness and conductivity of heat and electricity, are due to the presence of the free electrons. Thus, if an electric field is applied to the metal, the electrons can move (or drift) under the influence of the field, *i.e.* the metal conducts by electron transport. This applies to true metals (*e.g.* copper, nickel, iron, etc.); other metallic elements, however, such as aluminium, lead and zinc, possess the property in a smaller degree, since they have a more complex structure, with more than one type of bond between the atoms.

There are three typical ways in which the ions of most metals are packed together to form crystals, corresponding to three methods of stacking spheres of equal size. They are known as *cubic close-packing* (found, for example, in copper), *body-centred cubic packing (e.g.* iron at ordinary temperatures) and *hexagonal close-packing (e.g.* magnesium). These will now be described briefly. Some metals, as already mentioned, have more complex structures; further information should be sought in books such as R. C. Evans' *Crystal Chemistry.*

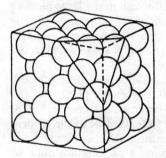 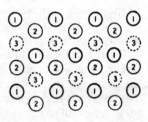

FIG. 7. Cubic Close-packing.

A Cubic close-packing. A close-packed layer of ions in a plane normal to one of the diagonals of the cube is shown with the ions above it removed.

B Diagram showing the relative positions of three successive diagonal layers, numbered 1, 2, 3.

When equal spheres are arranged in a single layer, the closest possible packing is obtained if the centres of the spheres lie at the corners of equilateral triangles, each sphere touching six others. Layers of this kind occur in both cubic and hexagonal close-packing.

(i) *Cubic close-packing.* The structure is shown in Fig. 7A, where the spheres represent metallic ions, all of equal size. The layers of ions which lie at right angles to a diagonal of the cube have the close-packing described above, with the ions arranged at the corners of equilateral triangles. Since there are four diagonals, there are also four directions in which layers of this type occur. In the complete structure, any one such diagonal layer fits against the next so that each sphere touches three in the layer above or below as shown diagrammatically in Fig. 7B. It is evident from the figure that a succession of three layers can be placed one on the other in different positions, as shown by the circles numbered 1, 2, 3, but a fourth layer would repeat the position 1. The pattern therefore repeats after every third layer.

The cube outline is added in Fig. 7A to show the cubic symmetry (see p. 97).

In a crystal of copper, movement or "gliding" may take place along the close-packed diagonal layers when the crystal is subjected to stress, and it is to this property that the metal owes its ductility and malleability. Although in the large number of crystals making up, say, a piece of copper wire the phenomenon is more complicated than in the case of a single crystal, gliding on the many diagonal planes takes place when the wire is stretched.

Other metals which have the cubic close-packed structure and, like copper, are ductile, include gold, silver, and platinum.

(ii) *Body-centred cubic packing*. This structure is shown in Fig. 8. The spheres again represent metallic ions; they lie at the corners of a series of cubes and at the centres of the cubes. Each sphere thus touches eight others, giving a packing not quite so close as in the previous type. The absence of close-packed layers of ions makes gliding much less easy, and a metal having this structure is in consequence harder and more brittle. The structure is found in iron at ordinary temperatures, but at higher temperatures iron has the cubic close-packed structure and becomes malleable. The importance of iron in metallurgy lies in this dual role, since the metal may be made to assume different properties according to the heat treatment it receives.

Other metals which have the body-centred cubic structure include tungsten, barium, vanadium and molybdenum.

(iii) *Hexagonal close-packing*. The third type of structure is illustrated in Fig. 9, which shows the stacking of a series of similar layers. In each layer the spheres are closely packed, with their centres lying at the corners of equilateral triangles. Adjacent layers are displaced relatively to one another so that any one sphere lies above or below three spheres in the layer next to it; but the spheres of alternate layers lie vertically above one another. The structure thus repeats after every second layer, instead of after every third layer as in cubic close-packing. Planes of easy gliding are fewer in the hexagonal close-packed structure, since they occur in one direction only; metals built up in this way are correspondingly less soft and ductile than those in the first group. Magnesium, titanium, one form of nickel, and calcium above

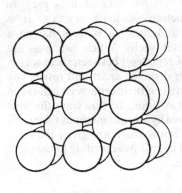

FIG. 8. Body-centred cubic packing.   FIG. 9. Hexagonal close-packing.

450°C. are among the metals whose crystals possess the hexagonal close-packed structure.

The Residual (van der Waals) Bond. This bond differs from the other three types of atomic linkage discussed above, in that it is not alone responsible for the coherence of any common substances. It occurs as a weak force of attraction between the ions or atoms of all solids, but its effect is completely masked in structures where ionic, homopolar, or metallic forces also occur. The only solids in which the atomic bonding is entirely of the residual type are the inert gases, *e.g.* argon, in the solid state. On the other hand, the residual bond is important in the realm of organic substances. The organic carbon compounds, for example, are formed of molecules containing carbon, hydrogen, and oxygen atoms arranged in many different ways; the bonding within these molecules is usually homopolar, but *between* the molecules it is in many cases residual. Thus, paraffin wax is built of molecules of a long chain hydrocarbon, and the molecules themselves are packed together, like a series of rods, to form crystals ('molecular crystals'). The soft nature of such a substance is due to the weakness of the residual bonding between the molecular crystals of which it is composed.

The relative sizes of Ions. The relative sizes of ions play an important part in the construction of ionic crystals. It is convenient to think of atoms and ions as spheres having a definite radius, that of their outer shells. When two ions are brought close together, a force of repulsion between them sets in abruptly when they are a certain distance apart and resists any closer approach. The distance between their centres is then taken as the sum of the radii of the two ions. In this way, the ions are treated as spheres in contact, and by various methods their radii can be measured. In the Table below there are given the *ionic radii* of certain elements arranged in groups according to the ionic charge or valency of the elements.

TABLE OF IONIC RADII IN ANGSTROM UNITS
(After Ahrens)

2—	1—	O	1+	2+	3+	4+
O 1·40	F 1·36	He	Li 0·68	Be 0·35		
S 1·84	Cl 1·81	Ne	Na 0·97	Mg 0·66	Al 0·51	Si 0·42
Se 1.91	Br 1·95	A	K 1·33	Ca 0·99	Ga 0·62	Ge 0·53
Te 2·11	I 2·16	Kr	Rb 1·47	Sr 1·12	In 0·81	Sn 0·71
		X	Cs 1·67	Ba 1·34	Tl 1·47	Pb 0·84

The radii of other metallic ions are as follows:—

Monovalent: Cu, 0·96   Ag, 1·26   Au, 1·37
Divalent: Fe, 0·74   Co, 0·72   Ni,   0·69   Mn, 0·80
                 Zn, 0·74   Cd, 0·97   Hg,   1·10   Pb, 1·20
Trivalent: Cr, 0·63   Fe, 0·64   Mn, 0·66

The relative sizes of ions important in rock-forming minerals are shown in Fig. 10 and should be carefully studied.

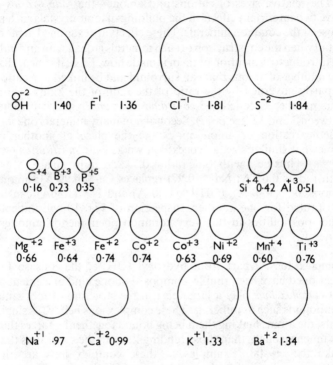

FIG. 10. Ionic Radii. In Ångstrom Units.

Oxygen is the commonest element of the earth's crust and most minerals are oxygen compounds. The size of the O^{2-} ion means that the crust is composed of over 90% by volume of this one element. The large oxygen ions are closely packed together in most minerals and the much smaller cations are situated in the interstices between the oxygens. Other negative ions (anions), e.g. F^{1-}, Cl^{1-}, S^{2-}, are similarly large (See Fig. 10).

The hydrogen ion is exceptional in being extremely small; when bound to an atom of oxygen it becomes embedded, as it were, in the oxygen, and the resulting (OH)-ion has virtually the same radius as that of oxygen. We can think of the H-ion as a centre of positive charge without dimensions.

The relative sizes of cations and anions—the *radius-ratios*—have two important effects in the building of ionic crystals such as those of the common minerals. These effects are examined in more detail when the construction of such minerals is dealt with and only an introductory mention of them is made now. First, it is clear that the number of anions that can surround and be linked to a cation is partly governed by the ratio of the radii of the kinds of ion. This number is called the *co-ordination number* and can range between 3 and 12 (see p. 77). Secondly, in many minerals one ion, whether cation or anion, may take the place of another of somewhat similar size, a process known as *ionic substitution* (see p. 355). Thus Fe^{2+} with ionic radius 0.74 Å readily replaces Mg^{2+} with radius 0.65 Å; Na^{1+} (0.97) replaces Ca^{2+} (0.99 Å); among anions, 0^{2-} (1.40 Å), $(OH)^{1-}$ (1.40 Å) and F^{1-} (1.36Å) can be substituted for one another. Inspection of Fig. 10 will indicate other possibilities in this very common phenomenon amongst minerals.

Complex Ions. Apart from hydroxyl $(OH)^{1-}$, the ions so far mentioned have been simple, composed of one kind of element—but *complex ions* play a large part in the structure of inorganic compounds such as minerals. These complex ions behave as single units, the atoms making them being bound together by forces that are much stronger than those binding them to the other ions that make the crystal. Examples of these complex ions are the carbonate ion CO_3^{2-}, the sulphate ion SO_4^{2-}, the silicate ion SiO_4^{4-} see Fig. 125a, p. 350) and the phosphate ion PO_4^{3-}.

The carbonate ion CO_3^{2-} has one carbon atom to which are attached three evenly-spaced oxygen atoms and it carries a double negative charge. The sulphate ion has four oxygens arranged at the corners of a tetrahedron around a central atom of sulphur. It will be seen later that the shapes of these complex ions affect the type of crystal structure of the minerals in which they occur, as well shown by calcite $CaCO_3$ (see p. 271).

The Periodic Classification of the Elements

When the elements are listed in the order of their atomic numbers, they may be divided into groups so that elements of similar chemical properties are brought together. This was first shown by Mendeleev, and modern views on the structures of the atoms have given a physical basis to his *Periodic Law*. In the Table below, each element is shown by its symbol and by its atomic number. The rows across the Table correspond to Mendeleev's original *periods,* and elements connected by lines running from top to bottom of the Table form the *groups*. Elements in the same group show similar chemical properties: they have the same orbital structure in their

PERIODIC TABLE OF THE ELEMENTS.

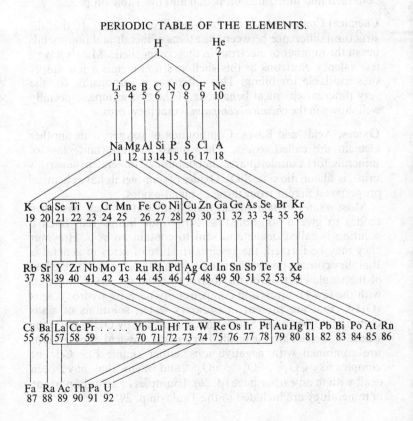

outer shell and thus the same main valency, and tend to replace one another in varying degree in minerals. Their compounds often crystallize in similar forms and they often occur together in nature. Thus, there are marked similarities between the corresponding minerals of Li, Na, and K; and of Ca, Sr and Ba. Elements in the 4th, 5th and 6th periods which are shown surrounded by a frame are of variable valency, having special features in their electronic structures, and are known as 'transitional'.

The Periodic Table is used as a basis for the classification of minerals described in Part II of this book. This classification is described and illustrated on p. 220 and the Table on p. 222.

Chemical Compounds. It has been noted earlier on p. 16 that the structural difference between the atoms of metals and non-metals lies in the number of electrons in their outer shells. Metals have a few valency electrons in this shell and non-metals a few empty sites available for filling. This difference is responsible for the very different chemical behaviour of the two groups, especially well shown in the *chemical compounds* that they form.

Oxides, Acids and Bases. Compounds of oxygen with another element are called *oxides*. They make an important class of minerals, for example, quartz, the commonest mineral in the earth's crust, is silicon dioxide, SiO_2. Oxides of non-metals have chemical properties different from those of metal-oxides.

Most *oxides of non-metals* react with water (itself hydrogen oxide) to give a solution that turns blue litmus red. Such a solution is called *acidic* and said to contain an *acid*. However they may be formed, the essential feature of acids in general is that they contain positive hydrogen ions (H^{1+}), formed by the loss of the single electron of the hydrogen atom (p. 13). Compared with the few H^{1+} ions and compensatory negative hydroxyl ions $(OH)^{1-}$ that are present in ordinary water, solutions of acids contain millions of H^{1+}-ions unaccompanied by OH^{1-}-ions; the acidic properties are the result. For a particular acid, the H^{1+}-ions are combined with negative ions, either simple like Cl^{1-} or complex like CO_3^{2-}, SO_4^{2-}, SiO_4^{4-}, and others that have been dealt with in an earlier page (p. 26). Examples of acids important in mineralogy are included in the Table on p. 29.

Most *metal oxides* do not dissolve in water; those few that do react to form *hydroxides* as, for example, sodium hydroxide NaOH, potassium hydroxide KOH and calcium hydroxide $Ca(OH)_2$. These hydroxides are *alkaline*, turning red litmus blue, and are termed *bases*. Their alkaline properties are related to the abundance of negative hydroxyl ions $(OH)^{1-}$. Many metal hydroxides, some of which occur as minerals, are obviously not formed in this way since their oxides are insoluble, but they lose their water on heating and form basic oxides. The term base has been extended to include the metal hydroxides as is noted in the next paragraph.

Salts. When an acid reacts with a base, part or the whole of the hydrogen of the acid is replaced by the metal of the base and the result is the formation of a *salt*. Thus, the action of hydrochloric acid HCl with the base, sodium hydroxide, NaOH, gives the salt sodium chloride NaCl, together with water H_2O, as shown in the equation below:—

$HCl + NaOH = NaCl + H_2O$

acid + base   = salt + water.

Expressed in terms of ions this equation becomes:—

$H^{1+} + Cl^{1-} + Na^{1+} + (OH)^{1-} = Na^{1+} + Cl^{1-} + H_2O$.

When all the H^{1+} ions from the acid have been partnered by the OH^{1-} ions from the base *neutralization* has taken place. Insoluble metal hydroxides act like the true soluble bases on neutralization by acids and form salts; on this account they are often included among the bases.

Many minerals are salts, and the names of some common acids and their corresponding salts are tabulated below:—

NAME OF ACID	NAME OF SALT	EXAMPLE OF SALT
Hydrochloric (HCl)	Chloride	Rock-salt (NaCl)
Hydrobromic (HBr)	Bromide	Bromyrite (AgBr)
Hydriodic (HI)	Iodide	Iodyrite (AgI)
Hydrofluoric (HF)	Fluoride	Fluorspar (CaF_2)
Nitric (HNO_3)	Nitrate	Nitre (KNO_3)
Sulphuric (H_2SO_4)	Sulphate	Barytes ($BaSO_4$)
Sulphuretted Hydrogen (H_2S)	Sulphide	Galena (PbS)
Carbonic (H_2CO_3)	Carbonate	Calcite ($CaCO_3$)
Tetraboric ($H_2B_4O_7$)	Borate	Borax ($Na_2B_4O_7$) Aq.
Phosphoric (H_3PO_4)	Phosphate	Apatite [$Ca_5(PO_4)_3$]

In the examples of mineral salts given in the table above, all the hydrogen of the acids has been replaced by metallic elements, and the resulting salts are called *normal salts*. When only a part of the hydrogen is replaced *acid salts* are produced. For example, K_2SO_4 is normal potassium sulphate, $KHSO_4$ is acid potassium sulphate. In *basic salts,* the whole of the base has not been neutralized by the acid portion; thus, the mineral malachite is a basic carbonate of copper and its composition may be written $CuCO_3.Cu(OH)_2$.

When certain salts are deposited from a saturated solution they form crystals that contain a definite and constant number of water molecules incorporated in the crystal structure. Such salts are *hydrates* and the combined water is called *water of crystallization*. Examples among common minerals are gypsum, $CaSO_4.2H_2O$ and natron (washing soda) $Na_2CO_3. 10 H_2O$. The water of crystallisation can be driven off at a moderate temperature, the *anhydrous* salt being left behind.

Oxidation and Reduction. A chemical change by which oxygen is added to an element or compound is called *oxidation*. The term *reduction* is applied to a change in which the oxygen or other non-metal is taken away from a compound. When oxidation occurs, reduction occurs as well and the reaction is called a *redox action*.

When metallic copper is heated in contact with air it is changed into a black oxide of copper, as in the following equation:—
$$2 Cu + O_2 = 2 (Cu^{2+}O^{2-})$$
The copper atoms in becoming Cu^2 ions have *lost* their external electrons on oxidation; the oxygen atoms have *gained* their electrons to fill the vacant sites in their outer orbit.

Again, in the reaction represented by
$$2Na + Cl_2 = 2(Na^{1+}Cl^{1-})$$
the sodium atoms have *lost* their single external electrons and the chlorine atoms have *gained* them. The general rule is thus:—if electrons have been lost by an atom the change is one of oxidation, if electrons have been gained the change is one of reduction. Oxidation and reduction are of great importance in the analysis of minerals and in the production of metals from their ores—these topics are now examined.

The Reactivity or Electrochemical Series of Metals. Metal elements can be arranged in an order which expresses the ease with which they part with electrons to form compounds. This order is called the *reactivity or electrochemical series* and for the commoner metals is stated below:—

K Na Ca Mg Al Zn Fe Sn Pb Cu Hg Ag Au

Potassium, the most active metal, begins the series and the least active, gold, ends it. The whole series shows a graduation in chemical and other properties from potassium to gold.

The recognition of the reactivity series ties together many chemical phenomena which are of importance in the occurrence and treatment of minerals. The more active metals are powerful reducing agents, combine readily with oxygen and corrode and tarnish readily. The stability of the precious metals illustrates the reverse of these properties.

The hydroxides and carbonates of the very active metals such as K and Na are soluble in water but calcium carbonate is insoluble and beginning with Mg neither hydroxide nor carbonate is soluble. With regard to acids, metals early in the reactivity series dissolve in hydrochloric and sulphuric acids with the evolution of hydrogen but with later metals the reaction is slower and it ceases altogether with silver and gold. All the metals listed dissolve in nitric acid except these precious metals.

The readiness with which compounds of metals decompose on heating increases from potassium to gold, a circumstance affecting the production of metals from their ores. Metals early in the series, that is potassium to aluminium, do not occur *native* or free in nature and these metals are obtained by electrolysis of their molten compounds. Metals in the middle of the series, zinc iron tin and lead, are rarely found native and are produced by reduction with carbon. Metals still less active, copper and mercury, often occur in the native state and these metals, as well as lead, can be readily obtained without a reducing agent by simple heating of their ores. The least reactive metals, silver and gold occur mostly native and uncombined.

CHEMICAL ANALYSIS

Qualitative and Quantitative Analysis. The splitting-up of a compound into its constituent elements is called *analysis*. The first step in analysis consists in determining the nature of the elementary substances contained in a compound, the next in determining the proportions of these constituents. The former is called *qualitative,* and the latter *quantitative* analysis.

In a *qualitative analysis* the recognition of the constituents hinges upon the fact that certain bases and certain acids produce well-marked phenomena in the presence of known substances or preparations termed *reagents.* The characteristic effect produced by a reagent is spoken of as a *reaction.* Thus hydrochloric acid is a reagent, and when added to clear solutions containing salts of lead, silver or mercury, it produces a dense white precipitate consisting of the chlorides of those metals—a reaction denoting the presence of one or more of them in the original solution. This reaction must be supplemented by others in order to determine which of the three metals is present in the salt.

The *quantitative analysis* of a mineral is conducted by a skilled chemist who reports his results as weight-percentages of the elements or the oxides constituting the mineral. The method of calculating the formula of a mineral from such analyses is illustrated by the following examples.

(1) The composition of the common copper ore *chalcopyrite* is given in weight-percentages of elements in Column 1 of the Table below. In Column 2 are given the atomic weights of copper, iron and sulphur and in Column 3 the atomic proportions of these three constituents, obtained by dividing the weight-percentage by the atomic weight for each constituent. The atomic ratio (Column 4) is derived from the atomic proportions and is seen to be 1 copper to 1 iron to 2 sulphur. The formula of chalcopyrite is thus $CuFeS_2$, copper iron sulphide.

ANALYSIS OF CHALCOPYRITE

	1. Weight percentage	2. Atomic weight	3. Atomic proportions	4. Atomic ratio
Cu	34.5	63.54	.5429	1
Fe	30.5	55.85	.5461	1
S	35.0	32.07	1.091	2

(2) The very common mineral *orthoclase* gives, if pure, the weight-percentages of its constituent oxides set out in Column 1, in the Table below. These percentages divided by the 'molecular' or formula-weights (see p. 15) of Column 2 produce the 'molecular' proportions of Column 3, from which the 'molecular' ratios are clearly 1 potassa, 1 alumina and 6 silica. The formula of orthoclase might be written $K_2O.Al_2O_3.6SiO_2$. This used to be the accepted way of formulating minerals, but it is misleading, because the oxides are not present as such in the mineral. The formulae of minerals are now written as far as possible in accordance with their atomic structure as revealed by X-ray studies, a matter of special importance in connection with the silicates. The formula of orthoclase, the commonest silicate, is thus $KAlSi_3O_8$.

ANALYSIS OF ORTHOCLASE

	1. Weight percentage	2. Formula weight	3. 'Molecular' proportions	4. 'Molecular' ratio
K_2O	16.91	94	0.180	1
Al_2O_3	18.35	102	0.180	1
SiO_2	64.74	60	1.079	6

Investigations conducted in solutions are called analysis *by the wet way*. There are, however, a number of *dry ways* of analysis, one of which, described immediately, is extremely convenient for the purposes of the mineralogist.

ANALYSIS BY THE BLOWPIPE

The Blowpipe. For examining a mineral in the dry way, the blowpipe is an invaluable instrument. It consists essentially of a tube bent at right angles, one extremity having a mouthpiece, the other being terminated by a finely perforated jet. The tube should bulge out between the two extremities into a cavity, in which the condensed moisture from the breath may lodge so as not to be carried through the jet or on to the *assay*, or portion of material being tested. It is important that the aperture of the nozzle of the blowpipe should be small and circular. This can be attained by nearly closing the apperture by gently tapping the nozzle on an iron surface and then inserting a square needle and by rotation producing a hole of the required size and shape.

In using the blowpipe the operator will probably experience at first some difficulty in keeping up a steady continuous blast. Practice will, however, soon enable him to use the instrument easily. While blowing, the cheeks should be kept inflated and the air expelled by their action only, fresh supplies of air being drawn in through the nose. Trial will teach far better than any description, and practice should be resorted to until a steady and uninterrupted blast can be kept up for some minutes. A gas flame is very convenient for blowpipe experiments, but the flame of a spirit lamp, an oil lamp, a paraffin-wax lamp, or a candle, will also answer the purpose. When a lamp or a candle is used, the wick should be bent in the direction in which the flame is blown. Many portable blowpipe lamps are on the market for use in the field.

The Two Kinds of Flame. In blowpipe analysis it is necessary to be able to produce and to recognise two types of flames, in one of

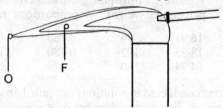

FIG. 11. Oxidizing Flame, showing Position of Blowpipe, and Points of Oxidation *(O)* and of Fusion *(F)*.

which oxidation of the substance under examination is brought about, and in the other, reduction.

1. *The Oxidizing Flame.* An oxidizing flame is produced when the nozzle of the blowpipe is introduced into the flame to about one-third of the breadth of the flame. It is advisable to blow somewhat more strongly than in the production of the reducing flame. The oxidizing flame is blue and feebly illuminating, and in it the air from the blowpipe is well mixed with the gases from the flame, and complete combustion ensues. There are two positions in this flame at which operations useful to the

experimenter are performed. The hottest part is just outside the inner blue cone, and is called the *point of fusion*. The best position for oxidation—the *point of oxidation*—is just beyond the visible part of the flame, for at this point the assay is heated surrounded by air, and hence oxidation takes place. The oxidizing flame is shown in Fig. 11, in which the positions of the point of fusion (F) and the point of oxidation (O) are seen.

2. *The Reducing Flame*. The reducing flame is produced when the nozzle of the blowpipe is placed some little distance from the flame. The reducing flame is bright yellow and luminous, ragged and noisy. In this flame the stream of air from the blowpipe drives

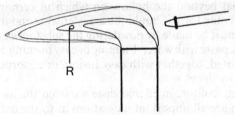

FIG. 12. Reducing Flame, showing Position of Blowpipe and Point of Reduction.

the whole flame rather feebly before it, and there is little mixing of air with the gases from the flame. The result is that these gases are not completely burnt, and hence they readily combine with the oxygen of any substance introduced into their midst. It will be seen, therefore, that the assay must be completely surrounded by the reducing flame, but care should be taken not to introduce the assay too far into the flame for a deposit of soot will then be formed and interfere with the heating of the substance. The reducing flame is shown in Fig. 12, and the *point of reduction* is at (R).

Supports. The portion of the substance under examination— the assay—may be supported in various ways according to the requirements of each particular case. After each experiment, all supports must be thoroughly cleaned before further use. The chief supports are charcoal blocks, platinum-tipped forceps and platinum wire.

Charcoal forms a good support by reason of its infusibility, feeble capacity for conducting heat, and its reducing action. The carbon of which the charcoal block consists readily combines with the oxygen which the assay may contain. The reduction of many metallic oxides to the metals may be effected by heating on charcoal in the reducing flame. In other cases charcoal may be used as a support in oxidation, provided that its reducing action does not materially interfere with the results which the operator desires to produce.

To obtain the maximum of information from the behaviour of an assay when heated on charcoal, it is necessary to observe several precautions. The assay should be placed in a small hollow scraped in the charcoal block, and there should be a large area of cool charcoal beyond the hollow on which in certain cases an encrustation may form. Should the assay crackle and fly about, a fresh assay must be made by powdering the substance and mixing it into a thick paste with water. Flaming or easy burning of the assay should be noted, together with easy fusion, or absorption of the fused assay by the charcoal.

The nature, colour, smell and distance from the assay of any encrustation are all important indications as to the nature of the elements present. Thus arsenic compounds give an encrustation far from the assay, whereas antimony compounds give one near the assay. White encrustations or residues, when moistened with cobalt nitrate and strongly reheated, give various colours characteristic of certain elements. Compounds containing some few metals—lead, mercury or bismuth—give characteristically coloured encrustations when heated on charcoal with potassium iodide and sulphur. Tables embodying these and other tests are given in later pages.

Platinum-tipped forceps are useful for holding small splinters of minerals in the blowpipe flame. When substances are examined in this way the colour of the flame should be noted, and the degree of fusibility of the mineral compared so far as is possible with the standard scale of fusibility described on p. 69.

A *platinum wire* may be used with excellent results for nearly all the operations usually carried out by the forceps. In the use of platinum, whether foil, forceps or wire, care should be taken that minerals suspected of containing iron, lead, antimony and other metals that form alloys with platinum are not supported by its means.

Platinum wire is used in the flame test and the bead tests. Several elements give distinctive colours to the blowpipe flame and this *flame test* is performed by introducing some of the finely powdered mineral, either alone or moistened with hydrochloric acid, into the flame on a platinum wire. The several very important *bead tests* are carried out by fusing the mineral with a flux in a small loop at the end of the platinum wire.

Fluxes. Certain substances are added to the assay for the purpose of effecting a more rapid fusion than could be obtained by heating the mineral by itself. Such substances are called *fluxes* and are especially useful when the constituents of the assay form a characteristic coloured compound with these substances. The most important fluxes are borax, microcosmic salt and sodium carbonate.

Borax is a hydrous sodium borate, $Na_2B_4O_5(OH)_4.8H_2O$. When used as a blowpipe reagent, the borax is finely powdered after having been deprived by heat of the greater part of its water. To make a borax bead the operator first bends one end of the platinum wire into a small loop. The loop is heated to redness in the blowpipe flame and immediately dipped into the powdered borax, some of which adheres to the wire. When the loop is heated again in the blowpipe flame, the powder froths up or intumesces, owing to the disengagement of the water still remaining in it, and gradually fuses to a clear transparent globule,—*the borax bead*. The powdered substance to be examined is touched with the hot bead so that a small quantity of the substance sticks to the bead. The bead is then heated by a well-sustained blast; its colour and other characters are noted, both when it is hot and cold; and these observation are made with both the oxidising and reducing flames. Some minerals should be added to the bead in very minute quantities, otherwise the reaction may be masked or rendered obscure, and difficulty experienced in determining the colour. Borax serves principally to reduce substances to the state of oxides, and it is by the colour and other properties which these oxides impart to the borax bead that we are able to ascertain to a certain extent the nature of the substance under examination.

Minerals containing sulphur or arsenic dissolve with difficulty in borax and the behaviour of such sulphides and arsenides in the borax bead differs materially from that of the oxides of the same metals. It is therefore advisable to roast the substance on charcoal in the oxidizing flame before making the usual borax bead test so that the sulphur or arsenic is volatilized.

Microcosmic salt is a hydrated sodium ammonium hydrogen phosphate ($NaNH_4HPO_4.4H_2O$). This substance is so fluid when it first fuses that it generally drops from the platinum wire. It is best, therefore, either to heat it on charcoal or platinum foil until the water and ammonia are expelled, when it can be easily taken up on the platinum loop which should be made rather small, or else to add the salt to the bead in small quantities at a time. The substance to be examined is added in just the same way as with the borax bead, and the whole fused in the blowpipe flame.

The action of microcosmic salt is to convert the oxides of metals into phosphates of a complex nature, imparting characteristic colours to the bead when hot and when cold; these colours often differ when the bead is produced in the oxidizing or reducing flame. Silica is insoluble in the microcosmic salt bead so that when silicates are dealt with, a silica skeleton appears in the bead.

Sodium carbonate has the chemical formula $Na_2CO_3.10H_2O$. It is used in the reduction of oxides or sulphides of metals to the metallic state. The mineral under examination is finely powdered and intimately mixed with sodium carbonate and powdered charcoal; the mixture is slightly moistened, placed in a hollow on charcoal and heated in the blowpipe flame. The powdered mineral should amount to about a third of the total mixture.

Sodium carbonate is valuable as a flux in the analysis of silicates as it then parts with carbonic acid and is converted into sodium silicate.

Manganese and chromium give charateristic colours when introduced into the sodium carbonate bead, owing to the formation of sodium manganate and chromate.

Sodium sulphide is formed from a sulphate by fusing the powdered mineral sulphate with sodium carbonate and charcoal on charcoal. The fused mass when placed on a silver coin and moistened gives a black stain of silver sulphide. Mineral sulphides give the same reaction but can be distinguished from sulphates by other tests.

Tube Tests. Reactions using the *closed* and *open tubes* are of great importance in blowpipe analysis. The closed tube consists of narrow soft tubing cut into 2–3 inch, 5–8 cm., lengths and sealed off at one end. In the closed tube the assay is heated practically out of contact with the oxygen of the air. In the open tube, which consists of hard glass tubing of 4–5 inch, 10–12 cm., lengths and open at both ends, heating takes place in a stream of hot air and oxidation results.

A small quantity of the powdered assay is introduced into the closed tube and heated. In many cases a deposit called the *sublimate* is formed on the cooler parts of the tube, and the colour and nature of this sublimate may indicate one or more of the elements present in the assay. Water driven out of the assay collects as drops towards the mouth of the tube. Again, the assay may be converted by heat into the oxides of the metals present, and some of these oxides have characteristic colours and properties. Thus brown limonite (hydrated iron oxides) is converted into black magnetic oxide by the expulsion of water which collects on the cooler parts of the tube.

With the open tube, the assay is placed towards one end and the tube inclined. The assay is thus heated in a current of air and is oxidized. Characteristic *smells* or *sublimates* are formed.

Reactions. The detection of several of the acid radicles present in minerals depends on the use of reagents, such as the usual acids, powdered magnesium, granulated tin, etc. For instance, carbonates give off carbon dioxide on being treated with hydrochloric acid, and some silicates gelatinize on being heated with the same acid. These and other reactions are given in the Tables of Blowpipe Analysis below.

TABLES OF BLOWPIPE ANALYSIS

FLAME TEST

Substance, either alone or moistened with HCl, HNO_3 or H_2SO_4, heated on a clean platinum wire, colours the outer part of the blowpipe flame.

Calcium	..	Brick-red
Strontium	..	Crimson
Lithium	..	Deep crimson
Sodium	..	Yellow
Potassium	..	Violet (masked by sodium, use blue glass filter to view flame)
Barium	..	Yellow-green
Copper	..	Emerald-green with HNO_3; sky-blue with HCl
Thallium	..	Bright green
Boron	..	Momentary yellow-green with H_2SO_4

Indefinite blue flames are given by lead, arsenic and antimony; and indefinite green flames by zinc, phosphorus and molybdenum. These elements are more satisfactorily detected by other tests.

BORAX BEAD TEST.

ELEMENT		OXIDIZING FLAME	REDUCING FLAME
Iron	..	Yellow hot, colourless cold	Bottle-green
Copper	..	Blue	Opaque red
Chromium	..	Yellowish-green	Emerald-green
Manganese	..	Reddish-violet	Colourless
Cobalt	..	Deep blue	Deep blue
Nickel	..	Reddish-brown	Opaque grey
Uranium	..	Yellow	Pale green

MICROCOSMIC SALT BEAD TEST.

ELEMENT, ETC.		OXIDIZING FLAME	REDUCING FLAME
Iron	..	Colourless to brownish-red	Reddish
Copper	..	Blue	Opaque red
Chromium	..	Red when hot, green cold	Green
Manganese	..	Violet	Colourless
Cobalt	..	Blue	Blue
Nickel	..	Yellow	Reddish-yellow
Uranium	..	Yellow when hot, yellow-green cold	Yellow-green hot, bright green cold
Tungsten	..	Colourless	Blue-green

Molybdenum	..	Bright green	Green
Titanium	..	Colourless	Yellow hot, violet cold
Silica	..	Remains undissolved in microcosmic bead	
Chlorine	..	Saturate microcosmic salt bead with copper oxide; if a powdered chloride is added, a rich blue flame surrounds the bead.	

SODIUM CARBONATE BEAD TEST.

ELEMENT		OXIDIZING FLAME
Manganese	..	Opaque blue-green
Chromium	..	Opaque yellow-green

REACTIONS ON CHARCOAL.

1. **Oxidation.** (a) Substance heated alone in oxidizing flame on charcoal:—

ELEMENT		ENCRUSTATION OR SMELL
Arsenic	..	White, far from assay; smell garlic
Antimony	..	White, near assay
Zinc	..	Yellow when hot, white when cold
Lead	..	Dark yellow when hot, yellow when cold
Bismuth	..	Dark orange when hot, paler when cold
Sulphur	..	Smell of sulphur dioxide
Tin	..	Yellow when hot, paler or colourless when cold
Molybdenum	..	Yellow when hot, yellow or colourless when cold; in reducing flame, blue

(b) White encrustations and residues, obtained from (a) above, moistened with cobalt nitrate and strongly reheated:—

ELEMENT, ETC.		COLOUR
Zinc	..	Encrustation grass-green
Tin	..	Encrustation blue-green
Antimony	..	Encrustation dirty green
Magnesium	..	Residue pink
Aluminium	..	Residue blue and unfused
Fusible silicates phosphates and borates	..	Residue blue fused and glassy-looking

(c) Substance heated in oxidizing flame with potassium iodide and sulphur:—

ELEMENT		ENCRUSTATION
Lead	..	Brilliant yellow
Bismuth	..	Scarlet; yellow near assay
Mercury	..	Greenish-yellow; and greenish-yellow fumes

2. **Reduction.** (d) Substance mixed with powdered charcoal and sodium carbonate and heated in oxidizing flame:—

ELEMENT		BEAD OR RESIDUE OBTAINED
Lead	..	Soft malleable metallic bead; easily fused; marks paper
Tin	..	Tin-white bead, soft and malleable; not marking paper
Silver	..	Silver-white malleable bead
Gold	..	Yellow bead, soft and malleable
Bismuth	..	Silver-white bead, brittle
Copper	..	Red spongy mass
Iron	..	Residue strongly magnetic
Cobalt	..	Residue feebly magnetic
Nickel	..	Residue feebly magnetic

(e) *Special Reduction Tests for Titanium and Tungsten:*—Substance fused with powdered charcoal and sodium carbonate; the residue boiled with hydrochloric acid and few grains of granulated tin:—

METAL		COLOUR OF SOLUTION
Titanium	..	Violet
Tungsten	..	Prussian blue

Substance fused with powdered charcoal and sodium carbonate; the residue dissolved in concentrated sulphuric acid with equal volume of water added; the solution is cooled, water added, then hydrogen peroxide added:—

METAL	COLOUR OF SOLUTION
Titanium ..	Amber

CLOSED TUBE TEST.

Assay heated in closed tube, either alone, or with sodium carbonate and powdered charcoal, or with magnesium:—

ELEMENT, ETC.		OBSERVATION
Sulphur	..	Orange sublimate
Arsenic	..	Black sublimate; smell of garlic
Mercury, with sulphur	..	Black sublimate, red on rubbing
Arsenic, with sulphur	..	Reddish-yellow sublimate, deep red while liquid
Antimony, with sulphur	..	Brownish-red sublimate, black while hot
Water	..	Colourless drops
Mercury	..	Heat with sodium carbonate and charcoal; globules of mercury as sublimate
Arsenic	..	Heat with sodium carbonate and charcoal; black mirror of arsenic, soluble in sodium hypochlorite
Phosphates	..	Heat with magnesium; add water. Characteristic smell of phosphoretted hydrogen

OPEN TUBE TEST.

Assay heated in open tube.

ELEMENT		OBSERVATION
Sulphur	..	Sulphurous fumes of sulphur dioxide
Arsenic	..	White sublimate, crystalline, volatile, far from assay; smell of garlic
Antimony	..	White sublimate near assay
Tellurium	..	Whitish sublimate fusible to colourless drops

REACTIONS FOR ACID RADICLE.

ACID RADICLE		TEST
Carbonate	..	With hydrochloric acid, carbon dioxide evolved, turning lime-water milky
Sulphides (some)	..	With hydrochloric acid, sulphuretted hydrogen evolved. Also indicated by closed tube, open tube and charcoal tests, q.v.
Fluoride	..	With strong sulphuric acid, greasy bubbles of hydrofluoric acid evolved, causing deposition of a white film of silica on a drop of water held at the mouth of the tube
Chloride	..	With sulphuric acid and manganese dioxide, greenish chlorine evolved. Also detected by microcosmic salt bead saturated with copper oxide, q.v.
Bromide	..	With sulphuric acid and manganese dioxide, brown bromine evolved
Iodide	..	With sulphuric acid and manganese dioxide, violet iodine evolved
Nitrate	..	With sulphuric acid, brown nitrous fumes evolved
Silicates (some)	..	With hydrochloric acid, gelatinize Silica skeleton in microcosmic salt bead
Sulphate	..	Heat substance on charcoal with sodium carbonate and powdered charcoal; place residue on silver coin and moisten. Black stain indicates sulphate (or sulphide)
Phosphate	..	Heat with magnesium in closed tube, add water; phosphoretted hydrogen evolved. Also detected by giving a fused blue mass when heated on charcoal, moistened with cobalt nitrate and strongly reheated
Telluride	..	Heat powdered mineral with a little strong sulphuric acid, reddish-violet solution, colour disappears on adding water to the cold solution and a grey precipitate is deposited.

SUMMARY OF TESTS FOR METALS.

Aluminium	..	Heated on charcoal, moistened with cobalt nitrate, strongly reheated, blue unfused residue
Antimony	..	Roasted on charcoal, white encrustation near assay. Heated in open tube white sublimate near assay. Heated in closed tube, red-brown sublimate, black when hot
Arsenic	..	Roasted on charcoal, white encrustation far from assay; garlic smell Heated in open tube, white volatile sublimate Heated in closed tube with sodium carbonate, black arsenic mirror, soluble in sodium hypochlorite
Barium	..	Flame test, yellow-green
Bismuth	..	Reduction on charcoal, brittle bead. Roasted with potassium iodide and sulphur, yellow encrustation near assay, outer parts scarlet
Calcium	..	Flame test, brick-red
Cadmium	..	Heated on charcoal with sodium carbonate reddish-brown sublimate
Chromium	..	Borax bead, green; microcosmic salt bead, green; sodium carbonate bead, yellow-green, opaque
Cobalt	..	Borax bead, deep blue; microcosmic salt bead, deep blue
Copper	..	Flame test, emerald-green with nitric acid, sky-blue with hydrochloric acid Borax bead, blue in oxidizing flame; opaque red in reducing flame Reduction on charcoal, red metallic copper
Gold	..	Reduction on charcoal, soft malleable gold bead
Iron	..	Borax bead, yellow hot, colourless cold, in oxidizing flame; bottle-green in reducing flame Reduction on charcoal, magnetic residue

Lead	..	Reduction on charcoal, malleable metallic bead, marking paper
		Roasted with potassium iodide and sulphur brilliant yellow encrustation
Lithium		Flame test, deep crimson, deeper than strontium flame
Magnesium	..	Heated on charcoal, moistened with cobalt nitrate, strongly reheated, pink residue
Manganese	..	Borax bead, reddish-violet in oxidizing flame; colourless in reducing flame
		Microcosmic bead, violet in oxidizing flame; colourless in reducing flame
		Sodium carbonate bead, blue-green, opaque
Mercury	..	Heated on charcoal with potassium iodide and sulphur, greenish-yellow encrustation and greenish-yellow fumes
		Heated in closed tube with sodium carbonate and charcoal, globules of mercury as sublimate
Molybdenum	..	Microcosmic salt bead, bright green in oxidizing flame; dirty green hot, fine rich green cold, in reducing flame
		Roasted on charcoal, yellow hot, yellow or colourless cold; in reducing flame, blue
Nickel	..	Borax bead, reddish-brown in oxidizing flame; opaque grey in reducing flame
Potassium	..	Flame test, violet, view through blue glass filter
Silver	..	Reduction on charcoal, silver bead
Sodium	..	Flame test, yellow
Strontium	..	Flame test, crimson
Tellurium		Heated in open tube, whitish sublimate, fusible to colourless drops
		Heated with strong sulphuric acid, reddish-violet solution
Thallium	..	Flame test, bright green
Tin	..	Reduction on charcoal, tin bead
Titanium	..	Microcosmic salt bead, yellow hot, violet cold, in reducing flame
		Reduction with tin, violet solution
		Hydrogen peroxide test, amber solution

Tungsten	..	Microcosmic salt bead, blue-green in reducing flame
		Reduction with tin, blue solution
Uranium	..	Microcosmic salt bead, yellow hot, yellow-green cold, in oxidizing flame; yellow-green hot, bright green cold, in reducing flame
Zinc	..	Roasted on charcoal, encrustation yellow when hot, white when cold
		Heated on charcoal, moistened with cobalt nitrate, and strongly reheated, grass-green encrustation

CHAPTER II

CERTAIN PHYSICAL PROPERTIES OF MINERALS

INTRODUCTION

Minerals possess certain physical properties that are considered in this chapter in the following order:

  (i) Certain characters depending upon light, such as colour, lustre, transparency, translucency, phosphorescence and fluorescence. Other optical properties especially valuable in the recognition of minerals in thin section under the micro-scope are dealt with in a later chapter.

 (ii) Characters depending upon certain senses, such as those of taste, odour and feel.

(iii) Characters depending upon the state of aggregation, such as form, pseudomorphism, polymorphism, hardness, tenacity, fracture, cleavage, and surface tension effects. Crystallo-graphy—the study of crystals—is considered in the next chapter.

 (iv) The specific gravity of minerals.

  (v) Characters depending upon heat, such as fusibility.

 (vi) Characters depending upon magnetism, electricity and radio-activity.

I.   COLOUR, LUSTRE, TRANSPARENCY, ETC.

Colour. Colour depends upon the absorption of some and the reflection of others of the coloured rays or vibrations which compose ordinary white light. When a body reflects light to so small an extent as not to affect the eye, it appears black, but when it reflects all the vibrations of the different colours which compose white light, it appears white. Again, if it reflects the red vibrations of ordinary light and absorbs all the other vibrations, it appears red. A blue mineral, such as sapphire, absorbs all the vibrations of white light with the exception of those that give the sensation of blueness to the eye.

The *colour* of a mineral is often its most striking property. Unfortunately for purposes of identification, however, the colours of minerals vary very greatly. Even in the same species specimens are found having very different colours. The mineral quartz, composed of silicon dioxide, is commonly colourless or white, but it is also found with pinkish-yellow, green, brown, amethystine and even black colours. Corundum, composed of alumina, varies in colour from pale brown to deep red and dark blue, the two latter varieties being the gemstones ruby and sapphire. The same crystal of a mineral may exhibit different colours, sometimes arranged in a regular fashion as in some crystals of tourmaline, at other times in patches as in certain specimens of fluorspar, calcium fluoride.

The *true colour* of a pure mineral depends on the nature and arrangement of the constituent ions. Thus minerals containing Al, Na, K, Ca, Mg, Ba as their main ions are generally colourless or light-coloured whilst those with Fe, Cr, Mn, Co, Ni, Ti, Va, Cu are coloured, often deeply. Different types of bonding of the carbon atoms are responsible for the colourless nature of diamond and the black opacity of graphite. With individual elements the valency can affect the colour of minerals, those with divalent iron, Fe^{2+}, for example being commonly green, with trivalent iron Fe^{3+} red, brown or yellow, with both, blue or deep green.

The *streak* of a mineral is the colour of its powder and may be quite different from that of the mineral in mass. For instance, black hematite gives a red powder. Streak is observed by producing a small quantity of the powdered mineral by scratching with a knife or file or by rubbing the mineral on a piece of unglazed porcelain or roughened glass called a *streak-plate*.

Some minerals, when turned about or looked at in different directions, display a changing series of prismatic colours, such as are seen in the rainbow or on looking through a glass prism. This is called a *play of colours*. It is shown by diamond and is produced by the splitting up of a ray of white light into its coloured constituents as it enters and emerges from the mineral. *Change of colour* is a somewhat similar phenomenon extending over broader surfaces, the succession of colours being produced as the mineral is turned. This phenomenon is excellently displayed by certain varieties of the mineral feldspar, the colours shown including blues, greens, yellows and reds. Such a feldspar is an abundant constituent of a rock from southern Norway, and polished slabs of this rock in which the feldspar crystals lie in various directions are used for ornamental purposes. The change of colour is caused by the interference of light reflected from thin plates of other minerals enclosed in parallel planes within the feldspar. *Schiller,* a nearly metallic lustre shown by certain surfaces of minerals such as hypersthene or schiller-spar, is due to a somewhat similar cause. Reflection takes place either from minute plates arranged on parallel planes, or from cavities due to chemical action along certain parallel planes within the mineral.

Opalescence is a somewhat pearly or milky appearance shown by opal and moonstone. *Iridescence* is a display of prismatic colours due to the interference of rays of light in minute fissures which wall in thin films of air or liquid. These fissures are often the result of incipient fracture. Iridescence may sometimes be seen in quartz, calcite and mica. The brilliant display of colours given by the precious opal is due to the presence of very thin curved or distorted layers with slightly different optical properties.

Some minerals *tarnish* on the surface when exposed to the air and sometimes exhibit iridescent colours. This tarnish may result either from oxidation, or from the chemical action of sulphur and other substances which are generally present in the atmosphere in minute quantities. Tarnish may be distinguished from the true colour by chipping or scratching the mineral, when the superficial nature of the tarnish is revealed. Copper pyrites often tarnishes to an iridescent mixture of colours. The mineral erubescite tarnishes readily on exposure to the air, and some varieties are called peacock ore.

Some crystals display different colours when viewed in different directions by transmitted light. This property, called *pleochroism,* is considered with the special optical properties on a later page.

Lustre. The lustre of minerals differs both in intensity and kind, depending upon the amount and type of reflection of light that take place at their surfaces.

There are several kinds of lustre, among them being the following:

Metallic, the ordinary lustre of metals. When feebly displayed this lustre is termed *submetallic*. Gold, iron pyrites and galena have a metallic lustre; chromite and cuprite have a submetallic lustre.

Vitreous, the lustre of broken glass. When less well developed, it is called *subvitreous lustre*. Quartz and rock-salt afford examples of vitreous lustre, calcite of subvitreous.

Resinous, the lustre of resin. Opal, amber and some kinds of zinc blende have a resinous lustre.

Pearly, the lustre of a pearl. It is shown by surfaces parallel to which the mineral is more or less separated into thin plates, reproducing to some extent the conditions of a pile of thin glass sheets, such as cover-glasses. Talc, brucite and selenite show pearly lustre.

Silky, the lustre of silk. This lustre is peculiar to minerals having a fibrous structure. The fibrous form of gypsum known as satin-spar, and the variety of asbestos called amianthus are good examples of minerals having a silky lustre.

Adamantine, the lustre of a diamond.

The lustre of minerals may be of different *degrees* of intensity, according to the amount of light reflected from their surfaces. Thus, when the surface of a mineral is sufficiently brilliant to reflect objects distinctly, as a mirror would do, it is said to be *splendent*. Certain varieties of hematite have a splendent lustre. When the surface is less brilliant and objects are reflected indistinctly, it is described as *shining*. When the surface is still less brilliant and is incapable of giving any image, it is termed *glistening,* and *glimmering* denotes a still more feeble lustre. Minerals with no lustre are described as *dull*.

As shown later, the various surfaces of a crystal may show different kinds and degrees of lustre.

Transparency and Translucency. A mineral is *transparent* when the outlines of objects seen through it appear sharp and distinct. Rock crystal—a variety of quartz—and selenite are good examples. Minerals are said to be *subtransparent* or *semitransparent* when objects seen through them appear indistinct. A mineral which, though capable of transmitting light, cannot be seen through is *translucent*. This condition is very common among minerals. When no light is transmitted the mineral is *opaque*, but it must be noted that this refers only to the appearance as usually seen. A large number of apparently opaque minerals become translucent when cut into very thin sections, and this property is of great importance, as shown in a later chapter, in the identification of minerals in rocks. Many minerals which are opaque in the mass are translucent on the sharply broken edges and in splinters, as in the case of the common black flint from the Chalk of the south of England.

Phosphorescence and Fluorescence. Phosphorescence is the property possessed by some substances of emitting light *after* having been subjected to certain conditions such as heating, rubbing, or exposure to electric radiation or to ultra-violet light. Some varieties of fluorspar, when powdered and heated on an iron plate, display a bright phosphorescence. Pieces of quartz when rubbed together in a dark room emit a phosphorescent light. Exposure to sunlight or even ordinary diffused light elicits a phosphorescence from many minerals, as may be observed by transferring them rapidly to a dark room. Diamond, ruby and certain other minerals show a brilliant phosphorescence after exposure to X-rays. Willemite, the zinc silicate from certain localities, phosphoresces when exposed to X-rays, a fact employed to make certain that this mineral has been completely extracted from its ore.

Some minerals emit light *whilst* exposed to certain electrical radiations. This phenomenon is best exhibited by fluorspar and for this reason is called *fluorescence*.

II. TASTE, ODOUR AND FEEL

Taste. The characters of minerals dependent upon taste are only perceptible when the minerals are soluble in water. The following are terms used in this connexion:—*saline*, the taste of common salt; *alkaline*, that of potash and soda; *cooling*, that of nitre or potassium chlorate; *astringent*, that of green vitriol; *sweetish astringent*, that of alum; *bitter*, that of Epsom salts, and *sour*, that of sulphuric acid.

Odour. Some minerals have characteristic odours when struck, rubbed, breathed upon or heated. Terms used are:—

Alliaceous—the odour of garlic, given when arsenic compounds are heated.

Horse-radish odour—the odour of decaying horse-radish, given when selenium compounds are heated.

Sulphurous—the odour of burning sulphur, given off by pyrites when struck, or by many sulphides when heated.

Foetid—the odour of rotten eggs, given by heating or rubbing certain varieties of quartz or limestone.

Argillaceous or Clayey—the odour of clay when breathed upon.

Feel. *Smooth, greasy* or *unctuous, harsh,* or *meagre* or *rough,* are kinds of feel of minerals that may aid in their identification. Certain minerals *adhere to the tongue.*

III. STATE OF AGGREGATION

Gases and Liquids. Oxygen, nitrogen and carbon dioxide are examples of natural gases; and water, mercury and petroleum are examples of natural liquids.

Solids. With the exception of mercury and the natural mineral oils, all the minerals with which we have to deal are found in the solid state, and the properties dependent on their state of aggregation are now considered.

Form. Under favourable circumstances minerals assume the definite geometrical forms of *crystals,* the recognition of which is a valuable aid in their identification. Crystallography or the study of crystals is dealt with in the next chapter. The following general descriptive terms are associated with the crystal characters of minerals:—

Crystallized—a term denoting that the mineral occurs as well-developed crystals. Most of the beautiful specimens in museums are of crystallized minerals.

Crystalline—a term denoting that no definite crystals are developed, but a confused aggregate of imperfectly formed crystal grains that have interfered with one another during their growth.

Cryptocrystalline—a general term to denote the possession of mere traces of crystalline structure. *Amorphous* is used to describe the complete absence of crystalline structure, a condition found in the natural glasses but rare in minerals.

Minerals assume various indeterminable forms that are not necessarily dependent on crystal character. These forms are described by the following terms, which have their customary meanings:—

Acicular—in fine needle-like crystals, as in natrolite.

Amygdaloidal—almond-shaped, as with the minerals known as zeolites which occupy the almond-shaped steam cavities of lavas.

Bladed—in forms shaped like a knife-blade or a lath, a form commonly exhibited by many museum specimens of kyanite.

Botryoidal—consisting of spheroidal aggregations, somewhat resembling a bunch of grapes, as with chalcedony.

Capillary—exhibiting a fine hair-like form as in millerite, nickel sulphide, whence the name capillary pyrites or hair pyrites for such varieties of this mineral.

Columnar—showing a form resembling slender columns, as in hornblende.

Concretionary and nodular—terms applied to minerals which are found in detached masses, the forms being spherical, ellipsoidal or irregular, as in the flint nodules found in the Chalk of the south of England.

Dendritic and arborescent—tree-like or moss-like forms, usually produced by the deposition of the mineral in very narrow planes or crevices, as with the dendrites of manganese oxide.

Fibrous—consisting of fine thread-like strands, as exhibited by the variety of gypsum called satin-spar, and by asbestos.

Foliated or, better, *foliaceous*—consisting of thin and separable lamellae or leaves, as with mica and other *micaceous* minerals.

Granular—in grains, either coarse or fine. Evenly granular aggregates of minerals, such as in marble, are often termed *saccharoidal* from their resemblance to lump sugar.

Lamellar—consisting of separable plates or leaves, as with wollastonite.

Lenticular—with the form of flattened balls or pellets, shown by many concretionary and nodular minerals.

Mammillated—displaying large mutually interfering spheroidal surfaces, as in malachite.

Radiating or divergent—showing crystals or fibres arranged around a central point, as in stibnite and in many cases of concretionary forms.

Reniform—kidney-shaped, the rounded surfaces of the mineral resembling those of kidneys and shown in perfection by the variety of hematite called kidney iron-ore.

Reticulated—in the form of cross-meshes like a net, as with the rutile needles found in some micas.

Scaly—in small plates as with tridymite.

Stellate—showing fibres radiating from a centre to produce star-like forms, as with wavellite.

Tabular—showing broad flat surfaces, as with wollastonite or tabular spar.

Tuberose—showing very irregular rounded surfaces often giving rise to gnarled, rootlike shapes as in the variety of aragonite called flos-ferri.

Wiry or filiform—in thin wires often twisted like the strands of a rope, and shown well by native silver and copper.

Pseudomorphism. Pseudomorphism is the assumption by a mineral of a form other than that which really belongs to it. Pseudomorphs may be formed in several ways:—

(1) A pseudomorph by *investment* or *incrustation* is produced by the deposition of a coating of one mineral on the crystals of another, for example, quartz on fluorspar.

(2) A pseudomorph by *infiltration* is formed when the cavity previously occupied by a certain crystal is refilled by the deposition in it of different mineral matter by the infiltration of a solution.

(3) A pseudomorph by *replacement* arises by the slow and gradual substitution of particles of new and different mineral matter

for the original particles which are successively removed by water or other solvents. This kind of pseudomorphism differs from the preceding in the circumstance that the new tenant enters before the old tenant has entirely evacuated its quarters.

(4) A pseudomorph by *alteration* is due to gradual chemical change which crystals sometimes undergo, their composition becoming so altered that they are no longer the same minerals, although they still retain the old forms. As an example may be instanced the common alteration of olivine to serpentine.

Pseudomorphs may often be recognised by a want of sharpness in the edges of the crystals, whilst their surfaces usually present a dull and somewhat granular or earthy aspect.

Polymorphism. It has already been mentioned that two minerals of markedly different physical properties, such as colour, hardness, crystal form, specific gravity, etc., may have identical chemical compositions. Such substances are said to be *dimorphous* and illustrate the general property of *polymorphism*. The minerals making up a polymorphous series are composed of the same atoms but have them arranged on different plans so that their physical properties differ.

As an example of dimorphism we may take the two forms of calcium carbonate occurring as the minerals calcite and aragonite. These two minerals form crystals of quite different types, their optical properties are different, and aragonite is harder and has a higher specific gravity than calcite. Again, the physically very dissimilar diamond and graphite are dimorphous forms of carbon; as illustrated in Fig. 6, p. 19, the arrangement of the atoms is radically different. In nature titanium dioxide, TiO_2, occurs in three forms or is *trimorphous*. The mineral anatase has a specific gravity of 3·9, brookite of 4·15, and rutile of 4·25, and their other physical characters are dissimilar, but in chemical composition they are all titanium dioxide. It is probable that the temperature, pressure, concentration, etc., operative at the time of formation of the mineral control what variety shall be produced.

Hardness. Hardness varies very greatly in minerals. Its determination is one of the most important tests used in the identification of minerals and may be made in several ways.

Hardness may be tested by rubbing the specimen over a tolerably fine-cut file and noting the amount of powder and the degree of noise produced in the operation. The less the powder and the greater the noise, the harder is the mineral. A soft mineral yields much powder and little noise. The noise and the amount of powder are compared with those produced by the minerals of the set used as standard examples for hardness tests. The scale in general use, and known by the name of *Mohs' Scale of Hardness,* is given below. The intervals on this scale are about equal except for that between corundum and diamond which is estimated to be thirty or more times as great.

MOHS' SCALE OF HARDNESS

Hardness	Standard Mineral
1	Talc
2	Rock-salt, or gypsum
3	Calcite
4	Fluorspar
5	Apatite
6	Orthoclase feldspar
7	Quartz
8	Topaz
9	Corundum
10	Diamond

Window-glass may be used in an emergency as a substitute for apatite, and flint for quartz.

The hardness test may also be made by endeavouring to scratch the specimens enumerated in the list with the mineral under examination. If, for example, the mineral scratches orthoclase feldspar but does not scratch quartz, it has a hardness between 6 and 7. A greater precision is sometimes attempted by giving the hardness as $6\frac{1}{4}$, $6\frac{1}{2}$, $6\frac{3}{4}$, according to whether the mineral in question approaches more nearly to feldspar or quartz in hardness.

Hardness may also be tested by means of a penknife or even the finger-nail, the former scratching up to about $6\frac{1}{2}$, the latter up to $2\frac{1}{2}$. Finger-nails, however, vary in hardness.

Several precautions are to be observed in testing hardness. A definite scratch must be produced in the softer mineral and this is best seen by blowing away (or licking away, if the observer cares to) the powder produced by scratching and then examining the place with a lens. A softer mineral drawn across a harder mineral often produces a whitish stripe which may be mistaken for a scratch in the harder mineral; in the same way an attempt to scratch harder minerals with the knife produces a steel mark on them. Granular specimens may give a kind of scratch by the breaking out of the mineral grains. Finally, it is of course necessary that a fresh surface, that is, one not coated with decomposition products or the like, of the mineral is subjected to the hardness test. During the hardness trial, the colour of the powder produced by the scratch is observed, this giving the *streak* of the mineral.

Hardness like other physical properties depends on the atomic structure of the mineral. It increases with the density of packing in the structure, with the valency of the ions and with the decrease in ionic size. The structural control results in hardness varying in different directions in the crystal. This difference is usually very small but in the mineral kyanite hardness varies between 7 and 5—an old name for kyanite is disthene, from the Greek for 'two strengths'.

Tenacity. Minerals possess certain properties dependent upon their tenacity, of which the following are the most important:—

(a) *Sectility*. A mineral is said to be sectile when it can be cut with a knife and the resulting slice breaks up under a hammer. Examples:—graphite, steatite, gypsum.

(b) *Malleability*. A mineral is malleable if a slice cut from it flattens out under a hammer. Examples:—native gold, silver and copper.

(c) *Flexibility* is the property of bending. In some minerals it can be observed by experimenting with thin plates or laminae only. A flexible mineral remains bent after the pressure is removed. Examples:—talc, selenite, etc.

(d) *Elasticity*, as the term is usually employed in mineralogy differs from flexibility in the fact that the portion bent springs back to its former position. Mica yields flexible elastic plates, whilst the somewhat similar mineral, chlorite, gives plates that are flexible but not elastic.

(e) *Brittleness* is a character common to many minerals and is shown by their crumbling or flying to powder instead of yielding a slice. Examples:—iron pyrites, apatite and fluorspar.

Fracture. It is very important to note the characters of the *fractures* displayed on the broken or chipped surfaces of minerals. It is equally important to distinguish between the smooth flat surfaces resulting from what is called the cleavage of a mineral,

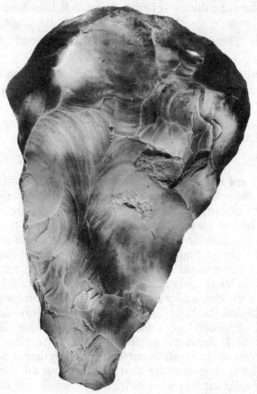

FIG. 13. Conchoidal Fracture shown by a flint implement.

and the irregular surfaces characterizing true fracture, these latter being totally independent of cleavage. Whilst the fracture is an important diagnostic character and, further, a recent fracture reveals the true colour of certain minerals, it is unwise to break or chip good crystals, as crystalline form is a far more valuable and constant a character by which to determine a mineral than its colour and, in many cases, than its fracture.

Fracture is said to be:—

(1) *Conchoidal.* The mineral breaks with a curved concave or convex fracture. This often shows concentric and gradually diminishing undulations towards the point of percussion, somewhat resembling the lines of growth on a shell. Conchoidal fracture is well shown by quartz, flint (Fig. 13) and natural glasses.

(2) *Even.* The fracture-surface is flattish or nearly flat, as in chert.

(3) *Uneven.* The fracture-surface is rough by reason of minute elevations and depressions. Most minerals have an uneven fracture.

(4) *Hackly.* The surface is studded with sharp and jagged elevations, as in cast-iron when broken.

(5) *Earthy.* As in the fracture of chalk, meerschaum, etc.

Cleavage and Parting. The tendency to split along certain definite planes—the *cleavage-planes*—possessed by many minerals is closely related to crystalline form and the internal structure of the crystal. In each cleavable mineral, the directions of the cleavage-planes are parallel to a certain face or to certain faces of a form in which the mineral may crystallize. In the plane of cleavage the atoms of the mineral are more closely packed together or the mutual electrical charges are greater than in directions at right angles to the cleavage-plane. This plane, therefore, is a plane of least cohesion and hence splitting or cleavage easily occurs along it. It is important, as already stated, to distinguish between fracture and cleavage, as the former is irregular and not connected with the crystalline structure of the mineral. Substances with no crystalline structure, that is, amorphous substances, show no cleavage. Certain rocks, such as slate, which split readily into thin sheets are said to be cleaved, but this property of slaty

FIG. 14. Cleavage: above galena, below mica.

cleavage, as it is best called, is the result of recrystallization under pressure and has no connexion with the cleavage which exists in minerals.

Minerals may show several cleavages but one is generally to be obtained with greater ease than the others. Cleavage is described by stating the crystallographic direction followed by the cleavage-planes and the degree of perfection shown by such planes. With regard to the latter, cleavage is described as perfect or eminent, good, distinct, poor, indistinct, difficult, etc. As examples of minerals with perfect cleavage, we may give fluorspar, galena, calcite, and mica. Fluorspar commonly crystallizes in cubes; if such a cube is taken and tapped with a hammer it will be found to cleave along planes truncating the corners of the cube, and if this cleaving is done in a regular way an octahedron is produced. Fluorspar is said, therefore, to have a perfect octahedral cleavage, and to give octahedra as its *cleavage-fragments*. Galena, which also crystallizes in cubes, cleaves parallel to the faces of the cube, so that its cleavage is cubic and its cleavage-fragments are cubes. Calcite, no matter what shapes its crystals are, produces rhombohedral cleavage-fragments on being crushed.

Graphite and mica illustrate how cleavage is controlled by crystal-structure. As described on p. 19 graphite is built up of sheets of carbon atoms that are tied to one another by strong covalent bonds, the sheets being loosely held together by weak residual forces. The perfect cleavage of graphite is parallel to the sheets. Writing with a 'lead' pencil is possible because of the rubbing-off of the cleavage-flakes of the graphite of which the 'lead' is made. In mica, the cleavage takes place along layers of potassium ions that have a much weaker bond than have the sheets of other ions that they separate (see Fig. 14 p. 61). The uses of mica in industry depend largely on the perfect mica cleavage Cleavage is a very important property in the recognition of minerals, both in the hand-specimen and, as is shown later, under the microscope.

Gliding-planes and *secondary twinning* are related to cleavage, and are produced in a mineral by pressure. For example, during the preparation of a thin slice of calcite for examination under the miscroscope, the pressure of grinding the mineral may cause it to show an excellent cleavage and some secondary twinning.

Twinning is discussed in later pages. The secondary twin-planes and the gliding-planes are often planes along which the mineral separates fairly readily—such planes are called *partings*.

Surface Tension Effects. The difference in adhesive power of various liquids to different minerals has formed the basis for numerous processes of ore separation and concentration. The surface tension between various metallic sulphides and a selected liquid is greater than that between the gangue minerals quartz, calcite, etc., and the same liquid. In the original Elmore Process a paste of sulphide and gangue was mixed with oil and water and agitated; the oil separated into a layer above the water and carried the sulphides with it. Somewhat the same principle underlies the method of extracting diamonds from their matrix, blue-ground, by causing them to adhere to grease upon shaking tables. The various Flotation Processes depend on surface tension. In these, bubbles of gas or air attach themselves to, say, fine-powdered zinc blende, agitated in a liquid containing oil or other suitable organic compounds, and float this mineral to the surface, leaving other sulphides and gangue material at the bottom of the liquid. By varying the conditions of flotation clean separations of various ore-minerals can be produced and in this way the working of mixed ores has been made economically possible.

IV SPECIFIC GRAVITY

Specific Gravity. The specific gravity of a body is the ratio of the weight of the body to that of an equal volume of water. This latter weight varies with the temperature, and this variation has to be considered in exact work. In the general practice of determinative mineralogy, however, this correction can be neglected. In selecting material for the determination of specific gravity it is necessary to obtain as pure a sample as possible and one free from alteration products, inclusions and the like.

The specific gravity of minerals depends on the atomic weight of the constituent elements and the way their atoms are packed in the crystal-structure. The first of these controls is illustrated by three mineral sulphates which have the same type of crystal-structure:—

Mineral	Formula	Atomic Weight of Cation	Specific Gravity
Celestine	$SrSO_4$	Sr, 87.63	2.9
Barytes	$BaSO_4$	Ba, 137.36	4.5
Anglesite	$PbSO_4$	Pb, 207.21	6.3

The influence of the style of packing is well shown by the two carbon minerals; diamond with close packing has a specific gravity of 2.54, graphite with loose packing 2.3.

Specific gravity is of great importance in the determination of minerals and the student should make himself familiar with the relative weights of pieces of approximately equal size of the common minerals. With the expected exception of minerals containing heavy metals such as Pb, Ba, etc., the 'sparry', non-metallic-looking minerals have specific gravity about 2·6–2·8, the metallic-looking minerals about 5.

The cardinal principle employed in most determinations of specific gravity is that the loss in weight of a body immersed in water is the weight of a volume of water equal to that of the body. If W_a is the weight of the body in air, W_w its weight in water, then $W_a - W_w$ is the weight of the water displaced by the body and the specific gravity of this is $\dfrac{W_a}{W_a - W_w}$

Methods of Determining Specific Gravity. The following are the chief methods of determining specific gravities in mineralogy, the particular method chosen depending usually upon the size and character of the specimen under examination.

(1) With the ordinary chemical balance, for fragments of a solid mineral about as big as a walnut.

(2) With Walker's steelyard, for large specimens.

(3) With Jolly's spring balance, for very small specimens.

(4) By measuring the displaced water, for the rapid determination of the approximate specific gravity of a number of specimens of a mineral.

(5) With the pycnometer or specific gravity bottle, for friable minerals, small fragments or liquids.

(6) With heavy liquids, used mainly for the separation of mineral mixtures into their pure components according to their specific gravities, but also for approximate determinations of specific gravity of mineral grains. For this latter determination, the diffusion column and Westphal Balance may be employed.

(1) *Determination of Specific Gravity with the Chemical Balance.*
The mineral is weighed on a good chemical balance. It is then suspended by thread or very fine wire from one arm of the balance and immersed in water contained in a beaker standing on a wooden bridge placed over the scale-pan. Bubbles of air sticking to the mineral are removed by a small brush, and the weight of the mineral immersed in water obtained. The specific gravity of the mineral is given by dividing its weight in air by the difference between its weights in air and water.

(2) *Walker's Steelyard.* This instrument is useful for determining the specific gravity of large specimens, and is shown in Fig. 15. The essential part of the apparatus is the long graduated beam which is pivoted near one end and counterbalanced by a heavy weight suspended to the short arm. The specimen is suspended and moved along the beam until it counterbalances the constant weight,

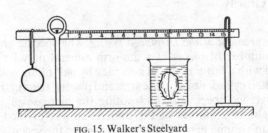

FIG. 15. Walker's Steelyard

the level position of the beam being observed by a mark on the up-right shown on the right of the figure. The reading (a) is taken. The specimen is then immersed in water and moved along the beam until the constant weight is again balanced and a second reading (b) is obtained. The readings (a) and (b) are inversely proportional to the weights of the body in air and in water respectively. Hence

$$\text{Specific Gravity} = \frac{\dfrac{1}{a}}{\dfrac{1}{a}-\dfrac{1}{b}} = \frac{b}{b-a}$$

whence the specific gravity is given by dividing the second reading by the difference between the second and first readings.

(3) *Jolly's Spring Balance*. This instrument consists of a spring suspended vertically against a graduated scale and is illustrated in Fig. 16. To the lower end of the spring are attached two scale-pans, one below the other, the lower scale-pan being always immersed in water. The reading (a) of the bottom of the spring on the scale is obtained. A small fragment of the mineral is placed in the upper pan, and a second reading (b) taken. The specimen is then transferred to the lower pan, and a third reading (c) taken.

Then (b—a) is proportional to the weight of the mineral in air, and (b—c) to the loss of weight in water, so that—

$$\text{Specific Gravity} = \frac{b-a}{b-c}$$

(4) *Measurement of the Displaced Water*. The specific gravity of a large number of pieces of a uniform mineral may be rapidly obtained with a fair amount of accuracy by half filling with water a graduated cylinder of suitable size, and placing therein the previously weighed specimens, and noting the increase of volume. The weight in grammes of the mineral in air, divided by the increase in volume in cubic centimetres, gives the specific gravity of the mineral.

(5) *The Pycnometer or Specific Gravity Bottle*. The pycnometer is used to obtain the specific gravity of liquids or of small fragments of minerals, gems, or porous or friable material. It is a small glass bottle fitted with a stopper through which is a fine opening. When filled up to a certain mark or to the top of the stopper, the bottle contains a known volume of liquid, so that by weighing

the bottle empty and then filled with liquid, the specific gravity of the latter can be obtained. If the volume of the bottle is not known, the specific gravity of a liquid may be determined by weighing the bottle empty, then filled with water, and finally filled with the liquid, when it is clear that the specific gravity of the latter is given by dividing the weight of the liquid by that of the water, since their volumes are the same.

In determining the specific gravity of mineral fragments, the mineral is first weighed. The bottle is filled with distilled water. Both the mineral and the filled bottle are placed in the same scale-pan and their combined weight obtained. The mineral is then put into the bottle from which it displaces an equal bulk of water, and the weight again determined. The weight of the water displaced is given by subtracting this last weight from the preceding. The specific gravity is obtained by dividing the weight of the mineral by the weight of the water it displaces.

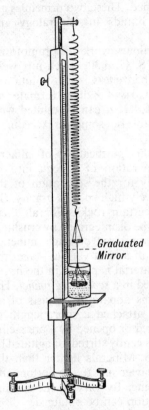

Graduated Mirror

FIG. 16. Jolly's Spring Balance.

(6) *The Use of Heavy Liquids.* If a mixture of two minerals of different specific gravities is placed in a liquid whose specific gravity lies between those of the minerals, the heavier mineral sinks in the liquid and the lighter mineral floats and thus a more or less complete separation of the two minerals can be effected. Further, by varying the specific gravity of a liquid a point can be reached when a given mineral placed in the liquid neither floats nor sinks; the specific gravity of the mineral and that of the liquid

are then the same and by determination of the latter the specific gravity of the mineral can be obtained. These two principles are the basis of the use of heavy liquids in mineralogy and petrology.

The liquids most suitable for ordinary work are bromoform, S. G. 2.89, and methylene iodide, S. G. 3.30, which mix well; they can be diluted with acetone or benzene. Clerici's Solution, a saturated solution in water of equal parts of thallium formate and malonate, has a specific gravity of 4.2; it can be diluted with water. Other heavy liquids are available but some are exceedingly poisonous or difficult to handle.

Heavy liquids are used for the purification of mineral material for analysis, for the separation of a rock into its component minerals and especially for the separation of the small amount of minerals of relatively high specific gravity, the *heavy residues* or *accessories,* in certain rocks. For all these purposes, the mineral or rock must be disintegrated by crushing, use of acids, etc., until particles composed of single minerals alone are present. Dust is washed off and at various stages the material is sieved. The prepared material is placed in the heavy liquid, usually bromoform, contained in a *separating funnel*. The simplest form, and the best, of this apparatus consists of an ordinary filter funnel to which is attached a short length of rubber tubing capable of being closed or opened by a press-clip. The mixture of material and liquid is gently stirred, or agitated by pressing the tubing above the clip. Minerals lighter than the liquid float to the top, and those heavier sink to the bottom and can be drawn off through the tubing. By varying the specific gravity of the liquid, a pure separation can be obtained.

In the determination of the specific gravity of a mineral by heavy liquids various methods are used. In the first, the heavy liquid is diluted until the mineral neither sinks nor rises in the liquid but remains suspended. The specific gravity of the liquid, and therefore of the mineral, is determined by means of the pycnometer (if there is a large amount of the liquid) or by using the *Westphal Balance*. In this, a sinker is immersed in the liquid and balanced by riders on a graduated arm. The arm is usually so graduated that the specific gravity of the liquid can be read off directly.

For testing the specific gravity of small samples the *diffusion column* is used. Two perfectly mixable liquids of different specific gravities are placed in a graduated tube without mixture, and allowed to stand for a day or more until regular diffusion of the two liquids has taken place. Thus is formed a column of liquid in which the specific gravity varies regularly from top to bottom. Small fragments of known specific gravity are placed in the liquid and, coming to rest at particular points in the column, serve as indices. A small quantity of the finely powdered sample is introduced, and its several constituents separate into bands with different specific gravities which can be told by their position with regard to the indices. The Berman torsion balance is available for obtaining the specific gravity of very small mineral particles.

V. CHARACTERS DEPENDENT UPON HEAT

Fusibility. The relative fusibility of certain minerals is a useful character as an aid in their determination by the blowpipe. A scale of six minerals, of which the temperature of fusion was supposed to increase by somewhat equal steps, was suggested by Von Kobell. These minerals are stibnite, natrolite, almandine garnet, actinolite, orthoclase, bronzite. All that can be said of this scale is that stibnite is easily fusible, whilst bronzite can hardly be fused in the ordinary blowpipe. The approximate melting points of the minerals of Von Kobell's scale are: stibnite 525°C, natrolite 965°, almandine garnet 1050°, actinolite 1200°, orthoclase 1300°, and bronzite 1400°.

VI. CHARACTERS DEPENDENT UPON MAGNETISM, ELECTRICITY AND RADIOACTIVITY

Magnetism. Magnetite, and in a less degree pyrrhotite, are the only minerals affected by an ordinary bar magnet, but a large number of minerals are attracted in varying degrees by the electromagnet. Minerals containing iron are generally magnetic, but not necessarily so, and the degree of magnetism displayed does not, in all cases, depend on the iron content. Minerals containing no iron may also be sufficiently magnetic to permit of their separation from non-magnetic materials, for example, monazite and some other cerium-bearing minerals. The electro-magnetic separation of minerals is an important ore-dressing process. By varying the strength of the electromagnet, minerals of varying magnetism can be separated from one another. Examples of such separations are the purification of magnetite from apatite, etc., the separation of pyrites from blende, siderite from blende, wolfram from tinstone, and monazite from magnetite and garnet. It is sometimes necessary to roast the ore in order to convert feebly magnetic materials, such as pyrites and siderite, into strongly magnetic material. A small electromagnet is used in the laboratory to separate the heavy residues obtained by the use of heavy liquids into magnetic and non-magnetic portions.

Highly magnetic. Magnetite, pyrrhotite ('magnetic pyrites').
Moderately magnetic. Siderite, iron-garnet, chromite, ilmen-ite, hematite, wolfram.
Weakly magnetic. Tourmaline, spinels, monazite.
Non-magnetic. Quartz, calcite, feldspar, corundum, cassiterite, blende.

Bodies of magnetite or pyrrhotite, the latter often accompany-ing nickel minerals, may be located by mapping the variations of the magnetic field of an area. This is done by a magnetometer, a glorified dip-needle, which for rapid results is often airborne. When certain rocks crystallize, the magnetic particles in them become orientated in the earth's magnetic field existing at that time and place. The direction of this 'fossil' magnetic field can be determined on samples of the rocks in the laboratory and this study, *palaeomagnetism,* is of great geological interest in the discussion of possible large-scale movements of the continents.

Electricity. Minerals vary in their capacity for conducting electricity. Good conductors are relatively few, mostly those with metallic lustre such as native metals and sulphides, with the notable exception of zinc blende. Good conductors can be separated in the laboratory from bad by inducing an electro-static charge on a glass rod by rubbing it with silk—the good conductors are attracted. The variation in conductivities is applied on a larger scale by dropping a finely crushed and dried ore on to a rotating iron cylinder which is electrically charged. Good conductors become charged and are repelled from the cylinder, bad conductors are repelled to a less degree, and hence the shower of ore is separated out into several minor showers which can be separately collected. For example, blende, a bad conductor, is separated from pyrites, a good conductor, in this way.

Certain minerals develop an electric charge when subjected to a temperature change, *pyroelectric minerals*, or to stress, *piezoelectric minerals*. Tourmaline is an example of a pyroelectric mineral; when heated a tourmaline crystal becomes negatively charged at its sharp end and positively charged at the blunt end (see p.132 for a figure of a tourmaline crystal). The most important piezoelectric mineral is quartz, suitably orientated thin plates of which respond to exceedingly small variations in directed pressure. Millions of such plates have been used for frequency control in radio and telephone systems.

Radioactivity. Many minerals containing elements of high atomic weight are radioactive and emit emanations or radiations which affect a photographic plate. The chief radioactive elements are radium itself, uranium and thorium, but certain isotopes of potassium and rubidium are of much greater importance in geology because, though their radiation is feeble, they occur in widespread and common minerals. Radioactive minerals are detected in the field by use of the Geiger Counter and the scintillometer.

Radioactive isotopes are subject to radioactive decay by which the unstable *parent elements* break down into stable *daughter elements*. These pairs of elements are given in the Table below in which the atomic weight of the isotope concerned is given above the symbol of the element and the atomic number below it:—

PARENT			DAUGHTER	
URANIUM,	235	$\longrightarrow$	LEAD,	207
	U			Pb
	92			82
URANIUM,	238	$\longrightarrow$	LEAD,	206
	U			Pb
	92			82
THORIUM,	232	$\longrightarrow$	LEAD,	208
	Th			Pb
	90			82
POTASSIUM,	40	$\longrightarrow$	ARGON,	40
	K			A
	19			18
RUBIDIUM,	87	$\longrightarrow$	STRONTIUM,	87
	Rb			Sr
	37			38

For each isotope the decay takes place at a determined and constant rate since it is a nuclear process. When the radioactive mineral is formed, the radiometric clock is started and, by spectrometry, the amount of the decay product can be measured and hence the age of the mineral obtained. The oldest mineral radiometrically dated is approximately 3,400 million years old.

CHAPTER III

THE ELEMENTS OF CRYSTALLOGRAPHY

INTRODUCTION

It was noticed by the ancient Greeks that a certain mineral, quartz, usually occurred in forms having a characteristic shape, being bounded by flat faces. From the transparency of this mineral and the occurrence in it of included material, it was thought quartz resulted from the freezing of water under intense cold, and hence the name *krustallos*—meaning *clear ice*—was given to the substance. There were, however, numerous other minerals known to the ancients which occurred in forms bounded by flat faces, and so, by a natural extension of the term, *krustallos* came to signify any mineral showing such forms. These forms are *crystals* and their study *crystallography*.

Crystals are bodies bounded by surfaces, usually flat, arranged on a definite plan which is an expression of the internal arrangements of the atoms. They are formed by the solidification of minerals from the gaseous or liquid states or from solutions,—a process known as *crystallization*.

From the definition of a crystal just given we see that the internal atomic structure is their fundamental property as illustrated by the structures of diamond and graphite already described and figured on p. 19. Though we could construct a model of a crystal in glass or some other amorphous material, such a model would not be a crystal since it would lack the essential atomic structure. In this book, however, we are chiefly concerned with the determination of minerals, so that for us the external form or *morphology* of crystals demands most attention. It is necessary, however, first to give a general account of *crystal-structure,* leaving the more detailed accounts of the structure of selected minerals, chief of which are the silicates, till their description in Part II of this book.

CRYSTAL-STRUCTURE

X-rays and Crystal-structure. The French mineralogist Haüy suggested in 1782 that crystals were built up of minute cleavage-fragments of the mineral, different modes of arrangements of the fragments producing different crystal forms. Later, Haüy's concept of solid cleavage-fragments was replaced by that of atoms or groups of atoms having a regular arrangement throughout a crystal. Within this arrangement a three dimensional pattern is repeated very many times, and the array of points at which the pattern repeats constitutes a *space-lattice* (Bravais, 1848). The unit of pattern is now called the *unit cell*. Although the geometrical aspects of space-lattices were extensively studied and speculations made relating internal atomic arrangements to the external crystal forms, it was not until 1912 that the true nature of crystal-structure was demonstrated when Laue and others used crystals as three-dimensional diffraction gratings for X-rays. The first structural analysis was made in 1913 by the British physicists, Sir W. H. Bragg and his son Sir Lawrence Bragg, on crystals of sodium chloride.

X-rays are somewhat like light waves but have much smaller wave-lengths (see p. 157), these being comparable to the distance between atoms in a crystalline solid. When a beam of X-rays falls on a crystal, it is scattered or diffracted by the parallel layers of atoms within the crystal, in the same way that light waves are diffracted by an optical grating. In the Laue experiments, the diffracted X-rays were allowed to fall on a photographic plate and the resulting photograph showed a series of spots which formed a symmetrical pattern (Fig. 17C). The interpretation of the Laue pattern is a complicated process and more direct methods of structure analysis are now available. Nevertheless, Laue photographs help in certain instances to determine the symmetry (p. 83) of the crystal under examination.

In the *rotation method,* a tiny crystal is rotated or oscillated about a vertical axis in a horizontal beam of monochromatic X-rays. As the crystal rotates, diffractions take place as successive layers of atoms are brought into suitable positions, and are recorded on a cylindrical film (Fig. 17B). From measurements made on such photographs, the arrangements of planes of atoms in the crystal and the distances between them can be deduced.

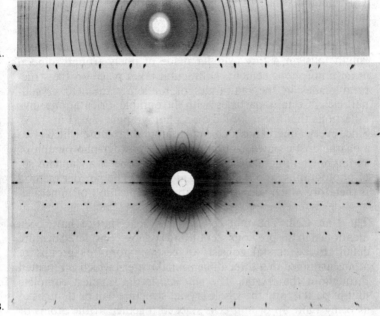

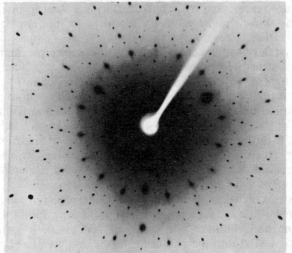

Fig. 17. Structural analysis of minerals; quartz structure revealed by three methods.
A. Powder diagram
B. Rotation diagram, crystal rotated about *c*-axis.
C. Laue diagram, looking down *c*-axis.

This method is useful in the determination of the size of the unit cell. Distances in the crystal-structure are expressed in Ångstrom units; one Ångstrom (Å) equals 10^{-8} cm.

In the *powder method* a beam of X-rays of one wavelength falls on a tiny cylinder made of finely-ground mineral in an amorphous cement. Diffraction takes place on the structural planes in the multitudes of randomly oriented crystalparticles. Certain particles with favourable orientations give reflections that are recorded on a photographic film (Fig. 17A). Since every mineral has a particular atomic structure which gives a characteristic powder-photograph, the powder-photograph of an unknown mineral can be matched with one of a standard set of photographs, and the mineral identified. The powder method is therefore much used for the rapid identification of minerals.

The Unit Cell. The unit cell and the space-lattice have been mentioned in the previous section and are now considered in more detail. Every crystal consists of certain atoms or groups of atoms arranged in a three-dimensional pattern which is repeated throughout the crystal. The *unit cell* is the smallest complete unit of pattern and the whole crystal-structure can be thought of as being built up of unit cells stacked together. If the atoms or ions are represented by points their arrangement in the crystal can be shown by the geometrical pattern or framework of the *space-lattice* or *point-system*.

Crystals of sodium chloride, NaCl, afford a simple example of crystal-structure and were, as already noted, the first to be investigated. As described on p. 18 and illustrated in Fig. 4, the Na^{1+} cations are bonded by ionic bonds to Cl^{1-} anions, in the formation of sodium chloride crystals. The Na and Cl ions are sited at the corners of a series of cubes and eight such cubes are stacked in the manner shown in Fig. 18A. This group of eight cubes is the *smallest unit* of structural pattern which by repetition can build up a crystal of sodium chloride and it is therefore the *unit cell* of the mineral.

Instead of points, ions of the unit cell can be represented by spheres whose radii are proportional to the ionic radii (p. 24) of sodium and chlorine. The radius of Na^{1+} is 0·97 Å, that of Cl^{1-} is 1·81 Å and the unit cell can be considered to be packed

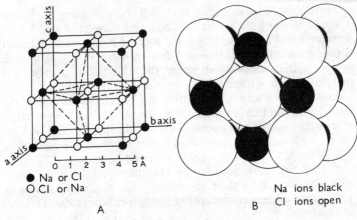

● Na or Cl
○ Cl or Na

Na ions black
Cl ions open

A

B

FIG. 18. Unit Cell of NaCl.

with spheres of these two sizes in appropriate orderly arrangement as shown in Fig. 18B. The edge of the unit cell of sodium chloride has been determined to be 5.6402 Å.

Some further aspects of the unit cell of sodium chloride of Fig. 18 are important as they illustrate some general points. First it should be noted that the central Cl^{1-} or Na^{1+} ion is in 6-fold *coordination*. (p. 26) with the surrounding Na^{1+} or Cl^{1-} ions. This coordination can be also described as *octohedral,* in that the six neighbours of a central ion form a regular octahedon about it, as indicated in Fig. 18A by the dotted lines. Second, the three edges of the unit cell that meet at a point can be used as *axes of reference* to denote the position of any plane in the crystal-lattice as a whole. These axes, lettered *a, b, c,* in Fig. 18A, are also the *crystallographic axes* that are of fundamental importance in the morphology of crystals, (p. 88).

Types of Unit Cells. There are seven shapes of unit cell (Fig. 34); they are called the cubic, tetragonal, hexagonal, trigonal or rhombohedral, orthorhombic, monoclinic and triclinic cells. The lengths of the edges of a unit cell may be different from one another and the angles between them may differ. The length of the cell-side along the a-axis (Fig. 18) is called a, that along the b-axis b and that along the c-axis, c. The angle between a and b is γ, between b and c is α and between c and a is β. The basic notation defining the unit cell is thus $a \wedge b = \gamma$, $b \wedge c = \alpha$, $c \wedge a = \beta$. In the sodium chloride structure the unit cell is a cube and therefore $a = b = c$ and $\alpha = \beta = \gamma = 90°$.

Every type of unit cell or lattice can be referred to one of six sets of axes and this grouping forms the basis of the classification of crystals into *crystal systems*. This classification is dealt with on pp. 94 – 96 and the types of unit cells involved are figured in Fig. 34 on p. 95 to which reference should be made.

As there are more space-lattices (14) than crystal systems, some unit cells allotted by their geometry to the same crystal system will contain different numbers or arrangements of atoms. This can be illustrated by the three cubic unit cells (Fig. 19)—the simple cube, the body-centred cube and the face-centred cube, the last being exemplified by the sodium chloride cell figured in Fig. 18A. The simplest possible unit cell in each crystal system is described as *primitive*.

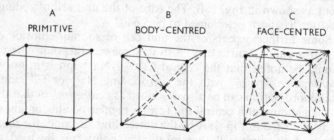

| A | B | C |
| PRIMITIVE | BODY-CENTRED | FACE-CENTRED |

FIG. 19. Cubic Unit Cells.

Crystal Structure and Morphology. In a space-lattice (defined on p. 76), the points which represent the corners of unit cells lie at regular intervals in parallel straight rows. Two sets of such parallel rows intersect to form a *net-plane* and three net-planes intersect to form a *unit parallelepiped*, that is repeated throughout the lattice. The intersecting rows and planes, an example of which is shown in Fig. 18A, are expressed in the *morphology* of crystals, i.e. the external shapes and the arrangements of crystal-faces (see p. 83).

EXTERNAL CHARACTERISTICS OF CRYSTALS

Faces. The plane surfaces or *faces* that are the most obvious feature of crystals are parallel to net-planes in the crystal-structure. The commonest faces are usually parallel to those net-planes containing the greatest number of lattice-points or ions. Wide spacing between the net-planes leads to the preferential development of faces in these planes,—the same factor that controls cleavage (p. 60).

Faces are of two kinds, like and unlike. Some crystals are limited by faces that are all alike. For instance, fluorspar commonly crystallizes in cubes, and any one face of the fluorspar cube is like all the other faces in its properties. Such faces that have the same properties are called *like faces,* whilst faces having different properties are *unlike faces*, the difference arising from the difference in the number and nature of the ions in the net-planes to which the unlike faces are parallel.

Forms. A crystal made up entirely of like faces is termed a *simple form*. For example, the cube and the octahedron are each of them simple forms, since all the faces of each have the same properties. The front face shown in the drawing of a cube in Fig. 20 can be

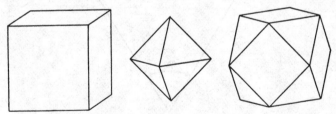

FIG. 20. Simple cube and simple octahedron. A combination of the cube and octahedron as found in crystals of galena.

replaced by any other of the cube faces without altering the drawing. A crystal which consists of two or more simple forms is called a *combination*. In Fig. 20, the cube and the octahedron are shown as simple forms and also as a combination such as occurs in crystals of galena.

Some simple forms occur by themselves in crystals as they can enclose space, but others can only occur in combinations, since they have too few faces to enclose space by themselves. Such latter forms are called *open*.

Edge. An *edge* is formed by the intersection of any two adjacent faces. It is therefore parallel to the *rows* of atoms occurring at the intersections of net-planes. The position in space of an edge depends, of course, upon the positions of the faces whose intersection gives rise to it.

Solid Angles. A *solid angle* is formed by the intersection of three or more faces.

Interfacial Angle. The angle between any two faces of a crystal is termed the *interfacial angle*. In crystallography, the actually

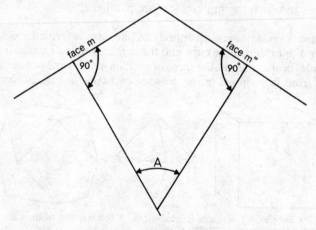

Interfacial Angle = A = 63° 48 ; mm^{III} = 63° 48

FIG. 21. The Interfacial Angle

measured angle is the angle between the normals, or perpendiculars, to the two faces. Thus, in Fig. 21, the interfacial angle between the two faces shown in section is A. Interfacial angles are of great importance in crystallography and are recorded in works of reference in the following way,—if the angle between the normals to two faces which we will call m and m''' is 63° 48′ it is recorded as $mm''' = 63° 48′$.

Measurement of Interfacial Angle. The interfacial angles of crystals are measured by the goniometer (or angle-measurer). Two types of this instrument are used, one termed the contact-goniometer, the other the reflecting goniometer.

The contact-goniometer consists of two straight-edged arms movable on a pivot or screw, and connected by a graduated arc, as shown in Fig. 22. These two arms are brought accurately

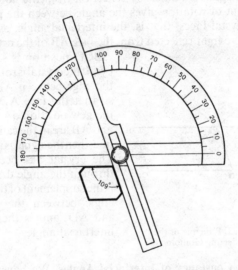

FIG. 22. Contact Goniometer.

into contact with adjacent faces of the crystal, and the angle between them read off on the graduated arc. In the illustration, the angle actually measured is the internal angle between the two faces, and this must be subtracted from 180° to give the interfacial angle used by the crystallographer.

Reflecting goniometers are rather elaborate instruments used with crystals possessing smooth or flawless faces. In general, the smaller the crystal, the more suitable for use with the reflecting goniometer will it be.

A common form of small reflecting goniometer consists of a vertical circle, graduated and capable of rotation, and a horizontal arm fixed at right angles to the plane of the circle. A mirror is fixed on the horizontal arm. The crystal is placed at the centre of the graduated circle with an edge parallel to the horizontal arm. The image of a distant signal is observed by reflection from the mirror, and also by reflection from the crystal face. By rotation of the graduated circle and with it the crystal, the two images are made to lie in the same straight line. The circle is then rotated until an image is obtained by reflection from the adjacent face. The amount of rotation gives the angle between the normals to the two crystal faces, that is, the interfacial angle, as shown in Fig. 23. Here light reflected from the face AB of the crystal in the

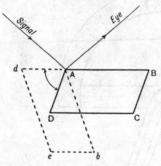

ABCD position is seen by the eye. If the crystal is rotated about the edge between AB and AD so that the face AD takes up the new position dA where dA and AB are in the same straight line, then the signal is again seen. The crystal has been rotated through the angle dAD, which is the supplement of the internal angle between the faces AB and AD, and is therefore the interfacial angle.

FIG. 23. Principle of the Reflecting Goniometer.

Law of the Constancy of Interfacial Angles. We have seen that the atomic structure for the crystals of any one mineral is fixed, so that it follows that the positions of the faces of such crystals are fixed also. This leads to the enunciation of the important law of the *Constancy of Interfacial Angles*. The corresponding interfacial angles are constant for all crystals of a given mineral, provided, of course, that the crystals have identical chemical compositions and that the measurements are made at the same temperatures.

Zones. Inspection of many crystals shows that their faces are so arranged that edges formed by the intersections of certain of the faces are parallel with one another. Such a set of faces constitutes a *zone,* and the line with which the edges are parallel is called the *zone-axis.* For instance, the common crystals of quartz or rock crystal such as are illustrated in Fig. 146, show six faces meeting in parallel edges, and terminated by a set of six usually triangular faces which do not meet in parallel edges; the first set of six faces forms a zone.

Symmetry. Examination of a crystal either with the eye or a goniometer shows that there is a certain regularity of position of like faces, edges, etc. This regularity constitutes the *symmetry* of the crystal and is of course dependent on the type of space-lattice and unit cell involved in its construction. For instance, the arrangement of the ions in the unit cells of sodium chloride (Fig. 18) controls the regularity of the external faces and edges. It has been noted already that structural analysis is also employed in the determination of crystal-symmetry. We are particularly concerned in this book, however, with the external manifestations of symmetry.

The degree of symmetry varies in different minerals and is employed, as elaborated later, in the classification of crystals. It is defined with reference to three *criteria of symmetry:*—

Plane of Symmetry.   Axis of Symmetry.   Centre of Symmetry.

Plane of Symmetry. A plane of symmetry divides a crystal into

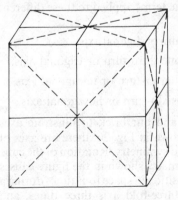

similar and similarly-placed halves. In other words, such a plane divides the crystal so that one half is the mirror-image of the other. Planes of symmetry can be illustrated by considering a cube. A cube has nine planes which divide it into two halves so that one half is the reflection of the other. The traces of these nine planes are indicated on the faces of the cube in Fig. 24 and the dissected planes are shown in Fig. 25.

FIG. 24. Cube, and Planes of Symmetry.

The geometrical symmetry of a matchbox or a brick is obviously lower than that of a cube for, as inspection shows, there are only three planes that divide the object into similar and similarly placed halves.

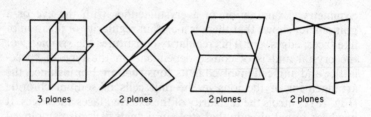

| 3 planes | 2 planes | 2 planes | 2 planes |

FIG. 25. Planes of Symmetry in the Cube

Axis of Symmetry. If a crystal, on being rotated, comes to occupy the same position in space more than once in a complete turn, the axis about which rotation has taken place is called an *axis of symmetry*. Depending upon the degree of symmetry, a crystal may come to occupy the same position two, three, four or six times in a complete rotation. The terms applied to these different classes of axes are as follows:—

Two times: two-fold, **diad**, half-turn or digonal axis.

Three times: three-fold, **triad**, one-third-turn or trigonal axis.

Four times: four-fold, **tetrad,** quarter-turn or tetragonal axis.

Six times: six-fold, **hexad**, one-sixth-turn or hexagonal axis.

We can again use the cube and our brick to illustrate axes of symmetry. In the cube, as shown in Fig. 26, there are axes of four-fold, three-fold and two-fold symmetry. Rotation of the cube about the axis of four-fold symmetry shown in the figure causes the cube to take up the same position in space four times during a complete rotation, about the three-fold axis three times, and about the two-fold axis twice. It is clear, moreover, that there are three axes of four-fold symmetry, four of three-fold symmetry and six of two-fold symmetry in the cube. This is expressed in the following way:—

Axes of Symmetry of the cube: $3^{iv}, 4^{iii}, 6^{ii}$.

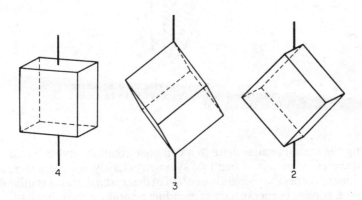

FIG. 26. Axes of Symmetry of the Cube.

In our brick there are only three axes of geometrical symmetry and these are of two-fold or diad type; they connect the middle points of the pairs of opposite faces of the brick.

Centre of Symmetry. A crystal has a centre of symmetry when like faces, edges, etc. are arranged in pairs in corresponding positions on opposite sides of a central point. The cube and brick obviously have centres of symmetry.

The Symmetry of Gypsum as an Example. A crystal of the common mineral gypsum may be taken to illustrate these definitions of symmetry. The usual form of such a crystal is shown in Fig. 27.

There is one plane which divides the crystal into two similar and similarly placed halves. This plane is the only plane of symmetry for this crystal. At right angles to this plane is an axis of symmetry. Rotation about this axis causes the crystal to take

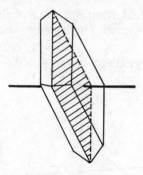

FIG. 27. A gypsum crystal, showing the one plane of
symmetry and the one axis of symmetry characteristic of this mineral.

up the same position *twice* in a complete rotation, and this axis
is therefore an axis of two-fold symmetry. Lastly, for every face,
edge or corner that occurs in one half of the crystal there is a similar
face, edge or corner in a corresponding position in the other half.
Therefore the crystal has a centre of symmetry.

Thus the symmetry of this gypsum crystal may be expressed in
the following way:—

Planes of symmetry, 1.

Axes of symmetry, 1^{ii}.

Centre of symmetry.

Crystallographic and Geometrical Symmetry. Crystallographic
symmetry must not be confused with geometrical symmetry.
Crystallographic symmetry
depends upon the internal
atomic structure of the crystal, and as the arrangement
of the atoms is the same for
parallel planes, it follows that
the angular position is the
only factor concerned, and
that the sizes of like faces

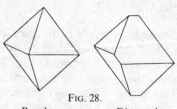

FIG. 28.

Regular   Distorted
Octahedron   Octahedron

and their distances from a plane or centre of symmetry are of no importance in this connection. This is illustrated in Fig. 28, which shows a regular octahedron with faces equally developed and a distorted octahedron with some faces larger than others. If such a distorted octahedron was examined with a goniometer it would be found that the interfacial angles were exactly the same as in the regular octahedron.

Crystals in which like faces are equally developed and are equal distances from the centre of the crystal are rare; but for convenience of study and of representation by diagrams, it is necessary to deal with crystals in their simplest and most intelligible shape, and that is when they have perfect geometrical symmetry.

Most crystals occur in *distorted forms,* having like faces not of the same size and not in the same geometrically symmetrical

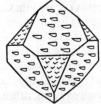

FIG. 29. Quartz Crystal showing diagrammatically similar etch-marks on like faces.

position. In many cases of distorted crystals the crystallographic symmetry has been made out from the fact that like faces have like properties. Etch-marks produced by solvents acting on the crystal faces, the behaviour towards heat and electricity, the hardness, lustre, etc., of the faces, have revealed the true symmetry of the distorted crystals. This is illustrated in the quartz crystal shown in Fig. 29, where the etch-marks are similar on like faces.

Distortion in crystals may be due to some restraint on growth of the crystal in certain directions or to a greater supply of material being available in one direction as compared with another.

Habit. The term *habit* is used to denote the characteristic shapes of crystals arising from variations in the number, size and shape of the faces; the distorted octahedron shown in Fig. 28 has a tabular habit; in Fig. 30 are shown two habits of apophyllite crystals. Variation in habit arises from many causes, such as speed and condition of crystal growth and the presence of impurities in the solution. Crystals from one locality may therefore be characterized by a particular habit.

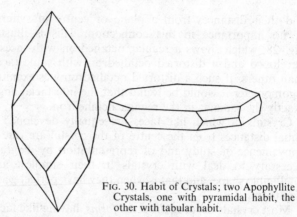

Fig. 30. Habit of Crystals; two Apophyllite Crystals, one with pyramidal habit, the other with tabular habit.

Crystallographic Axes. In solid geometry the position of a plane in space is given by the intercepts (or the lengths cut off) that the plane makes on three given lines called axes. This method of treatment is employed in crystallography, and the axes are termed the *crystallographic axes*. The crystallographic axes intersect at the *origin*. The three edges of the unit cell meeting at a point and extended front and back, sideways and up and down clearly form the most suitable crystallographic axes, as already mentioned (p. 77 Fig. 18).

Parameters. The *parameters* of a crystal face are the ratios of the distances from the origin at which the face cuts the crystallographic

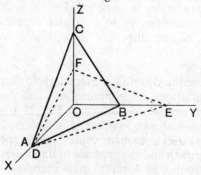

Fig. 31. Parameters.

axes,—that is, the parameters are the *ratios of the intercepts*. In Fig. 31, OX, OY, OZ, represent the crystallographic axes and ABC is a crystal face making intercepts of OA on OX, OB on OY, and OC on OZ. The parameters of the face ABC are given by the ratio of OA, OB, and OC. It is convenient to take the relative intercepts of this face as standard lengths for the purpose of representing the position of any other face, such as DEF. In the case of the face DEF, OD is equal to OA, OE is twice OB, and OF is half OC, and therefore $\frac{1}{1}$, $\frac{2}{1}$, $\frac{1}{2}$, are the parameters of DEF with reference to the standard face ABC.

The form whose face is taken as intersecting the axes at the unit lengths which are to be used for measuring the intercepts made by other forms on the same axes is called the *fundamental*, *parametral* or *unit* form. The selection of a suitable unit form depends on the properties and nature of the crystals. A form well developed, or commonly occurring, or parallel to which there is a good cleavage, was usually selected for this purpose, long before the advent of the X-ray analysis of crystal structure. It is not surprising, therefore, that the unit form of many crystal species has the same parameters as those of the corresponding unit cell or, if this is not the case, that there is a simple relationship between them.

Axial Ratio. The parameters of the unit form can be obtained by measurement and calculation, and can be expressed as multiples of one of their number. Take, for example, gypsum. It is found that the most commonly occurring form in gypsum crystals which makes intercepts with all three crystallographic axes does so in the ratio of $0\cdot374:1:0\cdot414$. This expression is called the *axial ratio*, and simply means that the standard or unit form cuts one axis at a distance represented by $0\cdot374$, the second axis at a distance represented by 1, and the third axis at a distance represented by $0\cdot414$. When we use this unit form to measure the intercepts, or to obtain the parameters, of any other form that cuts all three axes we shall do so by taking $0\cdot374$ as our unit of measurement along the first axis, 1 along the second axis, and $0\cdot414$ along the third axis.

Indices. The reciprocals of the parameters are called the indices and as we see later (p. 90) form the basis of crystallographic notation.

Lettering and Order of the Crystallographic Axes. There are certain conventions with regard to the lettering and order of the crystallographic axes. In the most general case, that in which the unit form cuts all three axes at unequal lengths and in which none of the axes is at right angles to any other, the crystallographic axis which is taken as the vertical axis is called c, that running from right to left is b, and that running from front to back is a. One end of each axis is positive, and the other end is negative, and the rule with regard to this illustrated in Fig. 32. The angle between $+a$ and $+b$ is called γ, that between $+b$ and $+c$ is called α, and that between $+c$ and $+a$ is β. This is, of course, the same nomenclature as that used for unit cells (p. 78).

In this most general case, the unit form cuts the three axes at unequal lengths from the origin, and this fact is often indicated loosely by stating that the crystallographic axes of this type of crystal are of unequal lengths.

In some crystals the unit form cuts two axes at an equal distance and the third at a different distance. In this case, the axes cut at equal distances are both called a and the third, placed vertical, is called c. It is customary to say here that the two axes are equal and the third different. Again, in other crystals, the unit form cuts all three axes at the same distance, so that all the axes are interchangeable; in this case the axes are all called a, and are loosely said to be equal.

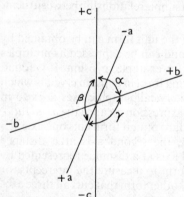

FIG. 32. Axial Conventions.

The position in space of the faces of most crystals are referred to *three* crystallographic axes, but in one group it is convenient to use *four* axes.

The planes in which two of the crystallographic axes lie are called the *axial planes*.

Crystallographic Notation. Crystallographic notation is a concise method of writing down the relation of any crystal face to the crystallographic axes. The most widely used systems depend upon either parameters or indices. Of these systems of notation, the chief are two—the Parameter System of Weiss, and the Index System of Miller (modified by Bravais).

Parameter System of Weiss. In this system of crystallographic notation, the axes are taken in the order explained above—that is, $a, b, c,$ for unequal axes, $a, a, c,$ for two axes equal, and $a, a, a,$ for three axes equal. The intercept that the crystal face under discussion makes on the a-axis is then written before $a,$ the intercept on the b-axis before $b,$ and the intercept on the c-axis before $c.$ These intercepts are of course measured in terms of the intercepts made by the unit form on the corresponding crystallographic axes. The most general expression for a crystal face in the Weiss notation is—

$$na, mb, pc,$$

where n, m, p are the lengths cut off by the face on the a, b, c axes as compared with the corresponding lengths cut off by the unit form. It is usual to reduce either n or m to unity.

If a crystal face is parallel to an axis, it can be imagined as cutting that axis at an infinite distance, and accordingly the sign of infinity, ∞, is placed as its parameter before the corresponding axial letter. Thus a face cutting the a-axis at a distance 1 unit,—that is at the same distance as the unit form cuts this same axis,—and cutting the b-axis at a distance 2 units or twice the distance cut off by the unit form along the b-axis, and running parallel to the c-axis has the Weiss symbol—

$$a, 2b, \infty c.$$

A face cutting the a-axis and parallel to the b-axis and c-axis obviously has the symbol

$$a, \infty b, \infty c.$$

Index System of Miller. We can use the Weiss method to explain the Miller System which is most commonly used. In this system of notation the indices, or reciprocals of the parameters, are used. They are written in the axial order, $a, b, c,$ and are always given in their most simple form by clearing of fractions. For example, consider the crystal face dealt with in the previous paragraph which has the Weiss symbol

$$a, 2b, \infty c.$$

The reciprocals of the parameters are—

$$1, \tfrac{1}{2}, 0.$$

Clearing of fractions and omitting the axial letters the Miller symbol is obtained—

$$210,$$

which is read as *two one nought*.

Similarly, the face parallel to the *b*-axis and the *c*-axis which has the Weiss symbol a, ∞ b, ∞ c will have the Miller symbol 100, which is read as *one nought nought*.

Several points are to be noted in connexion with Miller symbols. A face parallel to an axis will contain the symbol 0 (nought), the reciprocal of infinity, at the position in the symbol corresponding to the axis in question. Again, since the Miller symbols are based on reciprocals of the parameters, the larger the figure in the symbol the nearer to the origin will the face cut that particular axis. On the other hand, the smaller the figure in any given axial position in the symbol, the nearer does the face approach to parallelism with the axis in question; the limit is reached when the figure is zero and the face is parallel with the axis. The most general Millerian symbol is *hkl*.

Referring again to the unit form of gypsum, although this form cuts the three axes at unequal arithmetic distances, its Miller symbol is (111), since the unit form has been defined as cutting each axis at the standard or unit length for measurements of intercepts along the axes.

Consider the faces ABC, DEF shown in Fig. 31. p. 88. ABC is a face of the unit form and therefore has the symbols a, b, c, (Weiss) or 111 (Miller). For the face DEF, the Weiss symbol is a, 2b. $\tfrac{1}{2}$c: the reciprocals of these parameters give the indices 1, $\tfrac{1}{2}$, 2 which cleared of fractions, are the Miller symbols 214, read *two one four*.

Conventions in Notation. Several conventions in crystallographic notation must be considered. When it is required to indicate a crystal *form,* it is usual to enclose the symbols in a bracket thus (*hkl*), whereas, if the crystal *face* is indicated, the bracket is removed, thus—*hkl*. By some, however, the form is enclosed in a curly bracket or brace {hkl}, and the face in an ordinary bracket (hkl).

The convention, illustrated in Fig. 32, with regard to the signs of the ends of the crystallographic axes is of great importance in notation, since by adding the proper sign to the symbol it is possible to indicate any required face of a crystal form.

A face cutting the positive end of an axis is indicated by the corresponding index figure only, whilst one cutting the negative end has a negative sign placed above the index figure. This convention is illustrated in Fig. 33, which shows the form (111) made up of the eight faces 111, 1$\bar{1}$1, 1$\bar{1}\bar{1}$, 11$\bar{1}$, $\bar{1}$11, $\bar{1}\bar{1}$1, $\bar{1}\bar{1}\bar{1}$, $\bar{1}$1$\bar{1}$. Taking the face $\bar{1}\bar{1}$1 as an example, it is clear that this face cuts the negative

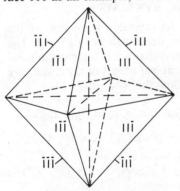

end of the first or *a*-axis, the negative end of the second or *b*-axis, and the positive end of the third or *c*-axis, and it must therefore lie in the back upper left-hand octant. Further, inspection of the figure illustrates the general rule that by changing the signs of the indices an opposite parallel face is indicated. Thus, in Fig. 33, the face opposite and parallel to the face 111 is clearly $\bar{1}\bar{1}\bar{1}$.

FIG. 33. The Form (111).

The Law of Rational Indices. The study of crystals has established the Law of Rational Indices which states that *the intercepts that any crystal face makes on the crystallographic axes are either infinite, or small rational multiples of the intercepts made by the unit form.* Thus symbols such as a, $\sqrt{2}$ a, ∞ a, or 2a, 1·736 . . . b, c are impossible. This law follows from the regularity of the space-lattice representing the atomic structure of crystals.

Classification of Crystals. It has been mathematically proved that there are possible among crystals only *thirty-two* types of symmetry, differing from each other in the degree and nature of the three elements of symmetry in each group. Of these thirty-two types, some are represented by artificially prepared compounds, some by very rare minerals, and some have no representative at all. The common minerals dealt with in this book are found in *eleven groups* only. The first classification of crystals is therefore into the *symmetry classes*.

We have seen (p. 78) that all types of unit cells and lattices can be referred to six sets of axes that are used as the crystallographic axes and that crystals belonging to different symmetry groups can be referred to the same set of crystallographic axes. All the crystal forms, of whatever symmetry, that can be referred to the same set of crystallographic axes fall in one *crystal system*. The second classification of crystals in therefore into *crystal systems*.

THE CRYSTAL SYSTEMS

Below is set out a synopsis of the crystal systems and of the symmetry classes in each that are dealt with in this book. The symmetry classes or types are named after a characteristic mineral belonging to the class and the technical names of the classes are given in the descriptions that follow this synopsis.

I. CUBIC SYSTEM. Axes, three equal, *a, a, a,* at right angles.

  1. *Galena Type* – symmetry: – 9 planes, 13 axes (6^{II}, 4^{III}, 3^{IV}), a centre.

  2. *Pyrite Type* – symmetry: – 3 planes, 7 axes (3^{II}, 4^{III}), a centre.

  3. *Tetrahedrite Type* – symmetry: – 6 planes, 7 axes (3^{II}, 4^{III}), no centre.

II. TETRAGONAL SYSTEM. Axes, *a, a, c,* two equal *horizontal*, one *vertical,* at right angles.

  4. *Zircon Type* – symmetry: – 5 planes, 5 axes (4^{II}, 1^{IV}), a centre.

III. HEXAGONAL SYSTEM. Axes, *a, a, a, c,* three equal *horizontal,* making angles of 120° with each other, a *vertical* axis at

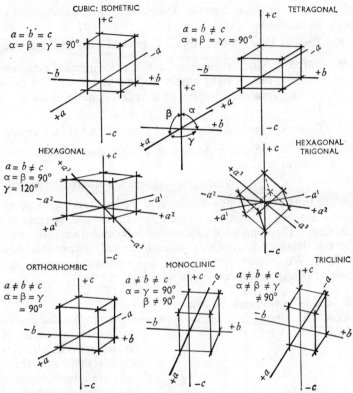

FIG. 34. The unit-cell types and the corresponding crystal system, showing the arrangement and nomenclature of the crystallographic axes. The top central small diagram gives the basic notation.

right angles to the plane containing the horizontal axes.

5. *Beryl Type* – symmetry: – 7 planes, 7 axes (6^{II}, 1^{VI}), a centre.

6. *Calcite Type* – symmetry: – 3 planes, 4 axes (3^{II}, 1^{III}), a centre.

7. *Tourmaline Type* – symmetry: – 3 planes, 1 axis (1^{III}), no centre.

8. *Quartz Type* – symmetry: – no planes, 4 axes (3^{II}, 1^{III}), no centre.

IV.   ORTHORHOMBIC SYSTEM. Axes, *a, b, c,* three unequal, all at right angles.

9.   *Barytes Type* – symmetry: – 3 planes, 3 axes (3^{II}), a centre.

V.   MONOCLINIC SYSTEM. Axes, *a, b, c,* three unequal axes, one vertical, one at right angles to the vertical axis, the third making an oblique angle with the plane containing the other two.

10.   *Gypsum Type* – symmetry: – 1 plane, 1 axis (1^{II}), a centre.

VI.   TRICLINIC SYSTEM. Axes, *a, b, c,* three unequal axes, none at right angles.

11.   *Axinite Type* – symmetry: – no planes, no axes, a centre.

The foregoing synopsis is illustrated in Fig. 34 and the relationships between the unit-cell types and the crystal systems are indicated. This condensed information is expanded and elaborated in the detailed descriptions of each crystal system that now follow. We begin with the most symmetrical system and work through to the least symmetrical.

I. Cubic or Isometric System

Axes. The cubic system includes all the crystal forms that can be referred to three axes of equal lengths which intersect at right angles. These *axes* are interchangeable and of equal value, so that they are all denoted by *a*. In order to be able to indicate any required face in a form, however, it is necessary to specify each axis. This is done by placing the crystal in what is known as the reading position, or the position in which the crystal is best studied, so that one axis, called a_3,

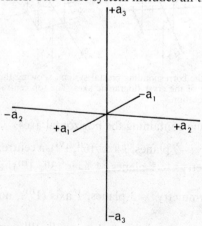

FIG. 35. The Cubic Axes.

is vertical, one a_2, runs right and left, and the third, a_1, runs front to back. The cubic axes are shown in Fig. 35, where also the positive and negative ends of the axes are indicated.

Three symmetry classes belonging to the Cubic System are dealt with here. They are:

  (i)   Galena Type or Hexoctahedral Class.
 (ii) / Pyrite Type or Diploidal Class.
(iii)   Tetrahedrite Type or Hexatetrahedral Class.

Pictorial representations of these three symmetries using the stereographic projection are given and explained in Fig. 50 p.113.

I. Galena Type or Hexoctahedral Class

This type takes its name from the common mineral galena, PbS, which belongs to it. It is often called the *Normal Type* of the Cubic System, since it shows the highest degree of symmetry of that system and, therefore, of all crystals. It is also called the Hexoctahedral Class from its *general form*, the hexoctahedron (Fig. 38); the meaning of general form is given on p.102.

SYMMETRY. The symmetry can be deduced from the typical form, the cube, as has been partly done on pp. 83–84. It is instructive to work out the symmetry by the aid of a model of a cube and a flat sheet of cardboard, celluloid or glass. There are clearly 3 planes of symmetry which each bisect the three sets of four parallel edges of the cube. As we shall see in the next section, these planes contain the crystallographic axes and are called therefore *axial* planes. In addition to these axial planes of symmetry, there are three pairs of planes which cut the faces of the cube diagonally; these we may call *diagonal* planes of symmetry. Thus, there are *nine planes of symmetry* in all, as illustrated in Figs. 24 and 25 on pp. 83–84.

With the model of a cube in front of us it is easy to demonstrate the axes of symmetry. Hold the cube between the thumb at the centre of one face and the first finger at the centre of the opposite parallel face. Rotation of the cube shows that in a complete turn a cube face appears four times, so that the axis of rotation, the line joining the centres of opposite parallel faces, is an axis of tetrad symmetry; there are three such axes, each joining the middle points of the three pairs of opposite parallel faces. These three axes are the crystallographic axes shown below. Secondly hold the cube by its opposite corners and rotate; the cube comes to occupy the same position three times during a complete turn, so that the line joining the opposite corners is an axis of three-fold symmetry and there are four such axes. Thirdly, hold the cube at the centres of opposite parallel edges; rotation gives two similar positions in a complete turn, and the six lines joining the centres of the six pairs of opposite parallel edges are axes of two-fold symmetry. There are thus in the cube three axes of four-fold symmetry, four of three-fold and six of two-fold—*thirteen axes of symmetry* in all (see Fig. 26, p. 85).

Lastly, the faces and edges of the cube occur in pairs on opposite sides of a central point, as may be tested by placing the cube on the table and noting that a face lies parallel with the table. The cube has a *centre of symmetry*.

The symmetry of the Galena Type may now be summarized as follows:—

Planes, 9 $\begin{cases} 3 \text{ axial.} \\ 6 \text{ diagonal.} \end{cases}$

Axes, 13 $\begin{cases} 3^{IV}. \text{ (the crystallographic axes).} \\ 4^{III}. \\ 6^{II}. \end{cases}$

A Centre of Symmetry.

FORMS. (i) *Cube*. As shown in Fig. 36A the cube is a six-faced solid bounded by faces which cut one axis and are parallel to the other two.

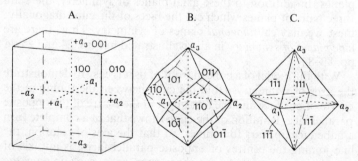

FIG. 36. Cube, Rhombdodecahedron and Octahedron, Axes, and Symbols of visible faces.

Each face is a square and is parallel to an axial plane of symmetry. The symbol of the form is obviously (100) and the form ·is made up of the faces 100 (front), $\overline{1}$00 (back), 010 (right), 0$\overline{1}$0 (left), 001 (top), and 00$\overline{1}$ (bottom).

(ii) *Rhombdodecahedron.* The rhombdodecahedron is a solid having twelve rhomb-shaped faces, each face cutting two axes at equal distances and being parallel to the third. The symbol of the form is therefore (110), and the symbols of several faces are shown in Fig. 36B.

(iii) *Octahedron.* The unit form of the Galena Type, the octahedron, is a solid having eight equilateral triangular faces, each face cutting all three axes at equal distances. The octahedron is shown in Fig. 36C. Its symbol is (111); the symbols of the front faces are given in Fig. 36C and of all the faces in Fig. 33, p. 93.

(iv) *Tetrahexahedron.* The tetrahexahedron is a solid having twenty-four faces, each face an isosceles triangle. For each face of the cube (or hexahedron) there appears a low pyramid of four (or tetra) faces, whence the name tetrahexahedron. Each face is parallel to one axis and cuts the other two axes at unequal lengths. The general symbol therefore is (hk0), *i.e.* (*h.k.* nought), which expresses the fact that each face cuts two axes at different distances and is parallel to the third. There can be many tetrahexahedra since the ratio of h and k or of the intercepts made by the face varies; on this account this form is said to be a *variable form*. Suppose either h or k becomes zero, the faces then become parallel to two axes and the form becomes a cube; suppose h is made equal to k, then the faces cut two axes at equal distances and are parallel to the third axis, and the form becomes a rhombdodecahedron. This relationship is expressed by saying that the *limiting forms* of the tetrahexahedron are the cube and the rhombdodecahedron.

We have seen that the general symbol of the tetrahexahedron is (hk0); a commonly occurring example is (210), others are (320), (310), (410), etc. The form (210), in which the faces cut one axis at a certain distance, the second at twice that distance and are parallel to the third, is shown in Fig. 37A, where also the symbols of several of the faces are given. Though the form has been called here the (210) tetrahexahedron, it must be realised that it might just as well have been called the (201), (120), etc., form; the ratio between the intercepts made by each face is, of course, of the same type throughout. Inspection of Fig. 37A shows moreover how by depressing the pyramids on the cube faces until their faces become parallel with two axes there can be obtained one of the limiting forms, the cube. Similarly, it can be noted how by steepening the pyramids until the intercepts on the two axes are equal there is produced the other limiting form, the rhomb-dodecahedron; the edges between the faces 210 and 120, between

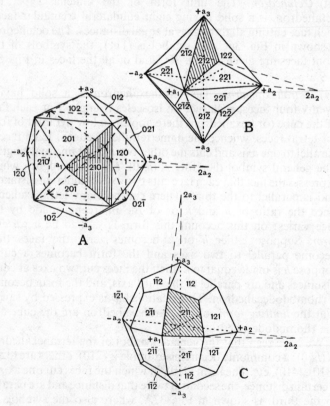

FIG. 37. Tetrahexahedron (210), Trisoctahedron (221), and Trapezohedron (211).

201 and 102, between 012 and 021, etc., disappear and each pair of faces becomes one face of the rhombdodecahedron, as 110, 101, 011, etc.

(v) *Trisoctahedron*. The trisoctahedron is a solid having twenty-four faces, each face an isosceles triangle. It has the appearance of being formed by a three-faced pyramid grown on each face of the octahedron. Each face cuts two axes at an *equal* length, and the third at a *greater* length (see Fig. 37B). The Weiss symbol for the form is therefore (*a, a, pa*), the Miller symbol is (*hhl*) and a commonly occurring form, illustrated in Fig. 37, is (221). Other trisoctahedra are (331), (332), etc. The symbols of the faces shown in the figure should be carefully studied, and the limiting forms—the octahedron (111) and the rhombdodecahedron (110) developed. Note, also, that the edges between the faces 221 and 212 and 122 form a Y upside down.

(vi) *Trapezohedron*. The trapezohedron or icositetrahedron has twenty-four like faces, each face a trapezoid. Each face cuts two axes at an *equal* length, and the third at a *smaller* length. The Miller symbol is therefore (*hll*) where *h* is greater than *l*, and a common form, shown in Fig. 37C, is (211). Other trapezohedra are (311), (411), (322), etc. Inspection of the figure shows that the limiting forms are the cube (100) and the octahedron (111), and further, that the faces 211, 121 and 112 meet in a Y which is right way up.

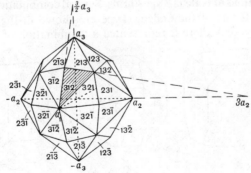

Weiss: $1a_1$, $3a_2$, $\frac{3}{2}a_3$: whence $\frac{1}{1} \cdot \frac{1}{3} \cdot \frac{2}{3}$ for Miller, which is (312) for the shaded face.

(vii) *Hexoctahedron*. The hexoctahedron is a solid having forty-eight like faces, each face a scalene triangle. Each face cuts the three axes at unequal lengths. The symbol is therefore (hkl), and an example is (321). This form is illustrated in Fig. 38, in which the symbols of the faces should be studied. By varying the relations between the intercepts made by the faces of the hexoctahedron, all the other forms of the Galena Type may be obtained. The hexoctahedron is therefore called the *general form* of this type of crystal; the six other forms belonging to this class are *special forms*.

Work with Models. The best method of studying the crystal systems is, of course, with the help of actual crystals, models and drawings. To the beginner models are exceedingly valuable. Fairly large glass or wooden models of the seven forms of the Galena Type should be examined by the student until he is satisfied that each form possesses the symmetry elements—9 planes, 13 axes and a centre—demanded by the Galena Type. The position of the crystallo-graphic axes and the symbols of the faces should be ascertained in each form. Lastly, it is instructive to start with the general form, the hexoctahedron, and to derive the other forms of the type from it by varying the intercepts; this can be demonstrated by placing models of the seven forms side by side in the reading positions.

Combinations of Galena Type Forms. Several combinations of two or more forms of the Galena Type are shown in Figs. 39 and 40. In Fig. 39A there is represented a combination of cube and

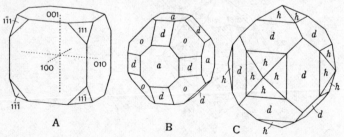

Fig. 39. A. Cube and Octahedron; B. Cube (a), octahedron (o), and rhombdo-decahedron (d); C. Rhombdodecahedron (d) and tetrahexahedron (h).

octahedron in which the solid angles of the cube are cut off by the faces of the octahedron. Three forms are combined in Fig. 39B, where the octahedral faces cut off the solid angles of the cube, and the faces of the rhombdodecahedron truncate the edges of the cube and at the same time those of the octahedron. In Fig. 39C the rhombdodecahedron and the tetrahexahedron are seen in combination. Figs. 40A and B show combinations of rhombdodecahedron and trapezohendron such as are commonly

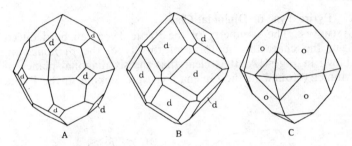

FIG. 40. Minerals of Galena Type. A, B, Garnet, showing combination of rhombdodecahedron (d) and trapezohedron. C. Spinel, showing combination of octahedron (o) and trapezohedron.

found in the mineral garnet, the relative developments of the two forms differing in the two figures, and Fig. 49C is a combination of octahedron with the trapezohedron as seen in crystals of spinel.

SOME COMMON GALENA TYPE MINERALS

Galena. Galena occurs commonly in crystals showing combinations of the cube and octahedron, as illustrated in Figs. 20 and 39A.

Fluorspar, Rock-salt and Sylvine. These minerals occur as simple cubes.

Leucite, Analcite. Leucite and analcite occur as simple trapezohedra.

Spinel, Magnetite. These minerals occur as simple octahedra, sometimes combined with rhombdodecahedra, or trapezohedra, as shown in Fig. 40C. The cleavage of *fluorspar* is parallel to the faces of the octahedron and the cleavage-fragments are octahedra.

Garnet. Common crystals of garnet are either the rhombdodec-ahedron or the trapezohedron, or combinations of the two, as shown in Fig. 40A and B.

Diamond. Diamond occurs in octahedral crystals, which on account of etching tests, etc., have been considered to be of lower symmetry than that of the Galena Type. X-ray examination shows, however, that diamond possesses the atomic structure of the Normal class of the Cubic System.

2. Pyrite Type or Diploidal Class

SYMMETRY. The symmetry of the Pyrite Type can be deduced from the characteristic form of this type, the *pyritohedron,* shown in Fig. 41A. It is clear that the six diagonal planes of symmetry found in the Galena Type are lacking in the Pyrite

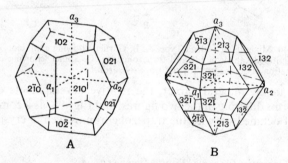

FIG. 41. A. The Pyritohedron, (210), with axes and symbols.
B. The Diploid (321), with axes and symbols.

Type, the only planes present in the latter type being the *three axial planes,* in which lie the three pairs of parallel edges of the pyritohedron. The three crystallographic axes are axes of twofold symmetry only. The four axes of three-fold symmetry of the Galena Type are also present in the pyritohedron, so that there are *seven axes* altogether. There is a *centre* of symmetry, since similar faces, edges, etc., occur on opposite sides of a central point. The student should work out this symmetry from a model of the pyritohedron. The symmetry of the Pyrite Type is therefore:—

Planes,   3 axial.

Axes,   7 $\begin{cases} 3^{\text{II}} \\ 4^{\text{III}} \end{cases}$ (the crystallographic axes).

A Centre of Symmetry.

FORMS. (i) *Pyritohedron*. This form receives its name from the fact that it is the common form in which pyrite crystallizes. It is also known as the *pentagonal dodecahedron,* since it is bounded by twelve pentagonal faces. These pentagons have one edge longer than the other four, which are alike: the long edges run in pairs parallel with the crystallographic axes, as shown in Fig. 41A. Each face cuts two axes at different lengths and is parallel to the third. The symbol is therefore $(hk0)$, and a typical form is (210), others being (310), (320), (410), etc. It will be recalled that the symbol of the tetrahexahedron of the Galena Type is also $(hk0)$, (210), (310), etc., and the relationship between the pyritohedron and tetrahexahedron is shown in Fig. 42, where it is seen that a pyritohedron is produced by the development of alternate faces of a tetrahexahedron. As there are no diagonal planes of symmetry in the Pyrite Type the tetrahexahedron itself is not possible. There will obviously be two pyritohedra corresponding to the same tetrahexahedron,—one, the (210) form, related to the shaded faces of Fig. 42, the other, the (201) form, related to the unshaded faces. The first form, (210), with its front edge vertical, is called the *positive pyritohedron;* the second form, (201), with its front edge horizontal, is the *negative pyritohedron.*

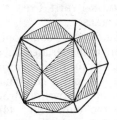

FIG. 42. Pyritohedron, showing its development from Tetrahexahedron.

(ii) *Diploid*. The diploid (Fig. 41B) is a solid bounded by twenty-four faces. Each face is a trapezium, and they are grouped in pairs, hence the name diploid. Each face cuts all three axes at different lengths, and a typical form is that having the symbol (321). It is the general form of the Pyrite Type and gives its name to the symmetry class. The diploid is related to the hexoctahedron (321) of the Galena Type in the same way as the pyritohedron is to the tetrahexahedron. It may be regarded as being produced by the development of alternate faces of the hexoctahedron. As with the pyritohedron, the diploid occurs as a *positive* form, *e.g.* (321), and a *negative* form, *e.g.* (231).

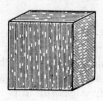

FIG. 43. Striated Cube of Pyrite.

The other five forms that satisfy the symmetry requirements of the Pyrite Type are the cube (100), rhombdodecahedron (110), octahedron (111), trisoctahedron (221), and trapezohedron (211). These are geometrically similar to corresponding forms in the Galena Type but have a lower structural symmetry. This lower symmetry is evident in the case of the common striated cube of pyrite shown in Fig. 43. It will be noticed that the striations on the three pairs of faces lie in three directions at right angles, parallel to the crystallographic axes. As can be seen from the diagram, any one of these axes is an axis of *two-fold* symmetry, whereas, in the cube of Galena Type, a crystallographic axis is an axis of four-fold symmetry. A striated cube of pyrite results from what is known as an *oscillatory combination*, both the cube and the pyritohedron having endeavoured, as it were, to assert their respective forms during crystallization.

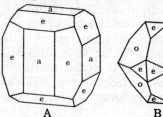

FIG. 44. Combinations seen in Pyrite. A. Pyritohedron (e) and cube (a). B. Pyritohedron (e) and octahedron (o).

COMMON PYRITE TYPE MINERALS

Pyrite. Pyrite commonly occurs in the striated cube (Fig. 43), the pyritohedron (Fig. 41), or in combinations of the pyritohedron with the cube (Fig. 44), with the octahedron (Fig. 44) or with the diploid.

Cobaltite, smaltite and *chloanthite* are other minerals that crystallize in the Pyrite Type with forms similar to those shown by pyrite itself.

3. Tetrahedrite Type or Hexatetrahedral Class

SYMMETRY. The form typical of the Tetrahedrite Type is the *tetrahedron* shown in Fig. 45A. Inspection of this figure or, better, manipulation of a model of a tetrahedron, reveals the following symmetry. The axial planes of symmetry of the Galena and Pyrite types are here absent, but the *six diagonal planes* of the former class are present in the tetrahedron. The three crystallographic axes are axes of two-fold symmetry, and there are, in

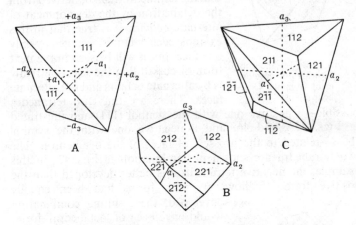

FIG. 45. A. Tetrahedron (111). B. Deltoid (221). C. Tristetrahedron (211).

addition, four axes of three-fold symmetry which join the centres of the faces to the apex opposite them. There are thus *seven axes* of symmetry. There is obviously *no centre* of symmetry, as can be shown by laying the tetrahedron on a face on the table when an apex comes above. The symmetry of the Tetrahedrite Type is therefore:—

Planes, 6 diagonal

Axes, 7 $\begin{cases} 3^{11} \quad \text{(the crystallographic axes)} \\ 4^{111} \end{cases}$

No Centre of Symmetry

FORMS. (i) *Tetrahedron*. The tetrahedron, as its name implies, is a four-faced solid. Each face is an equilateral triangle, and meets the axes at equal distances. The symbol of the form is therefore (111). It should be noted that the crystallographic axes of the tetrahedron join the centres of the opposite edges and that one axis is vertical; it is essential that the positions of the axes be realised with the help of a glass or cardboard model.

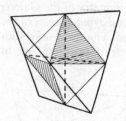

FIG. 46. Tetrahedron, showing its development from the Octahedron.

The symbol (111) of the tetrahedron shows that it is related to the octahedron (111) of the Galena Type. This relationship is shown in Fig. 46, where it is seen that the tetrahedron can be derived from the octahedron by the development of alternate *octants*. It is true that in this example each octant is occupied by only one face, but it will be seen in the next form discussed that the selection really is by alternate octants and not alternate faces. There are thus *two* tetrahedra possible—a positive one with the symbol (111) and illustrated in Figs. 45A and 46, and a negative one with the symbol (1$\bar{1}$1)—related to the *two* sets of faces of the octahedron. The two tetrahedra are shown in combination in Fig. 47. In this example, the negative tetrahedron is better developed than the positive form. If, however, both forms had been equally developed, the resulting combination would have been of 'octahedral' form. That all the faces of this 'octahedron' were not like faces and that two tetrahedra were present could be demonstrated by the arrangement of etchmarks, as indicated in Fig. 47. The faces of the two forms show very different etch-marks.

FIG. 47. Positive and Negative Tetrahedra in Zinc Blende showing different etchmarks on the faces of the two forms.

(ii) *Deltoid-dodecahedron*. The deltoid-dodecahedron is a solid bounded by twelve faces, each face being a trapezoid or deltoid in shape. The form is illustrated in Fig. 45B. Each face cuts two axes at an *equal* length and a third at a *greater* length. The general symbol is (*hhl*) and a typical form, shown in Fig. 45B, is (221). It will be recalled that the form with similar symbols in the Galena Type is the trisoctahedron (*hhl*, 221, etc.), and the deltoid-dodecahedron may be considered to be made up of the twelve faces of the trisoctahedron that occur in alternate *octants*. Note the upside-down Y-edges in the corresponding positions in the two forms. There are, of course, two deltoid-dodecahedra related to a trisoctahedron, just as the positive and negative tetrahedra are related to the octahedron.

(iii) *Tristetrahedron.* The tristetrahedron is a form bounded by twelve like triangular faces and has the appearance of having a low three-faced pyramid raised on each face of the tetrahedron, as shown in Fig. 45C. Each face cuts two faces at *equal* lengths and the third at a *smaller* length. The general symbol is therefore (hll) and a representative example is (211) as figured in Fig. 45C. The corresponding form in the Galena Type is the trapezohedron (211), etc., and the tristetrahedron shows the twelve faces of this form that occur in alternate octants. Note the right-way-up Y-edges in the corresponding positions in the two forms. There are again two tristetrahedra related to a trapezohedron.

(iv) *Hexatetrahedron.* The hexatetrahedron (Fig. 48) is a solid bounded by twenty-four triangular faces, these corresponding to the faces of the alternate octants of the hexoctahedron $(321$, etc.) of the Galena Type. Each face of the hexatetrahedron cuts the axes at unequal lengths, so that the general symbol is (hkl) and one example (321). Again there are two hexatetrahedra corresponding to each hexoctahedron. The hexatetrahedron is the *general* form of the Tetrahedrite Type and gives its name to the symmetry Class.

The other forms occurring in the Tetrahedrite Type are the *cube* (100), the *rhombdodecahedron* (110) and the *tetrahexadron* (210), etc., giving again seven forms in all. Though geometrically similar to the corresponding Galena Type forms, these Tetrahedrite Type forms have a lower structural symmetry, which finds expression in the distribution and character of etch-marks and striations, or by pyroelectricity.

It may be noted that the forms of the Tetrahedrite Type have the faces of alternate *octants* of the corresponding forms in the Galena Type, whilst the forms in the Pyrite Type have the alternate *faces* of the Galena Type. The three types of the Cubic System are compared in the Table on p. 111.

COMMON TETRAHEDRITE TYPE MINERALS

Tetrahedrite. Tetrahedrite occurs as the tetrahedron, and tristetra-

hedron, and as combinations of the tetrahedron and tristetrahedron and of the tetrahedron and cube.

Zinc Blende. The common crystals of blende are formed by the
combination of positive and negative tetrahedra, the two forms
being distinguished by etching tests as shown in Fig. 47.

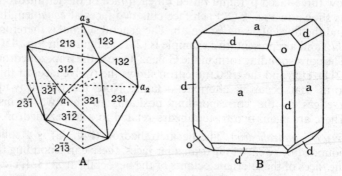

FIG. 48. A. Hexatetrahedron. B. Boracite; combination of cube (a), rhombdodecahedron (d), and tetrahedron (o).

Boracite. Boracite occurs in crystals showing a combination of
cube, tetrahedra and rhombdodecahedron, and a crystal of this
type is given in Fig. 48B.

STEREOGRAPHIC REPRESENTATION OF CRYSTAL SYMMETRY

By the use of the device known as *stereographic projection*,
crystal faces can be represented by a series of points on a circular
diagram, which is the projection of the upper or lower half of a
sphere on to a horizontal plane through its centre. Each point is
the intersection of the *normal* to a crystal face (assumed to lie
at the centre of the sphere) with the surface of the hemisphere, as
seen when projected on to the equatorial plane. The symmetry of
the crystal is then shown by the pattern of the points in the
diagram.

COMPARISON OF THE GALENA, PYRITE AND TETRAHEDRITE TYPES

GALENA TYPE OR HEXOCTAHEDRAL CLASS	PYRITE TYPE OR DIPLOIDAL CLASS	TETRAHEDRITE TYPE OR HEXATETRAHEDRAL CLASS
SYMMETRY:	SYMMETRY:	SYMMETRY:
Planes, 9 $\begin{cases}3 \text{ axial} \\ 6 \text{ diagonal}\end{cases}$	Planes, 3 Axial	Planes, 6 diagonal
Axes, 13 $\begin{cases}3^{\text{IV}} \\ 4^{\text{III}} \\ 6^{\text{II}}\end{cases}$	Axes, 7 $\begin{cases}3^{\text{II}} \\ 4^{\text{III}}\end{cases}$	Axes, 7 $\begin{cases}3^{\text{II}} \\ 4^{\text{III}}\end{cases}$
Centre of Symmetry	Centre of Symmetry	No Centre of Symmetry
FORMS:	FORMS:	FORMS:
(i) Cube (100)	(i) Cube (100)	(i) Cube (100)
(ii) Rhombdodecahedron (110)	(ii) Rhombdodecahedron (110)	(ii) Rhombdodecahedron (110)
(iii) Octahedron (111)	(iii) Octahedron (111)	(iii) *Tetrahedron* (111)
(iv) Tetrahexahedron (210) etc.	(iv) *Pyritohedron* (210) etc.	(iv) Tetrahexahedron (210) etc.
(v) Trisoctahedron (221) etc.	(v) Trisoctahedron (221) etc.	(v) *Deltoid-dodecahedron* (221) etc.
(vi) Trapezohedron (211) etc.	(vi) Trapezohedron (211) etc.	(vi) *Tristetrahedron* (211) etc.
(vii) Hexoctahedron (321) etc.	(vii) *Diploid* (321) etc.	(vii) *Hexatetrahedron* (321) etc.

The principle of the stereographic projection is shown in Fig. 49A. The normal to the face F at the centre of the sphere meets the surface of the sphere at P, which is called the *pole* of the face. The pole is projected on the equatorial plane at the point I where a line joining P to the south pole of the sphere cuts the equatorial plane. A plane passing through the centre of the sphere intersects the surface of the sphere in a *great circle* which projects on the equatorial plane in different ways according to its inclination to this plane. If the plane of the great circle is vertical it plots as a straight line, if it is horizontal it plots as the equatorial circumference, if it is inclined it plots as an arc whose chord is a diameter of the equatorial circle (Fig. 49B).

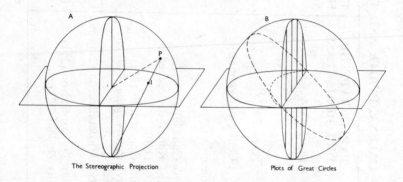

The Stereographic Projection   Plots of Great Circles

FIG. 49. A. Principle of the Stereographic Projection.
B. Great Circles.

Any detailed description of the use of stereographical projection in crystallography is beyond the scope of this book. Nevertheless, it may be used here to give a simplified *pictorial representation* of the symmetry elements of the different classes; this is now done in Fig. 50 for the three symmetry classes that have been described from the Cubic System. The projections show the crystallographic axes, the planes of symmetry and the axes of symmetry, the symbols of these latter being explained in the figure. The differences between the symmetry of the three classes are apparent; the symmetry element common to all three classes is the possession of four triad (or three-fold) axes of symmetry. If he so desires, the student can work out the symmetry from Fig. 50 and compare it with that listed in the Table on p. 111.

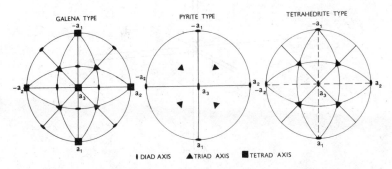

FIG. 50. Symmetry elements of the Galena, Pyrite and Tetrahedrite Types.

II.   TETRAGONAL SYSTEM

Axes. All crystals are classed in the Tetragonal System whose faces can be referred to two equal horizontal crystallographic axes and a third vertical axis which is either shorter or longer than the other two; all three axes are at right angles. The two equal horizontal axes have the same value and are interchangeable. They are both denoted by a, and for ease of reference that running from front to back is called a_1, and that from right to left a_2. The vertical axis is c. The axial plan and nomenclature is shown in Fig. 51. The reading position is c vertical, a_1 running front to back, a_2 right to left. Whilst we have, as the custom is, spoken of two equal horizontal axes and a vertical axis different from these, what is really meant is that the unit or fundamental form cuts the two horizontal axes at the same lengths and the vertical axis at a different length. The *axial ratio* is determined from these intercepts made by the fundamental form. The intercept on the c-axis is given in terms of that on a lateral axis. For

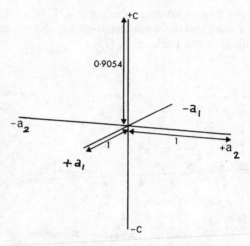

FIG. 51. *Tetragonal Axes,* showing lengths cut off by
the unit form of zircon, $c = 0.9054$.

example, the face belonging to what is chosen as the fundamental
form for the mineral zircon intersects the horizontal axes at a
distance taken as unity and the c-axis at a distance of 0.9054 units.
For zircon, therefore, the axial ratio is expressed as $c = 0.9054$.
The same ratio is expressed by the dimensions of the unit cell of
zircon, $a = 6.604$ Å, $c = 5.979$ Å.

One symmetry class, that characterized by *Zircon,* of the
Tetragonal System is dealt with here. The Zircon Type is
the Normal or highest symmetry class belonging to this
system; it is also called the Ditetragonal-bipyramidal class after
its general form, the ditetragonal bipyramid (*hkl*).

4.   Zircon Type or Ditetragonal-bipyramidal Class

SYMMETRY. The symmetry of the Zircon Type can be made out
from a consideration of the axes or by manipulation of models
of crystals belonging to this type. There is clearly a horizontal
plane of symmetry, containing the horizontal axes a_1 and a_2.
There are two vertical planes of symmetry, one containing the
vertical axis and the a_1 horizontal axis, the other containing the
vertical axis and the a_2 horizontal axis; these two planes are of

course at right angles. In addition, there are two other vertical planes of symmetry which contain the vertical axis and bisect the angles between the two horizontal axes. There are thus *five planes of symmetry* in all, one horizontal axial, two vertical axial, and two vertical diagonal. On account of the tetragonal symmetry, four of the diagonal planes of symmetry characteristic of the Normal or Galena Type of the Cubic System are absent.

The c-axis is clearly an axis of four-fold symmetry. There are four horizontal axes of two-fold symmetry, two of these being the horizontal crystallographic axes and the other two being diagonal axes bisecting the angles between these. There are thus *five axes of symmetry* in the Zircon Type.

Lastly, a *centre of symmetry* is present, corresponding faces, edges etc., being present in pairs on opposite sides of a central point.

The symmetry of the Zircon Type is therefore as follows:—

Planes, 5 $\begin{cases} \text{3 axial (1 horizontal, 2 vertical)} \\ \text{2 diagonal vertical} \end{cases}$

Axes, 5 $\begin{cases} 4^{\text{II}} \text{ (horizontal, 2 crystallographic axial, 2 diagonal)} \\ 1^{\text{IV}} \text{ (vertical crystallographic axial)} \end{cases}$

A Centre of Symmetry.

A pictorial representation of this symmetry is given in Fig. 52 and can be compared with those of the Cubic System given in Fig. 50, p. 113.

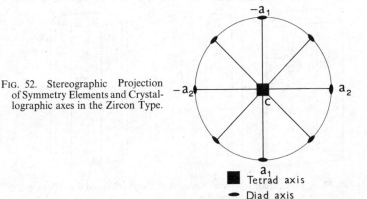

ZIRCON TYPE

FIG. 52. Stereographic Projection of Symmetry Elements and Crystallographic axes in the Zircon Type.

■ Tetrad axis
● Diad axis

FORMS. (i) *Basal Pinacoid*. We may begin with a face cutting the c-axis at its positive end and lying parallel with the horizontal plane of symmetry containing the horizontal axes a_1 and a_2. The symmetry is satisfied by the occurrence of an opposite parallel face cutting the c-axis at its negative end. The form is thus of two faces only, each cutting the c-axis and being parallel to a_1 and a_2. Such a form is called the *basal pinacoid*. The symbols are obviously (001) for the upper face and (00$\bar{1}$) for the lower face. Such a form is an example of an *open form* since it cannot occur alone in a crystal but must be combined with other forms to enclose space. It is shown in combination in Fig. 53.

(ii) *Tetragonal Prism* (100). Now consider a vertical face which is parallel to the c-axis and to one of the horizontal axes. The existence of the vertical diagonal planes of symmetry demands the presence of three other faces to complete the form. A horizontal crystallographic axis emerges at the *middle* of each of the four *faces*, as shown in Fig. 53A. Such a form is sometimes called a *Tetragonal Prism of the Second Order* and is clearly an open form. The symbols of the four faces are 100, 010, $\bar{1}$00, 0$\bar{1}$0, and the prism stands on a square base.

(iii) *Tetragonal Prism* (110). Suppose a vertical face cuts the two horizontal axes at equal distances and is parallel to the vertical axis. Four such faces satisfy the symmetry. The crystallographic axes emerge at the *centres* of the *vertical edges* as indicated in Fig. 53B. This form is an open form with a square base and is

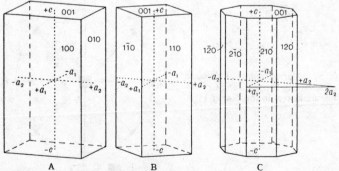

FIG. 53. A. Prism (100) and Basal Pinacoid. B. Prism (110) and Basal Pinacoid. C. Ditetragonal Prism (210) and Basal Pinacoid.

sometimes called a *Tetragonal Prism of the First Order*. Since each face cuts the two horizontal axes at equal distances and is parallel to the vertical axis, the symbols of the four faces are 110, $\bar{1}$10, $\bar{1}\bar{1}$0, 1$\bar{1}$0. The prism (110) and the (100) prism are geometrically alike; in crystals showing combinations of the two forms their naming depends on the position selected for the crystallographic axes.

(iv) *Ditetragonal Prism*. The prism just described meets the horizontal axes at equal distances; we will now consider a vertical face which cuts the horizontal axes at unequal distances. Such a face would be reflected by both the axial and diagonal vertical planes of symmetry so that eight vertical faces would result. This form is called the *Ditetragonal Prism* and is an open form with faces cutting the horizontal axes at unequal lengths and parallel to the vertical axis. There are accordingly a number of ditetragonal prisms depending upon the different intercepts made on the horizontal axes. The general symbol is (*hk*0) and a typical example, shown in Fig. 53C, is (210).

The relation between the three prisms is shown in Fig. 54, which is a plan of the horizontal plane of symmetry, with the a_1 axis running front to back, a_2 axis right to left, and the *c*-axis emerging at the intersection of these two axes. The student should complete the symbols of the various faces whose edges appear in the plan.

(v) *Tetragonal Bipyramid* (*h*0*l*). The prisms of the Zircon type with which we have dealt have faces which are parallel to the vertical

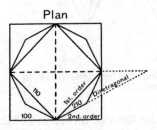

FIG. 54. Relations of the Tetragonal Prisms.

crystallographic axis. The three remaining forms—the bipyramids —now to be described are analogous to the prisms but cut the vertical axis; they are closed forms, that is, they can enclose space alone. The first is the Tetragonal Bipyramid (*h*0*l*) or *Bipyramid of the Second Order*. In this, each face cuts the vertical axis and one of the horizontal axes, and is parallel to the other horizontal axis. One such face in the upper half of the form is repeated three times by the diagonal planes of symmetry, and the resulting four upper faces are reflected in the lower half by the horizontal plane of symmetry. The form thus consists of a square double pyramid with eight isosceles faces. The horizontal crystallographic axes emerge at the *centres* of the horizontal edges. The general symbol is (*h0l*). If a face of this pyramid cuts the horizontal axis and the vertical axis in the same ratio as the axial ratio, then the symbol of the form is (101); such a form is illustrated in Fig. 55A. This form (101) clearly has the faces 101, 011, $\bar{1}$01, 0$\bar{1}$1 above, and 10$\bar{1}$, 01$\bar{1}$, $\bar{1}$0$\bar{1}$, 0$\bar{1}$$\bar{1}$, below. In any mineral, a number of pyramids of the second order may occur according to the variations in the value of *h* and *l* in the general symbol (*h0l*). Examples are (102), (103), (203), etc., which are flatter than the (101) form, and (201), (301), (302), etc., which are steeper. When the pyramid becomes, as it were, infinitely steep, that is when the intercept on the *c*-axis becomes infinite so that *l* = 0, the form becomes a second order prism (100).

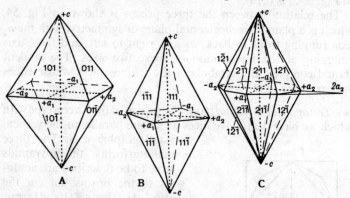

Fig. 55. Bipyramids in the Zircon Type: A. 2nd order bipyramid (101) B. 1st order bipyramid (111). C. Ditetragonal bipyramid (211).

(vi) *Tetragonal Bipyramid* (*hhl*). The faces of this form cut the horizontal axes at equal distances and also cut the vertical axis. The form consists of 8 isosceles faces giving a bipyramid similar in geometrical shape to that of the (*h0l*) bipyramid. The horizontal crystallographic axes emerge at the intersections of the horizontal

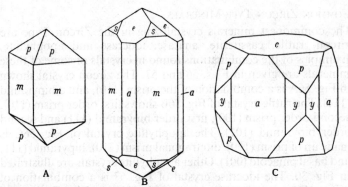

FIG. 56. Combinations in Zircon Type. A. Zircon, c 0·9054; m (110), p (111). B. Rutile, c 0·64; a (100) m (110), e (101), s (111). C. Apophyllite, c 1·761; a (100), c (001), y (310), p (111).

edges as shown in Fig. 55B. The general symbol for the unit or fundamental pyramid, which cuts the axes at the lengths taken as standard for each axis, is (111), illustrated in Fig. 55B. This form has the faces $\bar{1}11$, $\bar{1}\bar{1}1$, $1\bar{1}1$, 111 above, and $11\bar{1}$, $\bar{1}1\bar{1}$, $\bar{1}\bar{1}\bar{1}$, $1\bar{1}\bar{1}$ below. For each mineral there are a number of (hhl)-bipyramids or *bipyramids of the first order*. Pyramids flatter than the unit form (111) have symbols such as (112), (223), (114), etc.; pyramids steeper than the unit form are (221), (332), (441), etc. The limit to the flattening is the basal plane (001), to the steepening, the prism (110).

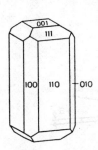

FIG. 57. Idocrase, showing the Forms, Prisms of First and Second Order, Bipyramid First Order, Basal Pinacoid.

(vii) *Ditetragonal Bipyramid*. This bipyramid has sixteen scalene faces which meet the axes at unequal distances. The symbol of this *general* form is (hkl) and an example is the form (211) shown in Fig. 55C. It should be noted that the essential feature of the ditetragonal bipyramid is that h and k are different. Thus, the forms (212), (211), etc., are clearly ditetragonal bipiramids. There are of course, a number of ditetragonal bipyramids possible for a given mineral.

COMMON ZIRCON TYPE MINERALS

The commonest minerals crystallizing in the Zircon Type are zircon, rutile, cassiterite, anatase, idocrase and apophyllite. Examples of the combinations found in crystals of some of these minerals are given in Figs. 56 and 57. The zircon crystal shown in Fig. 56A is a combination of the prism (110), and a bipyramid (111). The rutile crystal of Fig. 56B shows first order prism (110), second order prism (100), first order bipyramid (111) and second order bipyramid (101). The apophyllite crystal of Fig. 56C is made up of prism (100), ditetragonal prism (310), bipyramid (111) and basal pinacoid (001). Other apophyllite crystals are illustrated in Fig. 30. The idocrase crystal of Fig. 57 is a combination of prism (110), prism (100), bipyramid (111), and basal pinacoid (001).

III; HEXAGONAL SYSTEM

Axes. The Hexagonal System contains those crystals that can be referred to four axes,—three equal horizontal axes making angles of 120° with each other, and a vertical axis perpendicular to the plane containing the horizontal axes. The axial nomenclature and polarity are shown in Fig. 58, and should be carefully

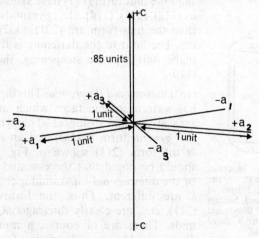

FIG. 58. Hexagonal Axes. Showing lengths cut off by the unit form of calcite, with $c = \cdot 85$.

studied. The three horizontal axes are lettered a_1, a_2, a_3, and the vertical axis is c. Note the positions of the positive ends of the horizontal axes, and that positive and negative ends of these axes alternate. Symbols of faces in the Hexagonal System thus have four numbers in them and the axial order is a_1 a_2 a_3 c. There is a simple relationship between the indices referring to the three horizontal axes. Their sum is always zero, so that two indices having been calculated the third is given by adding these two together and changing the sign. Examples of hexagonal symbols to show this relationship are $(21\bar{3}0)$, $(10\bar{1}1)$, $(3\bar{1}\bar{2}1)$, where it is noted that the sums of the first three figures are zero.

In the Hexagonal System, the unit form is bounded by faces that cut the vertical axes at the distance taken as the unit for that axis, and cut two horizontal axes at equal distances taken as the unit for the horizontal axes. It is clear that if a face cuts two horizontal axes at the same distances it must be parallel to the third horizontal axis. The unit form must therefore have a symbol like $(10\bar{1}1)$. From the unit form the *Axial Ratio* is obtained in the same way as in the Tetragonal System. It is the ratio of the intercept made by the unit form on the c-axis to that made on a horizontal axis.

The Hexagonal System can be divided into the *Hexagonal* and *Trigonal* divisions, in the first the vertical axis being of six-fold symmetry, in the second of three-fold. The *Beryl Type* possesses the highest symmetry possible in this system. Three other types dealt with here—the *Calcite, Tourmaline* and *Quartz* types— belong to the *Trigonal* division, regarded by some as a distinct system.

5.   Beryl Type or Dihexagonal-bipyramidal Class

The Beryl Type is not of importance except for the illustration of the normal hexagonal type of symmetry. It is exactly analogous with the Zircon Type of the Tetragonal System, and its symmetry can be easily deduced from that of that type, so that there is no need to consider it in detail here.

SYMMETRY. From a consideration of the axes there is clearly a horizontal plane of symmetry. Six vertical planes of symmetry, three axial and three diagonal, are comparable with the four vertical planes, two axial and two diagonal, of the Zircon Type. There are thus seven planes of symmetry in all. The vertical axis is an axis of six-fold symmetry, analogous with the four-fold axis of the Zircon Type. There are six axes of two-fold symmetry in the horizontal axial plane, three being the crystallographic axes and three diagonal; these compare with the four horizontal axes of symmetry of the Zircon Type. There are therefore seven axes of symmetry in the Beryl Type. There is also a centre of symmetry. The complete symmetry is therefore:

Planes, 7 { 4 axial (1 horizontal, 3 vertical)
{ 3 diagonal, vertical

Axes, 7 { 6^{II} (horizontal, 3 crystallographic axial, 3 diagonal)
{ 1^{VI} (vertical crystallographic axial)

A Centre of Symmetry.

The symmetry picture of the Beryl Type given in Fig. 73 should be compared with that of the Zircon Type in Fig. 52.

FORMS. (i) *Basal Pinacoid.* This is an open form consisting of two faces, each cutting the vertical axis and being parallel to the three horizontal axes. Symbols of the faces are (0001) and (000$\bar{1}$).

(ii) *Hexagonal Prism* (11$\bar{2}$0) *or of the Second Order.* This open form is a prism of six faces, the horizontal crystallographic axes joining the centres of opposite parallel faces, as shown in Fig 60. Each face is parallel to the vertical axis, and cuts all

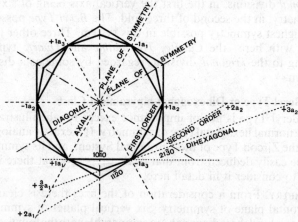

FIG. 59. Plan showing the Relations of the Prisms in the Beryl Type of the Hexagonal System.

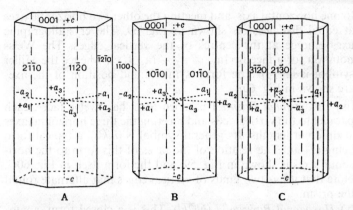

FIG. 60. Combinations of Prisms and Basal Pinacoid in the Beryl Type: A, 2nd
order prism and basal pinacoid. B, 1st order prism and basal pinacoid. C,
Dihexagonal prism and basal pinacoid.

three horizontal axes, one at the unit distance and the other two at
twice this distance. That this relation holds can be seen from
Fig. 59. The Weiss notation of a face is therefore $(2a_1, 2a_2. - 1a_3.$
$\infty c)$, whence the Miller symbol is $(11\bar{2}0)$. The form is represented
in Fig. 60A.

(iii) *Hexagonal Prism* $(10\bar{1}0)$ *or of the First Order*. This open form
is a prism of six faces, each face parallel to the vertical axis, and

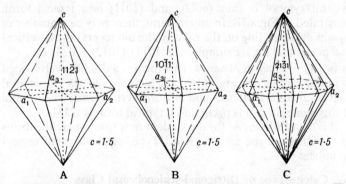

FIG. 61. Hexagonal Bipyramids. A. Hexagonal bipyramid of Second Order $(11\bar{2}1)$.
B. Hexagonal bipyramid of First Order $(10\bar{1}1)$. C. Dihexagonal bipyramid
$(21\bar{3}1)$. Axial Ratio, $c=1\cdot5$.

to one horizontal axis, and meeting the other two horizontal axes at equal distances, as shown in Fig. 59. The crystallographic axes emerge at the centres of the vertical edges. The Weiss notation for a face is $(1a_1, \infty\ a_2. - 1a_3\ \infty\ c)$, whence the Miller symbol is $(10\bar{1}0)$. The form, symbols and position of the axes are shown in Fig. 60B.

(iv) *Dihexagonal Prism.* This open form has twelve faces, each parallel to the vertical axis and meeting the three horizontal axes at different lengths. Its general symbol is $(hi\bar{k}0)$, a typical form being $(21\bar{3}0)$. The relation of the faces of this form to the horizontal axes is shown in Fig. 59, and the form itself in Fig. 60C. Note that it has two kinds of vertical edges that alternate round the prism.

(v) *Hexagonal Bipyramid* $(hh\bar{2}hl)$. This is a closed form, analogous with the Second Order Prism $(11\bar{2}0)$; each of its twelve faces cuts the vertical axis and all three horizontal axes, one at the unit distance and the other two at twice this distance. The general symbol is thus $(hh\bar{2}hl)$ and typical forms are $(11\bar{2}1)$, $(11\bar{2}2)$, $(22\bar{4}3)$, etc. A second order hexagonal bipyramid $(11\bar{2}1)$ is illustrated in Fig. 61A.

(vi) *Hexagonal Bipyramid* $(h0\bar{h}l)$. This closed form corresponds to the First Order Prism. It consists of twelve like faces each cutting two horizontal axes at equal distances and being parallel to the other horizontal axis, and cutting the vertical axis. The general symbol is thus $(h0\bar{h}l)$, and $(10\bar{1}1)$ is a typical form, illustrated in Fig. 61B. In any mineral, there may be a number of pyramids depending on the ratio of the intercepts on the vertical and horizontal axes. Examples are $(10\bar{1}1)$, $(10\bar{1}2)$, $(20\bar{2}3)$, etc.

(vii) *Dihexagonal Bipyramid.* This form is a double twelve-sided pyramid shown in Fig. 61C. Each face cuts the horizontal axes at unequal distances and also the vertical axis. The symbol of this general form is $(hk\bar{i}l)$ and a typical form is $(21\bar{3}1)$.

It is a useful exercise for the student now to compare the forms in the Zircon Type and the Beryl Type, especially with respect to indices.

6. Calcite Type or Ditrigonal-scalenohedral Class

SYMMETRY. The symmetry of the Calcite Type can be obtained by studying a typical form belonging to the type. The form we select for this purpose is the rhombohedron, described and figured below. It is a solid bounded by three pairs of rhomb-shaped faces, and it looks like a pushed-over cube. Inspection of the figures of the rhombohedron in the reading position or, better, manipulation of a model, shows that there are present three vertical planes of symmetry, which lie midway between the horizontal crystallographic axes. The vertical crystallographic axis, which joins the two solid angles formed by the obtuse angles of the rhomb faces, is an axis of three-fold symmetry, as may be tested by rotating the rhombohedron about this axis, when three similar rhomb faces occupy a similar position during a complete turn. The horizontal axes, which join the middles of the pairs of opposite edges, are axes of two-fold symmetry as may be demonstrated in the same way. The faces, edges, etc., of the rhombohedron occur in pairs on opposite sides of a central point. Lay the rhombohedron with one face on the table; a second face appears at the top parallel with the surface of the table. The rhombohedron therefore has a centre of symmetry.

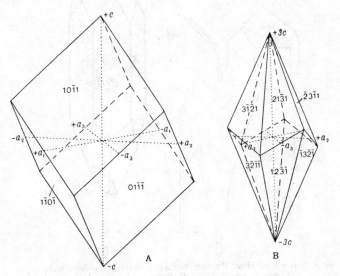

Fig. 62. A. Rhombohedron, (10$\bar{1}$1); axial ratio $c=2$. B. Scalenohedron (21$\bar{3}$1); axial ratio $c=\cdot85$, Calcite.

The symmetry of the Calcite Type is therefore:
Planes, 3 vertical diagonal

Axes, 4 $\begin{cases} 3^{II}, \text{ horizontal crystallographic axes} \\ 1^{III}, \text{ vertical crystallographic axis} \end{cases}$

A Centre of Symmetry.

FORMS. (i) *Rhombohedron*. The rhombohedron is figured in Fig. 62A. It has six like faces, each rhomb-shaped. The *c*-axis joins the two solid angles formed by the obtuse angles of the rhomb faces and the horizontal axes join the middles of the pairs of opposite edges. In the reading position, therefore, the rhombohedron shows three faces above, and three faces below. For each face above

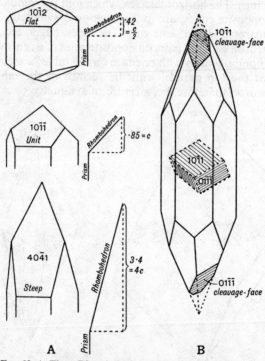

FIG. 63. A. Three Rhombohedra of Calcite, and inclinations of rhombohedral faces. B. Prism and scalenohedron of Calcite, with cleavage-faces of cleavage-rhombohedron (10$\bar{1}$1).

there is an edge below and for each edge above there is a face below. The lateral edges zig-zag round the crystal. Inspection of the rhombohedron will show that its faces are arranged in accordance with the symmetry of the Calcite Type. Instead of developing the symmetry of this type from an examination of the rhombohedron, we might have stated the symmetry and developed the form of the rhombohedron.

From the figure it is seen that the upper front face intersects the a_1 axis at a certain distance, is parallel to the a_2 axis, cuts the a_3 axis at a negative distance equal to the intercept on the a_1 axis, and cuts the c-axis. The general symbol is therefore $(h0\bar{h}l)$, and the unit rhombohedron, cutting the axes at the standard distances, has the symbol $(10\bar{1}1)$. The student should develop the indices of the faces of the unit rhombohedron shown in Fig. 62. There are clearly a number of rhombohedra possible in any given mineral. Rhombohedra flatter than the unit form have symbols such as $(10\bar{1}2)$, $(02\bar{2}3)$, etc.; rhombohedra steeper than the unit form have symbols such as $(20\bar{2}1)$, $(30\bar{3}2)$, etc. Examples are given in Fig. 63 where the profiles should be studied.

It will be recalled that the form having the general symbol $(h0\bar{h}l)$, or $(10\bar{1}1)$, etc., in the Beryl or Normal Type of the

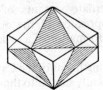

FIG. 64. Plus Rhombohedron showing its relation to the hexagonal bipyramid.

FIG. 65. Positive and Negative rhombohedra.

Hexagonal System is the hexagonal bipyramid of the first order. The relation between the rhombohedron and this latter form is shown in Fig. 64, where it is seen that the rhombohedron may be formed by developing the alternate faces (really, alternate sectants) of the hexagonal bipyramid. There are therefore possible two rhombohedra of similar geometrical form—the positive rhombohedron such as (10$\bar{1}$1) having a face uppermost, and a negative rhombohedron, such as (01$\bar{1}$1), having an edge uppermost, as shown in Fig. 65. The two rhombohedra may be distinguished by etch-marks or other properties.

(ii) *Scalenohedron.* The ditrigonal scalenohedron, shown in Fig. 62B, is a solid bounded by twelve faces, each face being a scalene triangle. The terminal edges are alternately blunt and sharp, and the lateral edges zig-zag round the crystal. Each face cuts the vertical axis, and also the three horizontal axes at unequal distances. The general symbol is thus (*hk$\bar{i}$l*) and an example is (21$\bar{3}$1) which is shown in the figure. This form is the *general* form and gives its name, ditetragonal scalenohedral, to the Calcite Type. There are, of course, many scalenohedra possible in the crystals of one particular mineral.

The form in the Beryl Type having symbols like (21$\bar{3}$1) is the dihexagonal bipyramid, and the relationship between the scalenohedron and this bipyramid is shown in Fig. 66. The development of the faces in alternate sectants of the pyramid and suppression of the others would lead to the formation of a scalenohedron.

The five remaining forms of the Calcite Type are geometrically similar to those of the Beryl Type, but have a lower structural symmetry. These forms are:—the *Basal Pinacoid* (0001), the three prisms— *Hexagonal Prism of the Second Order* (11$\bar{2}$0), *Hexagonal Prism of the First Order* (10$\bar{1}$0) and the *Dihexagonal Prism* (*hk$\bar{i}$0*), (21$\bar{3}$0), etc.—and, lastly, the *Hexagonal Bipyramid of the Second Order,* such as (11$\bar{2}$1).

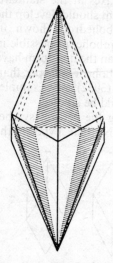

Fig. 66. The Scalenohedron (21$\bar{3}$1) showing its relation to the dihexagonal bipyramid.

COMMON CALCITE TYPE MINERALS

Calcite. Calcite occurs in various combinations of prisms, rhombohedra, and scalenohedra. Its crystals are of great beauty

(Fig. 1) and display a number of varied habits as shown in Figs. 63, 67. A common crystal of calcite is shown in Fig. 67, and consists of the prism m $(10\bar{1}0)$, a scalenohedron v $(21\bar{3}1)$ ter-

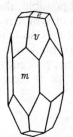

minated by a flat negative rhombohedron e $(01\bar{1}2)$. Calcite gives by cleaving the unit rhombohedron $(10\bar{1}1)$, and often the ends of scalenohedral crystals show one or more cleavage-faces, as illustrated in Fig. 63B, due to rough usage.

Siderite. Siderite occurs in rhombohedra the faces of which are sometimes curved.

Corundum. Corundum forms crystals of pyramidal or barrel-shaped habits; water-worn examples of

FIG. 67. Calcite showing the Forms, Prism (m), Scalenohedron (v) and Negative Rhombohedron (e).

these are shown in Fig. 123. The acute bipyramidal crystals show the form $(22\bar{4}3)$, a hexagonal bipyramid of the second order, dominant. The barrel-shaped crystals are combinations of various bipyramids of the second order, such as $(14.14.\overline{28}.3)$, $(22\bar{4}1)$, $(22\bar{4}3)$, terminated by the basal pinacoid (0001). Hexagonal prisms such as $(11\bar{2}0)$ and rhombohedra such as $(10\bar{1}1)$ are commonly seen in corundum crystals.

Hematite. Hematite occurs in thin tabular crystals, usually showing the basal pinacoid (0001) and a positive rhombohedron $(10\bar{1}1)$.

7. Tourmaline Type or Ditrigonal-pyramidal (hemimorphic) Class

SYMMETRY. Crystals belonging to the Tourmaline Type illustrate the phenomenon of *hemimorphism*. In the symmetry types already dealt with here we have found that the two ends of any crystallographic axis or axis of symmetry are similar with regard to the grouping of the crystal planes around them. In hemimorphic types, however, this is not the case, one half of the faces of a holohedral form being grouped about one end of the axis and none at the other end. Hemimorphic forms obviously cannot enclose space, and must occur in combination with other hemimorphic forms to produce a crystal differing markedly in appearance at the two ends of the axis in question. It is clear that there can be no centre of symmetry in hemimorphic crystals.

The symmetry of the Tourmaline Type can be derived from that of the Calcite Type by considering the consequences of the removal of the centre of symmetry from the latter type. The three vertical diagonal planes of symmetry are still possible; the horizontal crystallographic axes are no longer axes of symmetry since the forms differ at the two ends of the vertical axis, so that the axes of symmetry are reduced to one of three-fold symmetry,— the vertical crystallographic axis. There is, of course, no centre of symmetry. The symmetry of the Tourmaline Type is therefore:—

Planes, 3 vertical diagonal.

Axes, 1^{III}, the vertical crystallographic axis.

No Centre of Symmetry.

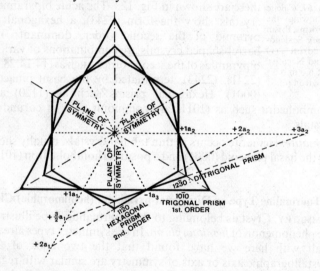

FIG. 68. Plan showing relations of prisms in the Tourmaline Type.

FORMS. (1a) and (1b). *Upper Basal Plane and Lower Basal Plane.* Since there is no centre of symmetry, the face cutting the *c*-axis and lying parallel to the horizontal axes fulfils the symmetry requirements in itself. The two faces 0001 and 000$\overline{1}$ of the basal pinacoid of the normal hexagonal crystal are two distinct forms, the upper basal plane or pedion (0001) and the lower basal plane or pedion (000$\bar{1}$).

(ii) *Hexagonal Prism* (11$\bar{2}$0) *or Prism of the Second Order*. The form (11$\bar{2}$0) satisfies the symmetry as seen in Fig. 68, and is a possible form in the Tourmaline Type. It is a hexagonal prism of the second order similar to that in the Beryl or Calcite Types.

(iii) *Trigonal Prism* (10$\bar{1}$0) *or of the First Order*. A form that has a face 10$\bar{1}$0 will be completed by the addition of two more faces, $\bar{1}$100 and 0$\bar{1}$10, as inspection of Fig. 68 shows. This form thus consists of three alternate faces of the hexagonal prism of the order, and is called the *Trigonal Prism of the First Order* (10$\bar{1}$0), or *Unit Trigonal Prism*. There are clearly two trigonal prisms, (10$\bar{1}$0) and (01$\bar{1}$0).

(iv) *Ditrigonal Prism*. It can be seen from Fig. 68 that a given face 12$\bar{3}$0 can be accompanied only by five other faces, by virtue of the diagonal planes of symmetry of the Tourmaline Type. Such a six-sided prism, with two alternating kinds of edges, is called the *Ditrigonal Prism,* and it clearly consists of alternate faces of the dihexagonal prism of the Calcite Type.

(v) *Hemimorphic Hexagonal Pyramid*. We have seen that the Hexagonal Prism (11$\bar{2}$0) is a possible form of the Tourmaline Type. A bipyramid such as (11$\bar{2}$1) corresponding to this form would agree with the vertical axis of threefold symmetry and the three diagonal planes of symmetry of the Tourmaline Type, but would not agree with the absence of a centre of symmetry. This last requirement limits the pyramid to the six upper faces only, so that in the Tourmaline Type there are two separate forms, one formed of the six upper faces of the hexagonal dipyramid and the other of the six lower faces. Such forms are called *Hemimorphic Hexagonal Pyramids*.

(vi) *Trigonal Pyramid*. Consider the face 10$\bar{1}$1, which, in the Calcite Type, is an upper face of the positive rhombohedron. In the Tourmaline Type, the symmetry is satisfied by the occurrence, in addition to the face 10$\bar{1}$1, of the faces $\bar{1}$101 and 0$\bar{1}$11, and these three faces are the three upper faces of the unit positive rhombohedron. Four such *trigonal pyramids* are related to one another so that one, *e.g.* (10$\bar{1}$1), may be considered to consist of the three upper faces of a positive rhombohedron, a second, *e.g.*

($\overline{1}01\overline{1}$), of the three lower faces of the same rhombohedron, a third, *e.g.* ($\overline{1}011$), of the three upper faces of the corresponding negative rhombohedron and the fourth, *e.g.* ($10\overline{1}\overline{1}$) of the three lower faces of the same negative rhombohedron.

(vii) *Ditrigonal Pyramid.* There are four *ditrigonal pyramids* which correspond to positive and negative scalenohedra in exactly the same way as the trigonal pyramids just described correspond to the two rhombohedra,—that is, one ditrigonal pyramid such as ($21\overline{3}1$) is made up of the upper faces of the positive scalenohedron, a second of the lower faces, a third of the upper faces of the negative scalenohedron and a fourth of the lower faces. This form is the *general* form and gives its name to the Tourmaline Type.

TOURMALINE CRYSTALS

Tourmaline commonly occurs in prismatic crystals having the two trigonal prisms ($10\overline{1}0$) and ($01\overline{1}0$) usually unequally developed. A

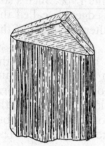

vertical striation is often seen due to oscillatory combination of the trigonal prism ($10\overline{1}0$) and the hexagonal prism ($11\overline{2}0$). Terminations of tourmaline crystals show various trigonal pyramids. A crystal of tourmaline is shown in Fig. 69.

8. Quartz Type or Trigonal Trapezohedral Class

FIG. 69. Tourmaline.

SYMMETRY. The commonest crystals of quartz occur in the form of an apparently hexagonal prism bounded above and below by apparently hexagonal pyramids. Thus it would appear that quartz belonged to the Beryl Type, and therefore possessed the highest symmetry of the Hexagonal System. However quartz is proved to have a symmetry much lower than that of beryl by several considerations. Etch-marks indicate that the pyramidal or prism faces are not all of the same kind, as shown in Fig. 29. The occurrence on some crystals of quartz of faces belonging to forms. called the trigonal bipyramid and the trigonal trapezohedron, having a much lower symmetry than the Beryl Type, shows that the characteristic forms are not simple hexagonal pyramids or prisms. These conclusions are confirmed by structural analysis.

The symmetry of quartz can be demonstrated with the general form, the trigonal trapezohedron, shown in Fig. 70. This is a six-faced solid bounded by trapezium-shaped faces and having the general symbol ($hk\bar{i}l$). Inspection of the drawing or of a model shows that there are no planes of symmetry present. The vertical crystallographic axis is an axis of three-fold symmetry, a similar face occupying a similar position three times during a complete revolution about this axis. The horizontal crystallographic axes are axes of two-fold symmetry. There is obviously no centre of symmetry. The symmetry of the Quartz Type is therefore:—

Planes, none

Axes, 4 $\begin{cases} 3^{II}\text{, the horizontal crystallographic axes.} \\ 1^{III}\text{, the vertical crystallographic axis.} \end{cases}$

No Centre of Symmetry.

FORMS. (i) *Trigonal Trapezohedron*. The Trigonal Trapezohedron is a solid bounded by six trapezium-shaped faces, and the relation

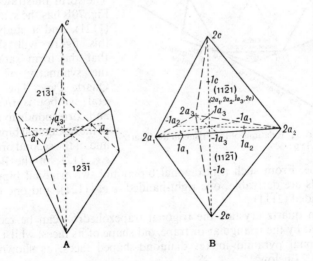

FIG. 70. Quartz Type. A. Trigonal Trapezohedron ($21\bar{3}1$). B. Trigonal Bipyramid ($11\bar{2}1$).

of its faces to the crystallographic axes is apparent from its general symbol ($hk\bar{i}l$). The horizontal crystallographic axes pass through the centres of opposite lateral edges. This form has one quarter of the faces of the dihexagonal bipyramid (*e.g.* 21$\bar{3}$1) of the holohedral Beryl Type, and is an example of so-called tetartohedrism. Four trapezohedra are related to each dihexagonal bipyramid, a right-handed positive form such as (21$\bar{3}$1), a left-handed positive form such as (3$\bar{1}$2$\bar{1}$), and two corresponding negative forms, ($\bar{1}$3$\bar{2}$1) and (12$\bar{3}$1). In Fig. 70, a right-handed positive trigonal trapezohedron is shown. A commonly occurring form in quartz is (51$\bar{6}$1).

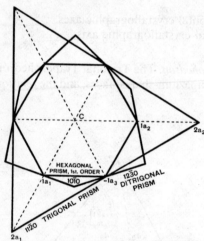

FIG. 71. Relations of the Prisms in the Quartz Type.

(ii) *Trigonal Bipyramid.* The trigonal bipyramid is a form bounded by the six faces of a double pyramid, the base of which is an equilateral triangle and the sides are isosceles triangles. The form illustrated in Fig. 70B has the symbol (11$\bar{2}$1), and a study of this figure will show that the form satisfies the symmetry of the Quartz Type. The general symbol is ($hh\overline{2h}l$) and corresponds to that of the hexagonal bipyramid of the second order, *e.g.* 11$\bar{2}$1, of the Beryl Type. From such a hexagonal bipyramid two trigonal bipyramids are derivable, one right-handed, *e.g.* (11$\bar{2}$1), and one left-handed (2$\bar{1}\bar{1}$1).

In quartz crystals the trigonal trapezohedron can be recognised by the triangular or trapezoid shape of its faces, whilst the trigonal pyramid makes diamond-shaped faces, as shown in Fig. 72 below.

(iii) *Rhombohedron.* A face such as 10$\bar{1}$1 would be repeated about the vertical axis of threefold symmetry to give three above, and the three horizontal axes of two-fold symmetry give three similar faces below. The resulting form is the ordinary *rhombohedron*, such as (10$\bar{1}$1). There are both positive and negative forms and, in some quartz crystals, they may be about equally developed, producing together a form like that of a hexagonal bipyramid, as shown in Fig. 146, p. 439.

(iv) *Trigonal Prisms (e.g.* 11$\bar{2}$0) *or Prisms of the Second Order.* By making the intercepts on the vertical axis of the faces of the trigonal bipyramid (11$\bar{2}$1) infinity, that is by making the faces parallel with the *c*-axis, a three-faced or *trigonal prism* (11$\bar{2}$0) is produced, as shown in plan in Fig. 71. A right-handed trigonal prism such as (11$\bar{2}$0) and a left-handed form (2$\bar{1}$$\bar{1}$0) are thus possible.

(v) *Ditrigonal Prism.* Similarly, by making the faces of the trigonal trapezohedron (21$\bar{3}$1) parallel with the vertical axis, a six-faced or *ditrigonal* prism (21$\bar{3}$0) is produced. The edges are sharper and blunter alternately, as can be seen from Fig. 71. The general symbol is (*hk$\bar{i}$*0).

(vi) *Hexagonal Prism* (10$\bar{1}$0) *or Prism of the First Order.* The hexagonal prism of the first order (10$\bar{1}$0) conforms to the symmetry demands of the Quartz Type and is a possible form in this type. It is geometrically similar to the corresponding form in the Beryl Type (see Fig. 71).

(vii) *Basal Pinacoid.* The horizontal axes of symmetry of the Quartz Type require two faces cutting the vertical axis and lying parallel with the horizontal axes. The symbols of the faces are of course (0001) and (000$\bar{1}$).

COMMON MINERALS OF THE QUARTZ TYPE

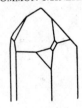

FIG. 72. Quartz Crystal.

Quartz and cinnabar are the commonest minerals crystallising in the Quartz Type.

Many crystals of quartz, as already mentioned, are combinations of the hexagonal prism (10$\bar{1}$0) terminated by the two rhombohedra. (10$\bar{1}$1) and (01$\bar{1}$1). When these two rhombohedra are equally developed, a crystal appearing to have the symmetry of the Beryl Type is produced. Often the two rhombohedra occur alone producing a combination having the outward form of a hexagonal bipyramid, as shown in Fig. 146, p. 439. Some crystals of quartz show, in addition to the prism and unequally developed rhombohedra, small faces belonging to the trigonal trapezohedron (51$\bar{6}$1) and the trigonal bipyramid (11$\bar{2}$1), faces of the former being trapezoid or triangular and of the latter diamond-shaped. In Fig. 72 there is given such a combination of hexagonal prism, positive and negative rhombohedra, and trigonal trapezohedron and trigonal bipyramid. Such a crystal is called right-handed.

Comparison of the Beryl, Calcite, Tourmaline and Quartz Types

The corresponding symmetries and forms of the four symmetry classes of the Hexagonal System dealt with here are shown in the Table opposite. In Fig. 73, there is given a pictorial representation of the symmetry elements of the four classes by use of the stereographic projection as explained on p. 110. The differences in symmetry between the classes are apparent in Fig. 73 and symmetries shown in the figure should be matched with their equivalents shown in the Table.

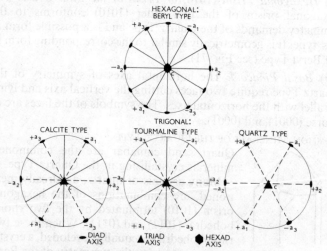

FIG. 73. Symmetry elements of the Beryl, Calcite, Tourmaline and Quartz Types of the Hexagonal System.

COMPARISON OF SYMMETRY TYPES OF THE HEXAGONAL SYSTEM

Beryl Type.	Calcite Type.	Tourmaline Type.	Quartz Type.
SYMMETRY.	SYMMETRY.	SYMMETRY.	SYMMETRY.
Planes, 7 $\begin{cases}4 \text{ axial.} \\ 3 \text{ diagonal.}\end{cases}$	Planes, 3 diagonal.	Planes, 3 diagonal.	Planes, none.
Axes, 7 $\begin{cases}6^{11} \\ 1^{v\,i}\end{cases}$	Axes, 4 $\begin{cases}3^{11} \\ 1^{111}\end{cases}$	Axes, 1^{111}.	Axes, 4 $\begin{cases}3^{11} \\ 1^{111}\end{cases}$
Centre of Symmetry.	Centre of Symmetry.	No Centre of Symmetry.	No Centre of Symmetry.
FORMS.	FORMS.	FORMS.	FORMS.
(i) Basal Pinacoid (0001).	(i) Basal Pinacoid (0001).	(i) Basal Planes (0001) and (000$\bar{1}$).	(i) Basal Pinacoid (0001).
(ii) Hexagonal Prism (11$\bar{2}$0).	(ii) Hexagonal Prism (11$\bar{2}$0).	(ii) Hexagonal Prism (11$\bar{2}$0).	(ii) *Trigonal Prism* (11$\bar{2}$0).
(iii) Hexagonal Prism (10$\bar{1}$0).	(iii) Hexagonal Prism (10$\bar{1}$0).	(iii) *Trigonal Prism* (10$\bar{1}$0).	(iii) Hexagonal Prism (10$\bar{1}$0).
(iv) Dihexagonal Prism (21$\bar{3}$0) etc.	(iv) Dihexagonal Prism (21$\bar{3}$0) etc.	(iv) *Ditrigonal Prism* (21$\bar{3}$0) etc.	(iv) *Ditrigonal Prism* (21$\bar{3}$0) etc.
(v) Hexagonal Bipyramid (11$\bar{2}$1) etc.	(v) Hexagonal Bipyramid (11$\bar{2}$1) etc.	(v) *Hemimorphic Hexagonal Pyramid* (11$\bar{2}$1) etc.	(v) *Trigonal Bipyramid* (11$\bar{2}$1) etc.
(vi) Hexagonal Bipyramid (10$\bar{1}$1) etc.	(vi) *Rhombohedron* (10$\bar{1}$1) etc.	(vi) *Trigonal Pyramid or Hemi-rhombohedron* (10$\bar{1}$1) etc.	(vi) *Rhombohedron* (10$\bar{1}$1) etc.
(vii) Dihexagonal Bipyramid (21$\bar{3}$1) etc.	(viii) *Scalenohedron* (21$\bar{3}$1) etc.	(vii) *Ditrigonal Pyramid or Hemi-scalenohedron* (21$\bar{3}$1) etc.	(vii) *Trigonal Trapezohedron* (21$\bar{3}$1) etc.

IV. ORTHORHOMBIC SYSTEM

Axes. All crystals whose faces can be referred to three unequal axes at right angles belong to the *Orthorhombic System*. The axes and their nomenclature are shown in Fig. 74. One axis, the *c* or vertical axis is placed vertical. One horizontal axis runs front to back and is the *a*-axis, the other horizontal axis runs right to left and is the *b*-axis. The *a*-axis was formerly called the *brachy*-axis (*short* axis) and the *b*-axis the *macro*-axis (*long* axis), but for some orthorhombic minerals such relative lengths do not hold for the axes chosen.

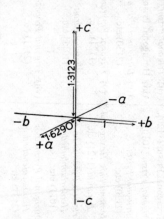

FIG. 74. Orthorhombic axes, showing lengths cut off by the unit form of Barytes, with a : b : c = 1·6290 : 1 : 1·3123.

As is the custom, we have spoken of the three axes of the Orthorhombic System as unequal, implying thereby that the intercepts made by the fundamental form on the axes are all different in length. The intercept that this form makes on the *b*-axis is taken as unity, and therefore the expression for the *axial ratio* is exemplified by that of barytes:

$$a : b : c = 1·6290 : 1 : 1·3123,$$

which states the observed fact that the form chosen as the standard or fundamental form and allotted the symbol (111) cuts the *a*-axis at 1·6290 units, the *b*-axis at unity, and the *c*-axis at 1·3123 units.

There is only one important symmetry class in the Orthorhombic System, namely the Normal Class of the system, the *Barytes Type*.

9. Barytes Type or Orthorhombic-bipyramidal Class

SYMMETRY. From a consideration of the axes of the orthorhombic system or from examination of characteristic forms, the symmetry can be made out. It is the geometrical symmetry of a matchbox or brick. There are three planes, which each contain two crystallographic axes, which divide a Barytes Type

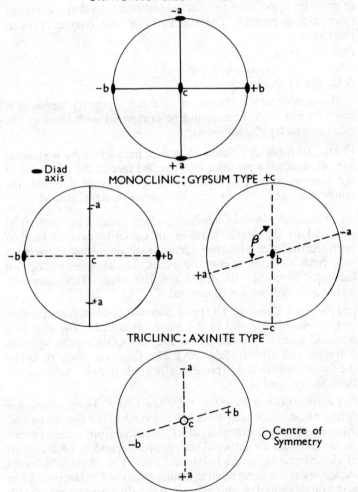

FIG. 75. Symmetry elements of the Barytes, Gypsum and Axinite Types.

crystal into two similar and similarly placed halves. Each crystallographic axis is an axis of two-fold symmetry, since rotation about them causes the crystal to take up the same position twice during a complete turn. The faces, edges, etc., run in pairs on opposite sides of a central point, so that a centre of symmetry is present. The symmetry of the Barytes Type is therefore:

Planes, 3 axial
Axes, 3^{II}, the crystallographic axes
A Centre of Symmetry.

The stereographic projection of these symmetry elements is given in Fig. 75A, and should be compared with those of the other symmetry classes shown.

FORMS. (i) *Basal or c Pinacoid.* A face parallel to the horizontal axes demands the presence of a parallel face at the other end of the vertical axis. This open form is the *basal pinacoid*, the symbols of the two faces being 001 above, and 00$\bar{1}$ below.

(ii) *Front or a Pinacoid.* Similarly, a face parallel to the vertical axis and to the *b*-axis requires an opposite parallel face to complete the form. This gives a pinacoid distinguished as the front or *a* pinacoid since it cuts the *a*-axis. The symbols of the two faces are obviously 100 front and $\bar{1}$00 back. This form was formerly called the *macropinacoid.*

(iii) *Side or b Pinacoid.* The third pinacoid has two faces parallel to the vertical axis and to the *a*-axis. It is called the side or *b* pinacoid since it cuts the *b*-axis and its faces have the symbols 010 right and 0$\bar{1}$0 left (Fig. 76A). This form was formerly called the *brachypinacoid.* All pinacoids are adequately indicated by their Miller symbol.

(iv) *Prism (hk0) or Prism of the Third Order.* These prisms are forms whose faces are parallel to the vertical (or *third*) axis and intersect the two horizontal axes. Prisms therefore consist of four faces. There are a number of such prisms depending on the ratios of the intercepts on the horizontal axes. The prism which cuts these axes at the same relative distances as the fundamental form is the fundamental or unit prism and clearly has the symbol (110) and its faces are 110, $\bar{1}$10, $\bar{1}\bar{1}$0, 1$\bar{1}$0. Other prisms are (210),

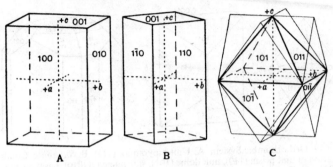

Fig. 76. Combinations in the Barytes Type. A. The three pinacoids. B. Unit prism and basal pinacoid. C. The domes (101) and (0$\bar{1}$1).

(120), (320), etc., and the general symbol is (hk0). The unit prism is illustrated in Fig. 76B.

(v) *Prism ($h0l$), Prism of the Second Order or Dome.* This is an open form of four faces which are parallel to the b (or *second*) axis, but cut the other two axes. It is, of course, of the same type of form as the prism just described and we may consider this form as a vertical dome or, conversely, the domes as horizontal prisms. The unit dome of the second order has the symbol (101) and faces 101, 10$\bar{1}$, $\bar{1}$0$\bar{1}$, $\bar{1}$01. Other domes of this kind are (201), (203), (103), etc., and the general symbol is ($h0l$). They were formerly called *macrodomes*.

(vi) *Prism ($0kl$), Prism of the First Order or Dome.* This form is an open one of four faces which are parallel to the a (or *first*) axis and cut the other two axes. The unit dome of the first order has the symbol (011) and the faces 011, 01$\bar{1}$, 0$\bar{1}\bar{1}$, 0$\bar{1}$1. Other domes of this kind are (012), (023), (031), etc., and the general symbol is ($0kl$). They were formerly called *brachydomes*. Domes of second and third orders are shown in Fig. 76C.

(vii) *Bipyramid.* A face of the bipyramid cuts all three axes. The symmetry requires that there should be eight such faces, so that a closed form results, bounded by faces each a scalene triangle. This is the *bipyramid*. The unit bipyramid has the symbol (111). Other bipyramids are (112), (213), (123), etc., depending on the

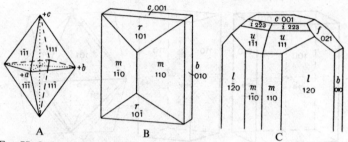

Fig. 77. Orthorhombic System. A. Unit bipyramid, (111). B. *Staurolite*. Combination of unit prism (110), unit dome (101), side pinacoid (010), basal pinacoid (001). C. *Topaz*: combination of prisms, m (110), l (120), bipyramids, u (111), i (223), dome, f (021), side pinacoid, b (010), basal pinacoid, c (001).

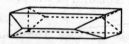

Fig. 78. Crystal of Barytes, showing the Forms, Prism (210), Side Pinacoid (010), Basal Pinacoid (001), and Dome (101).

ratio of the intercepts to those made by the unit form. The general symbol is (*hkl*) and this general form gives its name to the class. The unit bipyramid is illustrated in Fig. 77A.

COMMON BARYTES TYPE MINERALS

Many common minerals crystallize in the Barytes Type. Examples are barytes, celestite, olivine, enstatite, andalusite, topaz, anhydrite, aragonite, sulphur and staurolite. A crystal of barytes is illustrated in Fig. 78; the forms present are the prism (210), basal pinacoid (001), dome (101) and side pinacoid (010). The staurolite crystal shown in Fig. 77B has the forms prism (110), basal pinacoid (001), side pinacoid (010) and dome (101). The topaz crystal of Fig. 77C has two prisms, the unit prism (110) and another prism (120), and is terminated by the unit bipyramid (111), another bipyramid (223), the dome (021) and the basal pinacoid (001).

V. MONOCLINIC SYSTEM

Axes. The Monoclinic System includes all crystals that can be referred to three unequal axes, two of which are at right angles and the third makes an angle, not a right angle, with the plane containing the other two axes. Stated in another way, the Monoclinic System has one axis normal to the plane containing the other two axes which are not at right angles. The axes and their nomenclature are shown in Fig. 79. The *c*-axis is *vertical;* the *b*-axis, or *ortho-axis*, is at right angles to the *c*-axis; the axis inclined to the plane containing the *c* and *b* axes is the *a*-axis or *clino-axis*.

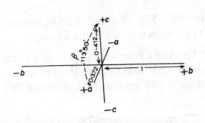

A monoclinic crystal is in the reading position when the clino-axis runs up and away from the observer, the ortho-axis runs right to left, and the vertical axis is vertical. The

FIG. 79. Monoclinic axes: Gypsum, a:b:c =0·372:1:0·412, β=113° 50'. Showing axial nomenclature and lengths cut off by the unit form (111).

obtuse angle between the clino-axis and the vertical axis is of importance and is of course fixed for each mineral species. It is given in the axial formula, thus:

Gypsum—a : b : c = ·372 : 1 : 412. β = 113° 50',

which registers the observed fact that the unit form of gypsum intersects the three axes in the ratio shown and that the obtuse angle between the vertical axis and the clino-axis is 113° 50'.

The commonest symmetry class of the Monoclinic System is the Normal or Holosymmetric Class characterised by the *Gypsum Type*, which is the only class considered here.

10. Gypsum Type or Prismatic Class

SYMMETRY. The symmetry of the Gypsum Type has already been worked out on p.85 and illustrated by Fig. 27. The inclination of the *a*-axis of this type has removed two of the three planes of symmetry found in the Barytes Type. The single plane of symmetry is that containing the clino-axis and the vertical axis. Similarly there is only one axis of symmetry, the *b*-axis or ortho-axis at right angles to the plane of symmetry. This is an axis of twofold symmetry. There is a centre of symmetry as shown by inspection of the crystal figured in Fig. 27. The symmetry of the Gypsum Type is therefore:—

Plane, 1, containing the clino-axis and the vertical axis.
Axis, 1^{II}, the ortho-axis.
A Centre of Symmetry.

This symmetry is illustrated in Fig. 75 and should be compared with that of the Barytes Type shown in the same figure.

FORMS. (i) *Basal Pinacoid*. This form consists of two faces, parallel to the clino-axis and the ortho-axis and cutting the vertical axis. The symbol of the upper face is 001 and of the lower face $00\bar{1}$.

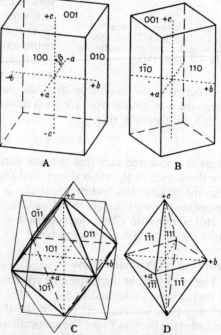

FIG. 80. Combinations in the Monoclinic System.
A. The three pinacoids. B. Unit prism and basal pinacoid. C. Clinodome and positive and negative hemiorthodomes. D. Positive and negative hemi-pyramids.

(ii) *Orthopinacoid*. This form has two faces, each parallel to the ortho-axis and the vertical axis, and cutting the clino-axis, the symbols of the front face being 100 and of the back face $\overline{1}$00. The form may be called the *front pinacoid*.

(iii) *Clinopinacoid*. In this form the two faces cut the ortho-axis and are parallel to the clino-axis and the vertical axis. The symbol of the right-hand face is 010 and of the left-hand face $0\overline{1}0$. The form may be called the *side pinacoid*.

The three pinacoids are shown in combination in Fig. 80A.

(iv) *Prism* (*hk*0). A face such as 110, which cuts the clino-axis and the ortho-axis and is parallel to the third axis, must be accompanied by the opposite parallel face $\overline{1}\overline{1}0$ by the virtue of the centre of symmetry, and these two faces must be accompanied by 1$\overline{1}$0 and

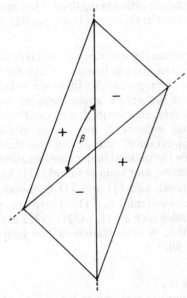

$\overline{1}$10 by virtue of the plane of symmetry. The resulting form is the prism *of the third order*. The unit prism which intersects the clino-axis and the ortho-axis in the same ratio as the unit form does has the symbol (110) Fig. 80B; other prisms are (210), (320), (130), etc., and the general symbol is (*hk*0).

(v) *Hemi-orthodomes* (*h*0*l*). No form which has faces parallel to the ortho-axis and cuts the clino-axis and the vertical axis can have *four* faces, since the orthopinacoid is not a plane of symmetry in the Gypsum Type. Only hemi-orthodomes, consisting of *two* faces, are therefore possible. The hemi-orthodomes are thus of two kinds, each consisting of two parallel faces

FIG. 81. Convention for sign of hemi-orthodomes in the Monoclinic System.

which are parallel to the ortho-axis and cut the other two axes. A hemi-orthodome is called *positive* if its faces lie in the obtuse angles between the vertical and clino-axis—that is in the obtuse angle β—and *negative* if they lie in the acute angles. This convention is shown in Fig. 81. The symbols of the faces of the unit negative hemi-orthodome are $\bar{1}01$ and $10\bar{1}$, and of the corresponding but independent positive form 101 and $\bar{1}0\bar{1}$. There are of course a number of hemi-orthodomes, the general symbols being positive $(h0l)$ and negative $(\bar{h}0l)$. The hemi-orthodome is also called a *pinacoid of the second order*, its faces being parallel to the *b* (or *second*) axis.

(vi) *Clinodome.* Consideration of the symmetry requirements shows that four faces are possible in a form whose faces are parallel with the clino-axis and cut the ortho-axis and the vertical axis. This form is the *clinodome*. The unit clinodome has the symbol 011 (see Fig. 80C). There are several clinodomes possible for any mineral species, the general symbol being $(0kl)$. They may also be called *prisms of the first order*, their faces being parallel to the *a* (or *first*) axis.

(vii) *Hemi-pyramids.* It follows from the symmetry of the Gypsum Type that a form cutting all three axes is limited to four faces only. Such forms are called *hemi-pyramids;* they are called *positive* if their faces occur in the obtuse angles between the plane containing the ortho-axis and vertical axis and that containing the clino-axis, and *negative* if they occur in the acute angles. The unit hemi-pyramid, intersecting the three axes at the standard lengths for the mineral under consideration, has the symbol (111). The positive unit hemi-pyramid (111) has the faces 111 and $1\bar{1}1$ up in front, and $\bar{1}\bar{1}\bar{1}$ and $\bar{1}1\bar{1}$ below and behind. The negative unit hemi-pyramid is $(\bar{1}11)$. There are of course a number of hemi-pyramids such as (112), (321), (132), etc., and the general symbol is (hkl). A combination of two hemi-pyramids is illustrated in Fig. 80D.

COMMON GYPSUM TYPE MINERALS

Gypsum. A typical crystal of gypsum is shown in Fig. 27, p. 86. It consists of the forms, clinopinacoid (010), prism (110), and negative hemi-pyramid $(\bar{1}11)$.

Orthoclase. Common crystals of orthoclase are combinations of clinopinacoid (010), basal pinacoid (001), prism (110) and negative hemi-orthodome (20$\bar{1}$), as shown in Fig. 82.

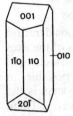

FIG. 82.
Ortho-
clase.

Augite. Common crystals of augite are illustrated in Fig. 137, p. 387, and are combinations of prism (110), orthopinacoid (100), clinopinacoid (010) and negative hemi-pyramid ($\bar{1}$11).

Hornblende. Hornblende often occurs in crystals showing the prism (110), clinopinacoid (010), clinodome (011) and negative hemi-orthodome ($\bar{1}$01), and shown in Fig. 139, p. 396.

Micas, *sphene* and *epidote* are other minerals crystallizing in the Gypsum Type.

VI. TRICLINIC SYSTEM

Axes. In the Triclinic System the crystallographic axes are all unequal and none is at right angles to another. One axis is placed vertical and is called the *c*-axis or the *vertical axis*. A second axis, the *a*-axis, runs up and away from the observer. The intercept made by the unit form on this axis is usually *smaller* than on the third axis (so that the *a*-axis was called the *brachy-axis* by analogy with the Orthorhombic System). The third axis is the *b*-axis (or *macro-axis*), and its slope depends upon how it was chosen in the original description of any particular mineral. The angle between the positive ends of *c* and *b* is called α, that between *a* and *c* is called β, and that between *a* and *b* is called γ. The axial nomenclature and conventions are shown in Fig. 83. There has been no standard way to orientate triclinic crystals and, even for the same mineral, many different orientations have been used—thus for axinite more than a dozen settings are on record, one such being that detailed in Fig. 83. The axial data consist of the ratios of the intercepts made on the axes by the unit form, and the angles between the axes, as—

Axinite.—a : b : c = 0·49 : 1 : 0·48
α = 82° 54′; β = 91° 52′; γ = 131° 32′.

The symmetry class described here is the Pinacoidal, usually also termed the Axinite Type though there is some doubt whether axinite actually belongs to the Pinacoidal Class.

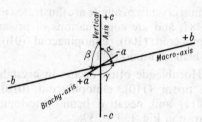

FIG. 83. *Triclinic Axes*. Axes for Axinite; a:b:c = ·49:1:·48; $a = 82°$ 54′, $\beta = 91°$ 52′, $\gamma = 131°$ 32′. The lengths of the axes shown correspond to the lengths cut off by the unit form of axinite.

II. Axinite Type or Pinacoidal Class

SYMMETRY. Consideration of the positions of the axes shows that the sole element of symmetry possible in the Triclinic System is the centre of symmetry. There can be no planes and no axes of symmetry. The symmetry of the Axinite Type is:—

Planes, none   Axes, none   A Centre of Symmetry.

The symmetry picture given in Fig. 75 is plotted on the plane perpendicular to the vertical crystallographic axis *c*.

FORMS. Since there is only a centre of symmetry in the Axinite Type, the presence of any one face necessitates only the presence of an opposite parallel face. Each form, therefore, consists of *two* faces and is thus a *pinacoid*. The forms were named by analogy with the Orthorhombic System, attention being paid of course to the limitation of the Axinite Type forms to two faces only. Nowadays they are usually indicated by their Miller symbols and are all called pinacoids.

(i) *Basal Pinacoid*. This form consists of two parallel faces each cutting the vertical axis and lying parallel to the other two axes. The symbol is (001).

(ii) *Front Pinacoid*. This form has two parallel faces each cutting the *a*-axis and lying parallel to the vertical axis and the *b*-axis. Its symbol is (100).

(iii) *Side Pinacoid*. The two parallel faces of this form cut the *b*-axis and are parallel to the vertical axis and the *a*-axis. The symbol is (010).

(iv) *Pinacoids* (*hk*0) *or Hemi-prisms*. These forms have two parallel faces cutting the *a* and *b* axes and being parallel to the vertical axis. One unit hemi-prism has the symbol (110) and the faces 110 and $\overline{1}\overline{1}0$; the other has the symbol (1$\overline{1}$0) and the faces 1$\overline{1}$0 and $\overline{1}$10. Other hemi-prisms have indices such as (210), (3$\overline{2}$0), (130), etc. They may be called *pinacoids of the third order* since they are parallel to *c*, the third axis.

(v) *Pinacoids* (*h0l*) *or Hemi-macrodomes.* The two faces of this form are parallel to the *b*-axis and cut the vertical and *a*-axes. A unit hemi-macrodome is (101) and others are (201), ($\overline{2}$03), etc. They may be called *pinacoids of the second order* as they are parallel to *b*, the second axis.

(vi) *Pinacoids* (*0kl*) *or Hemi-brachydomes.* The two faces here are parallel to the *a*-axis and cut the vertical axis and the *b*-axis. A unit hemi-brachydome is (011), and others are (021), (03$\overline{2}$), etc. They may be called *pinacoids of the first order* since they are parallel to *a*, the first axis.

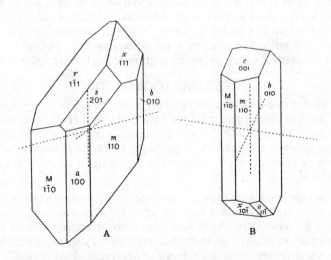

FIG. 84. A. *Axinite*: combination of hemi-prisms M (1$\overline{1}$0), m (110); front pinacoid a (100); side pinacoid b (010), quarter-bipyramids x (111), r (1$\overline{1}$1), hemi-macrodome s (201). B. *Albite*: combination of prisms M (1$\overline{1}$0), m (110), side pinacoid b (010), basal pinacoid c (001), quarter-bipyramid o (11$\overline{1}$), hemi-macrodome x (10$\overline{1}$).

(vii) *Pinacoids (hkl) or Quarter-bipyramids.* The two parallel faces of pinacoids which cut all three axes are quarter-bipyramids. There are obviously four unit quarter-bipyramids having the faces 111, $\overline{1}\overline{1}\overline{1}$; $\overline{1}11$, $1\overline{1}\overline{1}$; $\overline{1}\overline{1}1$, $11\overline{1}$; and $1\overline{1}1$, $\overline{1}1\overline{1}$. Quarter-bipyramids may be called *pinacoids of the fourth order*.

COMMON AXINITE TYPE MINERALS

Axinite. The axinite crystal given in Fig. 84A is a combination of the forms, two unit hemi-prisms (110) and $(1\overline{1}0)$, front pinacoid (100), side pinacoid (010), hemi-macrodome (201), and two unit quarter-bipyramids (111) and $(1\overline{1}1)$.

Plagioclase Feldspars. The very important mineral group of the plagioclase feldspars crystallizes in the Axinite Type. A crystal of albite, a member of the plagioclase series, is shown in Fig. 84B. It has the forms, two unit hemi-prisms (110) and $(1\overline{1}0)$, side pinacoid (010), basal pinacoid (001), hemi-macrodrome $(10\overline{1})$ and quarter-bipyramid $(11\overline{1})$. It will be noticed that the general shape of the albite crystal is like that of the orthoclase crystal of the Gypsum Type shown in Fig. 82. In albite, however, measurements show that the angle between the vertical axes and the macro-axis is $94°$, whereas the corresponding angle in orthoclase is of course $90°$.

CRYSTAL AGGREGATES

Two or more crystals associated together give a *crystal aggregate*. A mass of closely packed crystal grains without crystal forms is a *crystalline aggregate*. Crystal aggregates may be made up all of one mineral, and then the aggregate is said to be *homogeneous*, if two or more minerals occur in the aggregate, it is styled *heterogeneous*.

Heterogeneous Aggregates. There are various ways in which the two or more minerals making up a heterogeneous aggregate may be associated. They may show no arrangement and the orientation of any two individual crystals may show no relation to one another—this is an *irregular aggregate*. Two minerals of different natures and even belonging to different symmetry classes may sometimes form *regular growths* so that there is a partially similar arrangement of faces or axes. Examples are provided by the intergrowths of two types of feldspar known as perthite, or by the regular associations of the orthorhombic pyroxene called enstatite with the monoclinic pyroxene, augite. Another kind of heterogeneous aggregate is the *isomorphous growth,* in which occur concentric zones of various members of an *isomorphous series* of minerals of analogous chemical formulae and closely allied crystal structures; the important factor is the close similarity of sizes of the constituent anions and cations (p. 26). Isomorphous growths are found in many series of minerals, such as the plagioclase feldspars and the pyroxenes.

Homogeneous Aggregates. Here the aggregate consists of one mineral only. There are first the *irregular aggregates* in which there is no relation in the orientations of the crystals. The perfection of arrangement is seen in the *parallel growths,* in which the individuals have the same orientations, like edges and faces of the different crystals being parallel. The individuals may

FIG. 85. Group of Quartz Crystals.

be united on any crystal plane. In Fig. 85 several crystals of quartz are shown in parallel growth, the axes of the individual crystals being parallel. In a very important association of crystals there is an orientation that lies between the absolute absence of orientation of irregular aggregates and the perfect parallel orientation of parallel growths. In this association some crystallographic direction or plane is common to two or more individuals and this gives rise to what are called *twinned crystals.* The subject of twinning in crystals is considered in the next section.

TWIN CRYSTALS

Twins. Twinned crystals consist of two or more portions, of course consisting of the same substance, that are joined together in such a way that some crystallographic direction or plane is common to the parts of the twin. In twin crystals one part is in reverse position to the other part, or the second half of the twin may be conceived as produced by the rotation, about some line, of one half of the crystal through an angle of 180°. The plane dividing the twin so that one half is a reflection of the other half is called the *twin-plane,* and the axis about which rotation is necessary to restore the twin to its untwinned state is called the *twin-axis.* The twin-axis is usually perpendicular to the twin-plane. The conception of the twin-axis is merely a convenient way of describing twinning,—such a revolution has of course not occurred in the formation of twins. Twinning may be the result of growth on the two sides of a sheet of atoms, the growth keeping to the proper structural pattern but proceeding with different but appropriate orientations.

In Fig. 86 there is shown a twin of calcite. The form present is the scalenohedron. A plane parallel to the basal pinacoid (0001), indicated by re-entrant angles in the crystal, is a plane of symmetry for the twinned crystal—the upper half of the twin is reflected in this plane to produce the lower half. This is the *twin-plane* of the crystal and the crystal is said to be twinned on the basal pinacoid (0001). The *twin-axis* is the vertical crystallographic axis *c.*

The twin-plane is always a possible face of the crystal, and the twin-axis is always perpendicular to some possible face or parallel to a possible edge. A plane of symmetry cannot be a twin-plane, since it already divides the crystal into two halves one of which is a mirror-image of the other.

The plane along which the two halves of a twin are joined is called the *composition-plane.* It is usually but not necessarily coincident with the twin-plane, as in Fig. 86.

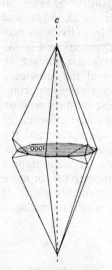

FIG. 86. Calcite Twin: Scalenohedron, twin-plane basal pinacoid 0001, twin-axis *c.*

Twin crystals may be recognised by the occurrence of re-entrant angles, and by the fact that all corresponding edges are not parallel and that the twin possesses a plane of symmetry which is absent in the two halves.

There are various types of twins, the following being the most important:—

Simple Twin. The twin of calcite of Fig. 86 is an example of a simple twin, consisting as it does of two halves symmetrical with respect to the twin-plane. It is also a *contact-twin.*

Penetration Twin. In this type, the two halves of the twin have grown so mixed together that the twin cannot be divided into two separate halves. Examples are supplied by the well-known "Iron Cross" twins of pyrites as shown in Fig. 87, interpenetrated cubes of fluorite (Fig. 119, p. 285) and the cross-shaped twins of staurolite (see Fig. 134, p. 370).

Repeated Twins; A repeated twin is produced by the repetition of twinning according to the same law. A crystal composed of three parts related to one another by the same twinning law is called a *trilling*, of four parts, a *fourling*. If the twin-plane in all parts of a repeated twin remains parallel, then the twinning is often called *polysynthetic*. Polysynthetic twinning is well seen in plagio-clase feldspar as described below and figured in Fig. 87. When the twin-plane does not remain parallel, the resulting twin approaches a curved form; this type of repeated twinning is called *cyclic*, and is illustrated by the twin of aragonite shown in Fig. 114, p. 276.

Compound or Complex Twin. A twin in which twinning has taken place on two or more laws is styled a *compound* or *complex* twin.

Examples of Common Twin Laws

CUBIC SYSTEM. The commonest type of twin in the Galena Type is on what is called the *Spinel Law,* in which the twin-plane is the octahedral face and the twin-axis is at right angles to this. *Fluorspar* commonly forms interpenetrated twins on this law. In the Pyrite Type, *pyrite* itself commonly twins in interpenetrated pyritohedra, giving a form known as the *iron-cross* (see Fig. 87A). The twin-plane is the rhombdodecahedral face and the twin-axis is normal to it. It may be noted that in the Normal or Galena Type, this plane is a plane of symmetry and this axis an axis of symmetry, so that the iron-cross twin of pyrite restores, as it were, the symmetry lost by the pyritohedron in its derivation from the tetrahexahedron.

TETRAGONAL SYSTEM. In the *Rutile Law,* the commonest type of twinning in this system, the twin-plane and composition-plane are a face of the pyramid of the second order. Twinning on the face 101 produces the knee-shaped or *geniculate* twins illustrated in Fig. 87B. A sharper 'knee' is given by twins on another pyramid (301).

HEXAGONAL SYSTEM. *Calcite* twins on the basal plane (0001), as shown in Fig. 86, or on rhombohedra, such as (01$\bar{1}$2), (10$\bar{1}$1) and (02$\bar{2}$1). Twinning on the rhombohedron (01$\bar{1}$2) is often due to gliding (see p. 62) and is observed in most thin sections of calcite under the microscope. The vertical axes of the two halves of the twin twinned on (10$\bar{1}$1) are about at right angles to one another, whilst in twins on (02$\bar{2}$1) they make an acute angle. *Quartz* commonly forms interpenetration twins in which the twin-

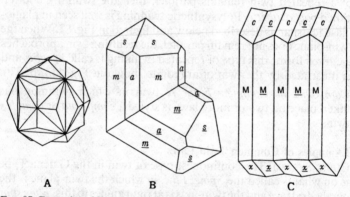

FIG. 87. Examples of Twins. A. Pyrite: Interpenetration Twin. B. Rutile: Geniculate Twin: twin-plane 101. C. Plagioclase: Repeated Twin: twin-plane 010.

axis is the vertical crystallographic axis and the twins may be regarded as consisting of two right-handed or two left-handed crystals, one of which has been rotated 180° about the *c*-axis. Such twins may be told by examining the positions and arrangements of the small trigonal trapezohedron and bipyramid faces. In another type of quartz twin, the crystal consists of a right- and left-handed interpenetrated pair; here the twin plane is a face of the prism (11$\bar{2}$0).

ORTHORHOMBIC SYSTEM. *Aragonite* forms cyclic twins, illustrated in Fig. 114, p. 276, by repeated twinning on the faces of the prism (110). The prism angle is nearly 60° so that twinning repeated five times produces a pseudo-hexagonal crystal. There is present, however, a re-entrant angle in the prism zone of the twin, and thus its true nature is detected. *Staurolite* forms two types of twins, illustrated in Fig. 134, p. 370. In the first type, the dome (032) is the twin-plane, and twinning on this law gives a 'Maltese Cross' twin (Fig. 134 right). The second type has a bipyramid (232) as twin-plane and gives a 'skew' twin (Fig. 134 left).

MONOCLINIC SYSTEM. *Gypsum* forms twins, sometimes called 'swallow-tails,' with the orthopinacoid (100) as the twin-plane. *Hornblende* and *augite* twins also have the orthopinacoid (100) as the twin-plane. It will be recalled from p. 147 that *hornblende*

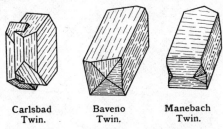

Carlsbad   Baveno   Manebach
Twin.   Twin.   Twin.

FIG. 88. Orthoclase Twins.

crystals commonly have three faces at the top and three at the bottom of the prism zone, as illustrated in Fig. 139, p. 396. By twinning on the orthopinacoid it often happens that, as it were, the orthodome face is transferred from the top of the crystal to the bottom, and the two clinodome faces from the bottom to the top,—so that many twinned crystals of hornblende show four faces at one end of the prism zone and two at the other. *Augite* twinned on (100) shows a marked re-entrant angle.

The feldspar, *orthoclase*, twins on three common laws, and drawings of crystals showing this are given in Fig. 88. The first type is called the *Carlsbad twin*, and in this the twin axis is the vertical crystallographic axis and the composition-plane is the clinopinacoid (010); interpenetration Carlsbad twins are common.

In the *Baveno twin,* the twin-plane and composition-plane are the clinodome (021), whilst in the *Manebach* twins they are the basal pinacoid (001).

TRICLINIC SYSTEM. Plagioclase feldspar is usually twinned on the *Albite Law,* in which the twin-plane is (010), the side pinacoid. Albite twinning is usually repeated and polysynthetic and the separate members of the twin, shown in Fig. 87C, are often exceedingly thin lamallae. This repeated twinning produces a striation on the basal planes of the crystal, and is also excellently seen in thin sections of plagioclase examined under the microscope, as illustrated in Fig. 144, right, p. 428, and dealt with on p. 424.

CHAPTER IV

THE OPTICAL PROPERTIES
OF MINERALS

Introduction. Certain characters of minerals dependent upon light have already been considered. We have now to deal with the optical properties that are especially important in the study of the group of minerals known as the *rock-forming minerals,* that is, those that make up the common rocks of the earth's crust. In this study, the rock-forming minerals are examined with the microscope and are either in the form of small grains or fragments or else in very thin slices. A microscope is also used in the examination of polished surfaces of ore-minerals. The description of the microscopes and of the methods of preparing the minerals for study with them is deferred until the principles of optics that underlie the study have been considered.

The Nature of Light. The optical behaviour of crystals is usually explained on the assumption that light consists of electro-magnetic vibrations. According to the undulatory or wave theory, the vibrations which take place when a ray of light is transmitted are in directions at right angles to the path of the ray. While this hypothesis does not explain all the phenomena connected with crystals (*e.g.* the interaction of light with atoms or electrons in a crystal), it provides a convenient basis for our discussion of the optical properties of crystals in this chapter. The transverse vibrations in a ray of light can be considered to take place in all possible directions perpendicular to the direction of propagation (Fig. 90). A *wave-length* is the distance between two points in exactly similar positions on a wave and moving in the same direction. Two waves are said to be in *phase* when they are of equal wave-length and their positions of zero amplitude occur at exactly the same instant. Thus, in the wave-form of Fig. 89, the points P_1 and P_3 are in the same phase, and P_1 and P_2 differ in phase by half a wave-length.

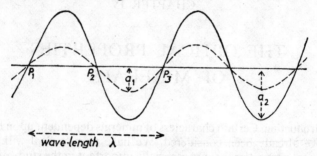

Fig. 89. Wave-forms, illustrating phase-differences and amplitudes, a_1 and a_2.

The time required for the disturbances to travel one wave-length is called the *periodic time* or *period*. If T is the period, λ the wave-length, then the velocity V is clearly $\frac{\lambda}{T}$.

The colour depends upon the wave-length of the light causing it. For violet light, the wave-length is 0·00039 mm. and for red light 0·00076 mm. White light, or ordinary light, consists of light with all the different wave-lengths between these extremes. Monochromatic light is light of one wave-length only.

Polarized Light. We have stated that the vibrations take place at right angles to the direction of propagation of the disturbance. In ordinary light, the vibrations take place in all directions in a plane at right angles to the ray, as shown in Fig. 90A. Light is said to be *plane polarized,* or, more usually, *polarized*, if the vibrations are confined to one direction in this plane. In Fig. 90B, all the vibrations lie in the plane of the paper and the light is polarized in this plane. Certain crystals, as we shall see, have the property of forcing, as it were, the complex vibrations of ordinary light to take place in two planes, at right angles to one another.

Isotropic and Anisotropic Substances. Suppose that a disturbance proceeds outwards in all directions from a point. At the end of a unit of time, say one second, the disturbance will have travelled

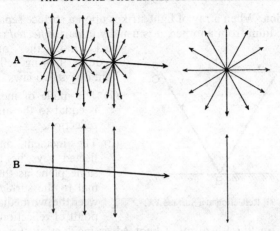

A

B

FIG. 90. Polarization of Light.

certain distances along all these directions, and we could construct some kind of surface that would pass through all the points reached in that time. This surface is called the *wave-surface*, and a section through it in any given plane passing through the point of origin is the *wave-front*.

The *shape* of the wave-surface depends upon the relative velocities of propagation of the disturbances along the different directions. If the velocity is the same for all directions, the wave-surface is a sphere and the wave-front is a circle. If the velocities differ in different directions the wave-surface is some surface other than a sphere—in many crystals it is an ellipsoid.

Two classes of substances can now be distinguished:

Isotropic. Substances in which light is transmitted with the same velocity in all directions. In these, the wave-surface is a sphere and the wave-front a circle. Glass and minerals crystallizing in the Cubic System are isotropic.

Anisotropic. Substances in which light is transmitted with different velocities in different directions. All crystals except those belonging to the Cubic System are anisotropic. In these, the wave-surface is an ellipsoid and the wave-front an ellipse.

Reflection. When a ray of light strikes upon a surface separating one medium from another, it is usually in part *reflected* or bent back into the original medium. During reflection light obeys two laws:

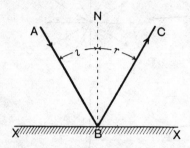

FIG. 91. Reflection at a Surface *XX*.

(1) The angle of incidence is equal to the angle of reflection.

(2) The incident and reflected rays lie in the same plane as the normal to the surface between the two media at the point of reflection.

Thus, in Fig. 91, a ray of light AB is incident on the surface XX at B. It is reflected along the path BC so that the angle between AB and the normal at B, BN, equals the angle between BC and BN, that is, so that the angle of incidence i equals the angle of reflection r. Further, the incident ray AB, the reflected ray BC and the normal BN all lie in the same plane, that of the paper, NXX.

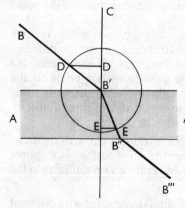

FIG. 92. Refraction.

Refraction. A part of the light of a ray striking the surface separating two media passes into the second medium. If the two media are of different properties the ray is bent or *refracted* on entering the second medium.

In Fig. 92 suppose AA to represent the section of a parallel-sided plate of glass, and BB' a ray of light incident at B'. The ray BB', instead of passing straight on, is bent or refracted to B'', and on emerging from the glass is again refracted to B''', parallel to the direction it pursued before entering the glass.

The two following laws govern simple refraction:—

(i) The sines of the angles made by the incident and refracted rays with a line perpendicular to the surface separating the two media always bear a definite ratio to one another.

(ii) The incident and the refracted ray are in the same plane, which is perpendicular to the surface separating the two media.

Refractive Index. If the angle of incidence is called i, and the angle of refraction r, we have seen that for the same two media $\dfrac{\sin i}{\sin r}$ is a constant. In Fig. 92, the lines DD and EE are perpendicular to the normal CB′, and DB′ = EB′ since they are radii of the circle with B′ as centre. Then $\sin i = \sin \text{DB}'\text{C} = \dfrac{\text{DD}}{\text{DB}'}$, and $\sin r = \sin \text{EB}'\text{E} = \dfrac{\text{EE}}{\text{DB}''}$, so that $\dfrac{\sin i}{\sin r} = \dfrac{\text{DD}}{\text{EE}}$ and is a constant.

This constant is called the *Refractive Index,* and is of fundamental importance in the optical investigation of minerals. In order to compare the refractive indices of different substances, we have to choose some medium for reference. The medium chosen is usually air, which is taken to have a refractive index of 1. Examples of other refractive indices are: water 1·33, fluorspar 1·43, crown glass 1·53, Canada balsam 1·54, garnet 1·77, diamond 2·42.

We have now to deal with the relation between velocities of light and refractive index. Suppose that the velocity of light in air is v_a and in a given medium is v_m. It can be shown by a simple geometrical construction that $\dfrac{\sin i}{\sin r} = \dfrac{v_a}{v_m}$ that is, that the Refractive Index is equal to the ratio between the velocities of light in air and in the medium. If light travels with velocities v_1 and v_2 in two given substances, then their refractive indices n_1 and n_2 are such that $n_1 = \dfrac{v_a}{v_1}$ and $n_2 = \dfrac{v_a}{v_2}$. Therefore $\dfrac{n_1}{n_2} = \dfrac{\frac{v_a}{v_1}}{\frac{v_a}{v_2}} = \dfrac{v_2}{v_1}$ and it follows that for a given colour of light, the refractive indices of two media are *inversely* proportional to the velocities of light in them.

It can also be shown that the refractive index increases as the wave-length of the light decreases. We have seen that the wave-length of red light is greater than the wave-length of blue light, so that the refractive index for red light is less than that for blue light. White light entering a medium is broken up into the colours of the spectrum, the blue colour occurring nearest to the normal since it has the greatest refractive index, and the red the farthest away. This breaking-up of white light is called *dispersion*.

It follows from the definitions given earlier that in *isotropic substances* the refractive index has a *constant* value no matter what direction the light is following, whilst in *anisotropic substances* the refractive index *varies* with the direction of transmission of the light.

Total Reflection and the Critical Angle. First examine the passage of light from a given medium into one of lower refractive index.

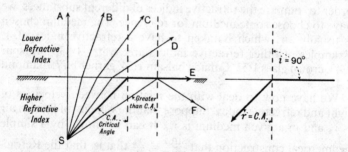

FIG. 93. The Critical Angle and Total Reflection.

In Fig. 93 a section through two media is shown, the lower having a higher refractive index than the upper. Rays of light travel from a source S and pass from the medium with higher refractive index into that with the lower. Ray A strikes the junction plane between the two media at right angles, so that its angle of incidence is zero and consequently its angle of refraction is zero too. It passes straight on unbent. Rays B, C, D are refracted away from the normal as they pass from the lower medium to the upper. There must be a position, as with ray, E, however, when the refracted ray just grazes the surface between the two media. A ray, such as F, meeting the junction plane at an angle greater than E does, is reflected back into the denser medium. A ray in the position of E is said to make the *critical angle* with the normal to the junction plane. Light falling on this plane at angles greater than the critical angle is *totally reflected*.

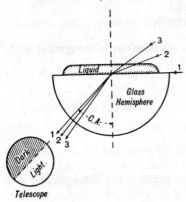

FIG. 94. Passage of Light through Refractometer.

Let us reverse the direction of the ray E as in Fig. 93, right. The angle of incidence i is 90°, the angle of refraction r is the critical angle (CA), whence the refractive index is given by $\frac{\sin i}{\sin r} = \frac{\sin 90°}{\sin CA} = \frac{1}{\sin CA}$, that is, the sine of the critical angle is equal to the reciprocal of the refractive index. The determination of the critical angle as a means of finding the refractive index is performed with one type of the instruments called *refractometers*, which are briefly described.

Refractometers. The principle of these instruments is illustrated in Fig. 94. A hemisphere of glass having a high refractive index, 1·8 to 1·9, is used. The mineral whose refractive index is to be determined is placed on the hemisphere, with a film of liquid, of a refractive index intermediate between those of the hemisphere and the mineral, between them. If the refractive index of a liquid is to be determined, as is usually the case, a drop is placed on the upper surface of the hemisphere. Light is thrown by a mirror at grazing incidence so that it is refracted at the critical angle. A telescope placed to observe the emergent rays shows a field that is half light and half dark, as shown in the figure. The value of the critical angle is thus obtained.

The refractive index of the hemisphere is determined for air, from the equation R.I. $= \dfrac{1}{\sin CA}$ where CA is the critical angle for air. Let this refractive index be N, and the refractive index of the mineral be n. Then, as in Fig. 93, $\dfrac{\sin i}{\sin r} = \dfrac{\text{velocity in mineral}}{\text{velocity in hemisphere}} = \dfrac{N}{n}$, whence $n\sin i = N\sin r$. But $i = 90°$, and $\sin i = 1$, and r is the critical angle, so $n = N \times$ sine of the critical angle. Therefore the refractive index of the mineral is found by multiplying the sine of the critical angle by the refractive index of the hemisphere.

The chief instrument of this type is the *Abbe Refractometer*. Smaller portable instruments suitable for the rapid determination of the refractive indices of gems are the *Herbert Smith* and the *Tully* refractometers:—in these the boundary of the light and dark fields is thrown on to a scale and read directly through an eyepiece.

Outline and Relief of Minerals in a Mount. For examination under the microscope minerals may be permanently embedded in a mount, such as Canada balsam, or for temporary purposes they may be immersed in an oil. The surfaces of the slice or fragments of the mineral are made up of tiny elevations and depressions which reflect and refract light at various angles, and so cause the surfaces to appear pitted. The greater the difference between the refractive index of the mineral and of the mount the rougher the surface appears.

A mineral of high refractive index embedded in a mount of lower refractive index acts as a little lens, and rays of light coming from the bottom surface of the mineral appear to come from a slightly higher point. Such minerals therefore seem to stand out in relief from their surroundings.

The distinctness of the outlines of minerals depends also upon the *difference* in refractive index between them and the mount. If the refractive indices of the mineral and the mount are the same, the mineral is invisible. The mineral cryolite has a mean refractive index of 1·339, and when immersed in water, with refractive index of 1·335, it can scarcely be seen. Minerals with refractive index markedly different from that of the mount have well-marked dark borders due to the production of shadows by total reflection. It is important to realise that this phenomenon is seen with minerals with a much lower refractive index as well as with those with a much higher refractive index than the mount.

Becke Effect. The relative refractive indices of two minerals in contact, or of a mineral and the mount, can be observed by studying the very important *Becke Effect*. In Fig. 95 a mineral of lower refractive index on the left is in contact with one with higher refractive index on the right. Of a bundle of rays thrown on to the contact between the two minerals some are refracted and some totally reflected so that they are concentrated just within the mineral with higher refractive index. Under the microscope a narrow line of light—the *Becke Line*—appears in this position, and as the microscope objective is raised, the Becke line appears to travel into the mineral with the higher refractive index. In practice, the Becke line is best seen by using a high-power objective and cutting off some of the light passing through the mineral by a diaphragm below the microscope stage. The Becke Effect is so constantly used that the student must remember the rule:—*as the objective is raised, the light band travels into the mineral of higher refractive index.*

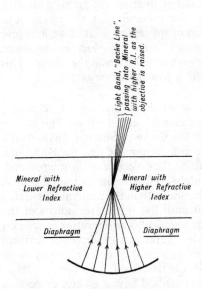

FIG. 95. The Becke Effect.

Shadow Method with Inclined Illumination. Another method of determining the relative refractive indices especially of minerals immersed in oils is to use inclined illumination; this can be effected by placing the finger or a card below the microscope stage so that a portion of the light passing through the mineral is cut off at one side. The rays then strike the contact surface between the mineral and the mount obliquely; if they pass from a mineral of higher refractive index into a mount of lower refractive index they are concentrated by refraction and form a light band; if they pass from the mount of lower refractive index into the higher mineral, they are spread out by refraction and so produce a shadow. Thus, if the finger is put in from the right, a shadow appears on the left side of the mineral. But the microscope objective reverses the position of the image, so that we may give the rule:—*If the shadow appears on the same side as the finger, the mineral has a higher refractive index than the mount.* This phenomenon is best seen with a low-power objective.

Immersion Method. For accurate determination of refractive index of mineral grains the immersion method is used. The grain is placed in an oil and by using the Becke or Shadow methods the relative refractive indices are noted. Suppose that the Becke line passes into the mineral on raising the objective of the microscope. The mineral therefore has a higher refractive index than the oil. A second oil, of higher refractive index than the mineral, is selected and mixed with the first oil until the refractive indices of the mixture and of the mineral are identical. The operation is done on the usual glass microscope slide. The refractive index of the mixture of oils, and therefore of the mineral, is then determined by using one of the refractometers already described.

Since the liquids used in this method have a greater dispersion than minerals, that is, they have a greater difference in refractive indices for lights of different colours, a stage is reached when the mineral and liquid have the same refractive index for yellow light, but the refractive index for red light of the mineral is greater than that of the liquid for the same light, and the refractive index for blue light of the mineral is less than that for blue light of the liquid. At this point, *colour fringes* appear at the edge of the mineral grain when the Shadow method is applied. One edge of the grain is blue and the other red, indicating that for some intermediate colour, such as yellow, the refractive indices of liquid and mineral are the same. For accurate work, mono-chromatic light is then used and the Becke Effect applied.

Suitable immersion oils are: kerosene 1·448, clove oil 1·53, α-monobromnapthalene 1·658, methylene iodide 1·740, and methylene iodide saturated with sulphur 1·778. Sets of immersion liquids whose stated refractive indices increase by regular steps can be purchased.

Double Refraction. We have seen that in *isotropic* substances the refractive index has the same value for all directions. A ray of light entering such a substance remains a single ray, though bent from its course. Isotropic substances are thus *singly refracting*.

It is different with *anisotropic* substances. A ray of light passing from an isotropic to an anisotropic medium forms *two* refracted rays. This phenomen is called *double refraction*.

Double refraction is shown by all anisotropic minerals but especially well by the colourless transparent variety of calcite called *Iceland Spar*. The student should obtain a small cleavage-fragment of this mineral. As will be recalled the cleavage-fragment is a rhombohedron in shape. If the rhomb is placed over a dot, it will be found that two images of the dot are seen. If the rhomb is rotated, one of these images remains stationary whilst the other moves round the stationary dot. The image which does not move is called the *ordinary image* since it is

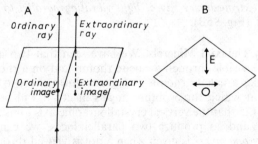

FIG. 96.
A. Paths of Ordinary and Extraordinary Rays.
B. Directions of vibrations of Ordinary Ray, O, and Extraordinary Ray, E.

formed by the *ordinary ray* which has passed through the calcite as if this were an isotropic medium. The other image is the *extra-ordinary image* formed by the *extraordinary ray*. The paths of the two rays are shown in Fig. 96A where it is seen that, though the light is incident perpendicular to the lower surface of the rhomb, still the extraordinary ray is refracted there, and is again refracted where it leaves the rhomb. That the ordinary and extraordinary rays travel with different velocities can be demons-trated by looking through two opposite pinholes in paper pasted on opposite faces of the rhomb. Images are seen in two positions of the eye, so that the ordinary and extraordinary rays are differently refracted when they emerge and therefore have different velocities.

We can examine the character of the light of the ordinary and extraordinary rays by means of a *tourmaline plate*, which has the property of transmitting light vibrating in a single plane, that is, polarized light. If the tourmaline plate is placed over the calcite rhomb so that the known vibration-direction of light passing through the tourmaline is parallel with the long axis of the rhomb face, it is observed that the extraordinary image disappears and only the ordinary image is seen. Similarly, if the tourmaline plate is rotated through 90°, the extraordinary image is seen and the ordinary image disappears. This experiment demonstrates that the light of the ordinary and extraordinary rays is polarized at right angles and that the *ordinary ray consists of light vibrating parallel to the long diagonal of the rhomb face and the extraordinary ray of light vibrating parallel to the short diagonal*. (Fig. 96B).

Optically Uniaxial Minerals. We have seen that two images of a dot are visible through a calcite rhomb laid on a rhomb face. If we take a calcite rhomb and grind down the two opposite corners at which three obtuse angles meet—it will be recalled from p. 125 that the vertical crystallographic axis joins these two corners—and so produce two parallel faces, we can observe that only *one image* is given when a dot is viewed through these faces. Examination of calcite rhombs cut in numerous other directions would show that the vertical crystallographic axis was the only direction along which only one image appeared. Along this direction the ordinary and extraordinary rays have the same velocities and no double refraction occurs. This direction is called the *Optic Axis*, and since there is only one such direction, crystals like calcite are said to be *Uniaxial*.

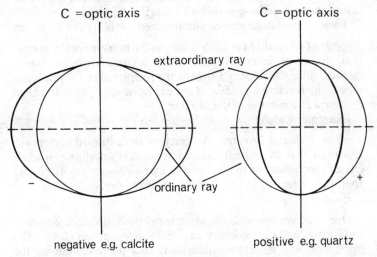

FIG. 97. Wave Fronts in Uniaxial Crystals

Experiments with innumerable sections of calcite have demonstrated that the ordinary ray travels with a constant velocity and has a constant refractive index no matter what its direction may be. Its wave-front is a sphere and a section of this a circle. On the other hand, it is found that the velocity of the extraordinary ray varies with its direction. Along the optic axis it has the same velocity as the ordinary ray; at right angles to the optic axis it has a maximum velocity greater than that of the ordinary ray. In intermediate positions it has intermediate velocities. The wave-front of the extraordinary ray is an ellipsoid of rotation with its short axis equal to the radius of the sphere representing the wave-front of the ordinary ray. A section of the extraordinary wave-front is an ellipse. The wave-fronts of calcite are represented in Fig. 97 left, the circle of the ordinary ray being inside the ellipse of the extraordinary ray. Minerals like calcite which have the velocity of the extraordinary ray greater than that of the ordinary ray are said to be *negative*. The opposite condition is seen in another group of uniaxial minerals, of which quartz is one, where the extraordinary ray is slower than the ordinary ray, so that the diagram of the wave-fronts shows an ellipse within a circle (Fig. 97 right).

These conventions can be summarized as follows:

Negative Uniaxial Minerals. Velocity of extraordinary ray greater than that of ordinary ray; wave-fronts a circle inside an ellipse; further, since the velocity varies as the reciprocal of the refractive index, the refractive index ε of the extraordinary ray is less than the refractive index ω of the ordinary ray.

Example. Calcite.

Positive Uniaxial Minerals. Velocity of the extraordinary ray is less than that of ordinary ray; wave-fronts an ellipse inside a circle; the refractive index ε of the extraordinary ray is greater than the refractive index ω of the ordinary ray.

Example. Quartz.

Our experiments with the cleavage-rhomb of calcite demonstrate also that the ordinary ray vibrates perpendicular to the optic axis, and the extraordinary ray in a plane containing the optic axis and lying at right angles to the direction of vibration of the ordinary ray.

Uniaxial minerals belong to the Tetragonal and Hexagonal Systems, so that the following classification has so far been established:

Isotropic minerals. Cubic.
Anisotropic minerals.   (1) Uniaxial—Tetragonal and Hexagonal Minerals.
                               (2) Other minerals.

The Nicol Prism. It is essential for the examination of minerals under the microscope that polarized light should be available. One means of producing this is by the *Nicol Prism* or *Nicol*. This apparatus depends for its action upon the double refraction of calcite. Long transparent cleavage-rhombs of Iceland spar are employed. The ends are ground down till they make an angle of 68° to the long edge. The rhomb is cut into two longitudinally by a plane running through the two corners which have three obtuse angles, and the two halves are recemented with Canada balsam. The inclination of this film of balsam is such as to cause the total reflection of the ordinary ray while allowing the extraordinary ray to be transmitted. Thus, in Fig. 98, a ray of light IR entering one end of the rhomb is doubly refracted into the

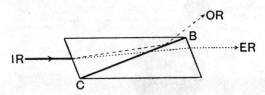

FIG.98. Nicol Prism, showing Ordinary Ray totally reflected by the Layer of Balsam whilst the Extraordinary Ray is transmitted.

ordinary ray OR and the extraordinary ray ER. The ordinary ray meets the film of balsam at an angle greater than the critical angle, undergoes total reflection and is driven on one side, so that it does not emerge at the other end of the rhomb. The extraordinary ray passes through the rhomb, so that *polarized light* emerges. It is important to note that since it is the extraordinary ray that is transmitted, the direction of vibration of the polarized light from a Nicol is parallel to the *short* diagonal of the rhomb face at the end of the Nicol.

Another device for producing polarized light is by means of special filters. We may call these devices of whatever type the *polars*. In the following pages where we examine the behaviour of minerals in polarized light, we refer to Nicol prisms but this behaviour is of course the same whatever type of polar is used.

The Petrological Microscope. A type of microscope used in the study of *thin sections* of minerals and rocks is shown in Fig. 99. The essential parts of the instrument are indicated in this figure. Two Nicol prisms are fitted—one below the stage called the *polarizer* and another called the *analyser* either above the eyepiece or, as shown in the type of microscope figured here, in the

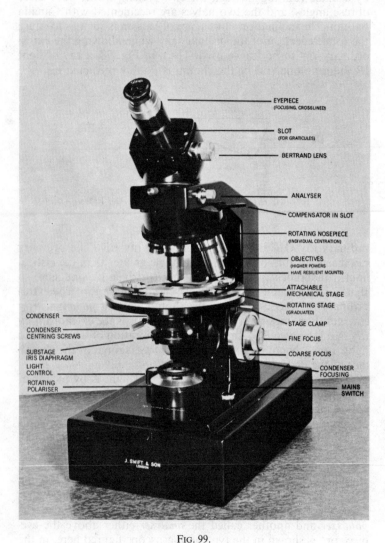

EYEPIECE
(FOCUSING, CROSSLINED)

SLOT
(FOR GRATICULES)

BERTRAND LENS

ANALYSER

COMPENSATOR IN SLOT

ROTATING NOSEPIECE
(INDIVIDUAL CENTRATION)

OBJECTIVES
(HIGHER POWERS
HAVE RESILIENT MOUNTS)

ATTACHABLE
MECHANICAL STAGE

ROTATING STAGE
(GRADUATED)

STAGE CLAMP

FINE FOCUS

COARSE FOCUS

CONDENSER
FOCUSING

MAINS
SWITCH

CONDENSER

CONDENSER
CENTRING SCREWS

SUBSTAGE
IRIS DIAPHRAGM

LIGHT
CONTROL

ROTATING POLARISER

J. SWIFT & SON
LONDON

FIG. 99.

tube between the eyepiece and objective. In some modern microscopes, discs of 'polaroid' are used instead of the usual Nicol prisms. A condensing lens can be placed between the lower nicol or polarizer and the stage. The mirror is used to throw a beam of light up through the polarizer if this is in use, through the thin slice, etc., of the mineral on the stage and hence through the objective, and the mineral is viewed by the eyepiece which is fitted with two cross-wires at right angles. The polarizer, analyser, condenser, and Bertrand lens can be inserted into or removed from the optical system at will. The student is recommended to make himself thoroughly familiar with the parts of the microscope that he uses.

Crossed Nicols. The two nicols of the microscope are said to be *crossed* when the shorter diagonal of one is at *right angles* to the shorter diagonal of the other. Consider a ray of light entering the lower nicol or polarizer. Only the extraordinary ray leaves the polarizer and vibrates parallel to the *short* diameter. When this extraordinary ray enters the top nicol or analyser it is vibrating parallel to the *long* axis of this, since the nicols are crossed. This is the direction of vibration of the ordinary ray of the analyser so that the ray is thrown aside and no light whatever gets through the analyser. A dark field of view results. The polars of the microscope used should be tested for accurate crossing.

Isotropic Minerals Between Crossed Nicols. It will be recalled that isotropic minerals are single refracting. A plate of such a mineral placed on the stage between crossed polars allows the polarized light leaving the lower polar to pass through it unaltered, so far as its vibration-direction is concerned. The dark field of the crossed polars therefore remains undisturbed. It is clear, too, that rotation of the mineral plate on the stage cannot affect this condition. The important rule is: *isotropic substances —that is, minerals of the Cubic System, natural glasses and a few minerals such as opal—give blackness between crossed polars at all positions of the stage.*

Anisotropic Minerals Between Crossed Nicols. Anisotropic minerals are doubly refracting, so that a ray of light entering a plate is broken up into two rays, vibrating at right angles and travelling with different velocities. The directions of vibration of the two rays are called the *vibration-directions* and one ray is the *fast* ray and one the *slow*.

Suppose a parallel-sided plate of an anisotropic mineral is placed on the microscope stage and examined between crossed nicols. A ray of *monochromatic light*—that is, light all of one wave-length—leaving the polarizer is vibrating parallel to the short diagonal of this nicol. This ray, meeting the lower surface of the mineral plate which has its vibration-directions at an angle to the nicol planes, as shown in Fig. 100, is doubly refracted into two rays which pass through the plate with different velocities. These two rays therefore differ in phase when they leave the mineral plate and enter the analyser. In the analyser each of the two rays is again split into two. One ray of each pair—the ordinary ray vibrating parallel to the long diagonal of the analyser—is thrown aside, and the two extraordinary rays, both vibrating parallel to the short diagonal, leave the analyser. These two polarized rays vibrate in the same plane, and have the same wave-length since monochromatic light is used, but they have travelled different distances so that there is a phase-difference between them. Under these circumstances the two rays *interfere*.

The two rays traversing the mineral plate differ in phase by some part of a wave-length. Suppose first of all they differ by 1, 2, 3 or any number of whole wave-lengths. Certain considerations dealing with the amplitudes of the vibrations along the vibration-directions of the mineral and their resolution along the vibration-direction of the analyser—these considerations are beyond the scope of this book—show that the vibrations are opposed in action and cancel one another out. Blackness results and this occurs in all positions of the vibration-directions, that is, throughout a revolution of the microscopic stage.

In the next place consider a phase-difference of $\frac{1}{2}$, $\frac{3}{2}$, $\frac{5}{2}$ wave-lengths, that is, any odd number of half wave-lengths. Here the vibrations are found to help one another, and the greatest amount of light gets through. This condition, however, does not hold throughout a rotation of the microscope stage. Consider the conditions when the vibration-planes of the mineral are parallel with the nicol planes, that is, with the vibration-direc-

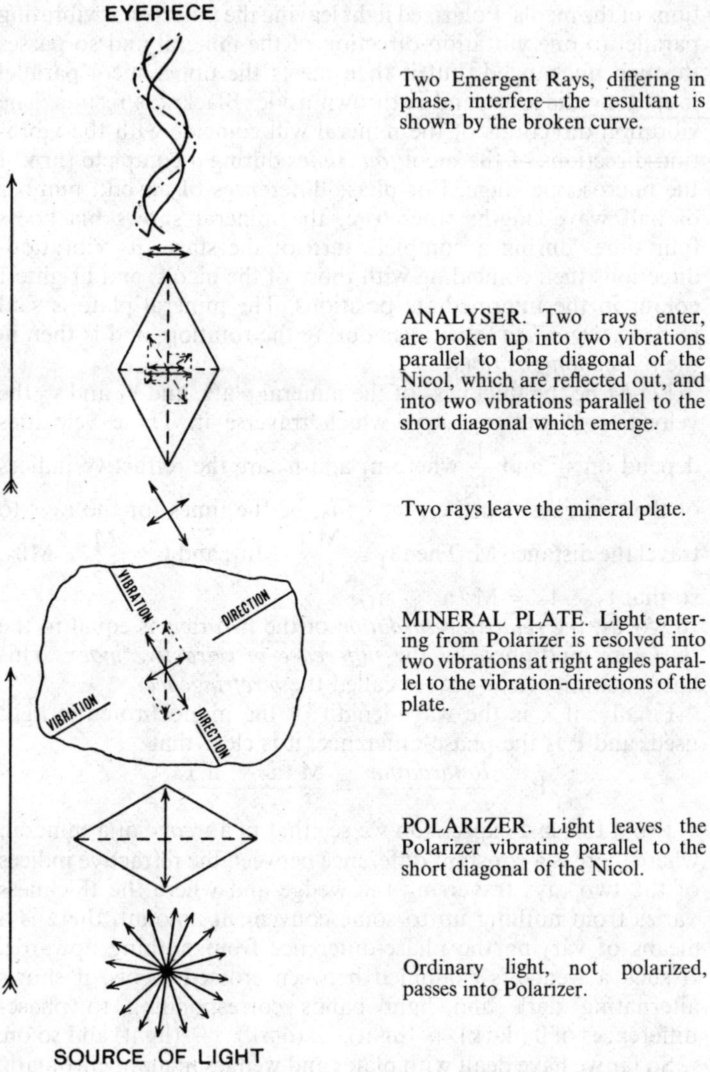
EYEPIECE
Two Emergent Rays, differing in phase, interfere—the resultant is shown by the broken curve.
ANALYSER. Two rays enter, are broken up into two vibrations parallel to long diagonal of the Nicol, which are reflected out, and into two vibrations parallel to the short diagonal which emerge.
Two rays leave the mineral plate.
VIBRATION
DIRECTION
VIBRATION
DIRECTION
MINERAL PLATE. Light entering from Polarizer is resolved into two vibrations at right angles parallel to the vibration-directions of the plate.
POLARIZER. Light leaves the Polarizer vibrating parallel to the short diagonal of the Nicol.
Ordinary light, not polarized, passes into Polarizer.
SOURCE OF LIGHT

FIG. 100. Diagram representing the Happenings in a Mineral Plate viewed between Crossed Nicols

tions of the nicols. Polarized light leaving the polarizer is vibrating parallel to one vibration-direction of the mineral and so passes through unchanged, but it then meets the upper nicol parallel to its long diagonal and is thrown aside. Blackness results. The vibration-directions of the mineral will coincide with the vibration-directions of the nicol *four times* during a complete turn of the microscope stage. For phase-differences of an odd number of half wave-lengths, therefore, the mineral shows blackness four times during a complete turn of the stage, its vibration-directions then coinciding with those of the nicols, and brightest colour in the intermediate positions. The mineral plate is said to be *extinguished* four times during the rotation, and is then in the *extinction-position*.

Let M be the thickness of the mineral plate, and v_1 and v_2 the velocities of the two rays which traverse it. These velocities depend on $\frac{1}{n_1}$ and $\frac{1}{n_2}$ where n_1 and n_2 are the refractive indices of the two rays. Further, let t_1, t_2, be the times for the rays to travel the distance M. Then $t_1 = \frac{M}{v_1} = Mn_1$, and $t_2 = \frac{M}{v_2} = Mn_2$, so that $t_2 - t_1 = M(n_2 - n_1)$.

In words, the *relative retardation* of the two rays is equal to the *thickness* multiplied by the *difference in refractive index*. This latter quantity $(n_2 - n_1)$ is called the *birefringence*.

Finally, if λ is the wave-length of the monochromatic light used, and P is the phase-difference, it is clear that

$$P = \frac{Retardation}{\lambda} = \frac{M(n_2 - n_1)}{\lambda}.$$

From this last expression we see that in a *wedge* of a mineral, where there is a constant difference between the refractive indices of the two rays traversing the wedge and where the thickness varies from nothing up to some convenient amount, there is a means of varying the phase-difference from nothing upwards. If such a wedge is examined between crossed nicols, it shows alternating dark and light bands corresponding to phase-differences of 0 (dark), $\frac{1}{2}\lambda$ (light), 1λ (dark), $1\frac{1}{2}\lambda$ (light) and so on.

So far we have dealt with plates and wedges in monochromatic light, and we must now deal with them in *white light* consisting of a number of rays of different wave-lengths. For a given thickness of a *plate*, the two rays leaving the polarizer have a certain phase-difference. If this phase-difference corresponds to 1, 2, 3, etc., wave-lengths of any one colour of light, that light will be extinguished, whilst if it corresponds to $\frac{1}{2}$, $\frac{3}{2}$, $\frac{5}{2}$, etc., wave-lengths of any colour, that colour will be strongest. The colour produced is called the *interference colour* or *polarization colour* of the plate. The colour does not change during rotation of the stage but simply alters in intensity.

Now consider the *wedge* between crossed nicols in *white light*. The various components of different wave-lengths that make up white light produce darkness and maximum colour at different positions for each light along the wedge. The overlapping of the various darknesses and brightnesses combines to form a series of colours known as *Newton's Scale of Interference Colours*. A familiar example of this scale is seen in the colours formed by thin films of oil on the surface of water. Newton's Scale is illustrated in the larger text-books. The colours are divided into *orders*, and a crude description of the scale is as follows:

First Order	Dark Grey Light Grey Greyish White Yellow Orange Red	↑ Stronger
Second Order	Violet Blue Green Yellow Pinkish Red	Fainter
Third Order	Blue Green Yellow Pink	↓
Higher Orders	Pale greens and brownish pinks.	

The student should examine a wedge of quartz between crossed nicols and familiarize himself with the colours.

The interference colour depends on the phase-difference which varies, as we have seen, with the thickness of the slice and the birefringence. It is clear that thickening the slice gives a colour higher in Newton's Scale. Similarly, if the birefringence, i.e. the difference between the refractive indices of the two rays traversing the plate, is higher, the interference colours are higher. The birefringence depends, in any particular mineral, on the direction of the slice.

Accessory Plates: *Quartz-wedge; Gypsum-plate; Mica-plate.* These three simple accessories to the petrographical microscope are of considerable use. The *quartz-wedge* provides Newton's Scale of Colours and can be used, as explained below, to estimate the birefringence and to determine the optical sign of uniaxial minerals. The wedge has marked on it the direction of the *slow* or *fast* vibrations. A wedge in which the slow vibration is parallel with the length of the wedge is called a *slow-wedge* or *slow-along wedge*. A second accessory is the gypsum- or selenite-plate which is a plate of gypsum of such a thickness that it gives the *sensitive tint*, the purple at the end of the First Order, between crossed nicols. When placed over a mineral, the gypsum-plate gives blue when the phase-difference is increased and red or yellow when it is decreased, so that phase-differences are easily told. The *mica-plate* is a thin plate of mica of such a thickness that for yellow light it gives a retardation of a quarter of a wave-length. The gypsum-plate and mica-plate have the character, fast or slow, of the vibration parallel to their lengths marked on them.

Compensation, and the Determination of Interference Colour. We have already noted that one of the two rays transmitted by a mineral plate is slow, and the other at right angles is fast. If a mineral plate is placed between crossed nicols and an accessory plate put above it so that the slow direction of the accessory plate coincides with the slow direction of the mineral plate, it is clear that the effect is one of thickening the slice, or increasing the retardation and so raising the interference colour. If, on the other hand, the slow of the accessory plate coincides with the fast of the mineral, then the effect is one of thinning, decreasing the retardation and lowering the interference colour. If the accessory plate is of the proper thickness, the gain in the mineral plate is just neutralized by the loss in the accessory plate so that the phase-difference is zero and blackness is produced. This is called *compensation*.

The quartz-wedge is of changing thickness, and is used to produce compensation. The mineral plate is placed in the position of greatest brightness, half-way between the positions of extinction, and the quartz-wedge inserted in the slot of the microscope between the nicol prisms. Two cases occur. When the slow of the wedge coincides with the slow of the mineral no compensation can be produced, and the colours in the wedge will be raised to those of higher orders. If the mineral plate is then turned through 90° the slow direction of the wedge is over the fast direction of the mineral and for some thickness of the wedge the gain in the plate is equal to the loss in the wedge, or vice versa —so that a *compensation band* appears in the Newton's Scale given by the wedge. This compensation band corresponds to the *interference colour* of the mineral so that this latter is determined.

Minerals are often consistently longer in one direction than another. The orientation of fast or slow rays with respect to this elongation is often an important optical feature. This *sign of elongation*, or simply the *elongation*, is determined with the quartz-wedge or other plate. The elongated mineral is put in the 45° position, and the character of the ray, either fast or slow, vibrating parallel with the elongation determined as in the previous paragraphs. *If the slow ray vibrates parallel to the elongation, this is said to be positive;* shortly, *slow long positive, fast long negative.* It should be carefully noted that sign of elongation is not the same as the optical sign.

Determination of the Optical Sign of Uniaxial Minerals where the C-Axis is known. The birefringence—the difference between the refractive indices of the two rays traversing a section— depends on the direction of the section with respect to the optic axis in *uniaxial minerals*. Light travelling perpendicular to the optic axis has two rays with the ordinary and the extraordinary refractive indices, and so birefringence is the greatest for the mineral. In a direction oblique to the optic axis, the two rays are the ordinary ray and an extraordinary ray which has a refractive index intermediate between those of the ordinary and extraordinary values—such a section has a lower birefringence. Along the optic axis, the ordinary ray travels with the same velocity as the extraordinary ray, the refractive indices are the same, there is no phase-difference, and the section is isotropic.

We may illustrate these remarks by reference to sections of quartz. A section of the usual thickness of quartz cut parallel to the vertical crystallographic axis, which is, of course, the optic axis, shows between crossed nicols yellow of the First Order in Newton's Scale. This is the highest polarization colour of quartz for that thickness of section. A second section cut obliquely to the *c*-axis, say parallel to a rhombohedral face, shows between crossed nicols a lower polarization colour, possibly pale grey of the First Order. A basal section is isotropic between crossed nicols, remaining black in all positions.

If the position of the *c*-axis or optic axis of a uniaxial crystal is known in the section, it is possible to find the *optic sign*. In uniaxial minerals, as already stated, the ordinary ray vibrates perpendicular to the optic axis or *c*-axis, the extraordinary ray perpendicular to the ordinary ray. Take calcite, a negative mineral, as an example. Remember the circle inside the ellipse of Fig. 97, and so state that the *extraordinary ray* is *fast, ordinary ray slow*. Hence the ray vibrating parallel to the *c*-axis is fast. In general terms:

The · *vertical crystallographic axis is fast, uniaxial mineral optically negative.*

The *vertical crystallographic axis is slow, uniaxial mineral optically positive.*

Whether the *c*-axis is a direction of fast or slow vibrations is determined in the manner just described with an accessory plate between crossed nicols. Unfortunately, the *c*-axis is recognisable only in a few cases in practice.

Vibration-Directions and Optic Orientation. In all anisotropic minerals there are fast and slow directions of vibration at right angles and also a third direction, at right angles to the other two, along which the vibrations have an intermediate speed. In uniaxial crystals, which we have mainly considered so far, this third speed becomes equal to that of either the ordinary or extraordinary rays—the vibration-directions are only two.

The three vibration-directions at right angles are called the *principal vibration-axes* or *axes of the optical ellipsoid*. The vibration-axes are denoted by fast, X, or α; intermediate, Y, or b, and slow, Z or c. In every section of an anisotropic crystal there are two vibration-directions at right angles, one slow and one fast, but not necessarily the slowest or the fastest for the mineral.

The *optical orientation* of a mineral is the relation between the vibration-axes and the crystallographic axes. The following scheme shows the optical orientations in the different crystal systems:

CUBIC SYSTEM. Three equal vibration-axes, wave-front a sphere, all directions alike.

TETRAGONAL AND HEXAGONAL SYSTEMS. The optic axis is the vertical crystallographic axis; the vibrations taking place perpendicular to this axis are all equal.

ORTHORHOMBIC SYSTEM. The vibration-axes X, Y, Z coincide with the crystallographic axes, any one of the first with any one of the second.

MONOCLINIC SYSTEM. One vibration-axis coincides with the crystallographic axis b, the ortho-axis, and the other two lie in two rectangular directions in the plane of symmetry.

TRICLINIC SYSTEM. The vibration-axes are in any position whatever, but, of course, all at right angles to one another.

Position of Extinction and Extinction Angle. We have seen that most sections of anisotropic minerals extinguish or show darkness between crossed nicols four times during a complete revolution of the stage, and that extinction occurs when the vibration-planes of the mineral are parallel to the nicol planes. From the previous paragraph we see that the optical orientation characteristic of each crystal system shows where extinction shall occur with reference to the crystallographic directions. In sections of many minerals there are seen cleavages or crystal edges of determinable crystallographic orientation, and the position of extinction with regard to these can be observed. The *extinction-angle* is the angle between a vibration-plane and a crystallographic direction in a given section of the mineral. It is measured by placing the mineral in the extinction position and reading the microscope stage; the stage is then rotated till the cleavage

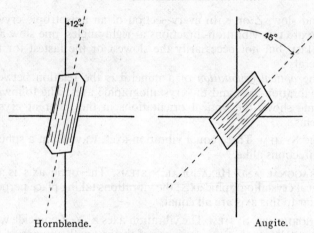

Hornblende.   Augite.

FIG. 101. Hornblende: Extinction Angle, 12°. Augite: Extinction
Angle, 45°

or edge is parallel with a cross-wire in the eyepiece and therefore
parallel with a nicol plane, and the stage reading taken again.
The difference between the two readings gives the extinction-
angle.

There are various positions of extinction. In *straight* or *parallel
extinction*, extinction takes place when the crystallographic
direction, cleavage, etc., is parallel with a cross-wire. *Inclined*
or *oblique extinction* occurs when the crystallographic direction
makes an angle with the cross-wires in the position of extinction
of the section. In *symmetrical extinction* the cross-wires in the
position of extinction bisect the angles between two sets of
cleavages or edges.

It must be noted that extinction-angles can be measured from
the fast or slow vibration-direction in the crystal plate to the
cross-wire, so that it is necessary to determine with the help of
an accessory plate the character of the vibration-direction dealt
with. Extinction-angles are often of great diagnostic value, and
this is illustrated in Fig. 101 where the extinction-angles of corresponding sections of hornblende and augite are shown.

In summary are now given the types of extinction characteristic
of each crystal system:

Cᴜʙɪᴄ Sʏsᴛᴇᴍ. All sections isotropic.

Tᴇᴛʀᴀɢᴏɴᴀʟ ᴀɴᴅ Hᴇxᴀɢᴏɴᴀʟ Sʏsᴛᴇᴍs. Basal sections, being perpendicular to the optic axis, are isotropic. Vertical sections give straight extinction.

Oʀᴛʜᴏʀʜᴏᴍʙɪᴄ Sʏsᴛᴇᴍ. Pinacoidal sections give straight extinction.

Mᴏɴᴏᴄʟɪɴɪᴄ Sʏsᴛᴇᴍ. Orthopinacoidal and basal pinacoidal sections give straight extinction; clinopinacoidal sections, as in Fig. 101, give oblique extinction.

Tʀɪᴄʟɪɴɪᴄ Sʏsᴛᴇᴍ. All sections give oblique extinction.

Anomalous Polarization Colours. Several minerals show interference colours not found in Newton's Scale, such colours being called *anomalous*. For instance, the body colour of a mineral, for example, biotite, may be strong enough to mask a delicate interference colour. Again, in some minerals, such as idocrase, zoisite and chlorite, queer Berlin or inky blues and washed-out browns are given between crossed nicols. This phenomenon depends fundamentally on dispersion and will not be considered further.

Pleochroism and Absorption. A mineral is said to be *pleochroic* when it shows change in quality or quantity of colour when rotated in polarized light. Pleochroism is due to the unequal absorption by the mineral of light vibrating in different planes. As an example, consider a longitudinal section of the dark mica, biotite. If the polarized light vibrates parallel with the cleavages, the mineral is a deep brown, and sometimes almost black in colour; if the polarized light vibrates across the cleavages, the mineral is a pale yellow. Pleochoism is best observed under the microscope by rotating the polarizer and watching for changes of tint in the mineral under examination; the analyser, of course, is not in position.

In *isotropic* substances, the absorption must be the same in all directions so that in one slice all sections of isotropic minerals show the same colour and are non-pleochroic. In *uniaxial minerals*, basal sections, since all the rays are ordinary rays, show no pleochroism, whilst vertical sections may show greatest difference. The third great class of minerals, *biaxial minerals*, may show three tints or colours according to the direction of the light. This pleochroism is described by stating the colour of the light vibrating parallel to the vibration-axes X, Y, Z. Thus, for some specimens of hornblende, the pleochroism is X yellow, Y blue-green, C blue, and the absorption X < Y < Z.

Around minute inclusions in some minerals, for example cordierite and biotite, small areas are more strongly pleochroic than the main part of the mineral. These pleochroic spots are called *pleochroic halos*, and are due to bombardment and alteration of the host mineral by radioactive emanations from the inclusion.

The property of pleochroism is of special value in the determination of gems, for which purpose use is made of a little instrument called a *dichroiscope*, which consists of a cleavage rhomb of Iceland Spar contained in a tube provided with a square aperture at one end, and a lens at the other. On looking through the lens at a transparent crystal placed over the aperture at the other end, the observer sees two images of the aperture side by side. One image is formed by the ordinary ray, the other by the extraordinary ray and the vibrations of these are at right angles, so that in the case of a pleochroic mineral different colours are seen, according to the direction in which the crystal is placed.

Convergent Light. So far we have dealt with the examination of minerals in ordinary light and parallel polarized light. We now have to consider briefly the application of convergent light. A converging lens system—the condenser—is placed below the microscope stage, the nicols are crossed and a high-power objective used. Under suitable conditions an *interference figure* is produced which can be rendered visible in three ways, (1) by inserting the Bertrand lens, (2) by placing a lens above the eyepiece and (3) by removing the eyepiece. Possibly, the last means is best in practice.

The kind of interference figure depends upon the optical character of the mineral, that is, whether it is uniaxial or not, the position of the section in the crystal and the type of light, monochromatic or not, that is used. We deal here only with the more simple cases.

Interference Figures in Uniaxial Crystals. We shall consider only the interference figure produced by a basal section of a uniaxial mineral, that is, a section perpendicular to the optic axis or *c*-axis. Figures given by other sections are too difficult for a beginner. The condenser throws a cone of light up into the mineral plate, as shown in Fig. 102.

The ray striking the plate at normal incidence travels along the optic axis and undergoes no double refraction; rays oblique to this direction are doubly refracted and the resulting rays have a phase-difference. This phase-difference is zero along the optic axis and gradually increases outwards from this as the rays become more inclined. There are thus produced exactly the same conditions as in a quartz-wedge between crossed nicols, so that we can consider our plate in convergent light to give the same result as a quartz-wedge rapidly rotated, or as a shallow quartz dish with a double wedged-shaped cross-section as shown in Fig. 102, examined in plane-polarized light between crossed nicols. Applying what has been said on pp. 177–8 on the quartz-wedge between crossed nicols, we see that the interference figure must consist of a series of coloured rings showing Newton's Scale and a black cross, the two arms intersecting at the centre of the microscope field.

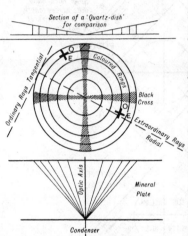

FIG. 102—Uniaxial Interference Figure given by a section cut perpendicular to the Optic Axis.

Determination of Optical Sign of Uniaxial Minerals by the Interference Figure. We have already seen that in a uniaxial mineral the extraordinary ray vibrates in the plane passing through the ray and the optic axis, and the ordinary ray at right angles to this and to the optic axis. Examples of the vibration-directions are shown by E and O in Fig. 102, whence the general rule, *extraordinary ray vibrates radially, ordinary ray vibrates tangentially*, is derivable. Note, in passing, that the black cross is produced when these vibration-directions are parallel to the vibration-planes of the nicols.

In the uniaxial interference figure of this kind, therefore, the vibration-directions of the extraordinary and ordinary rays are known so that the optical sign can be determined by using an

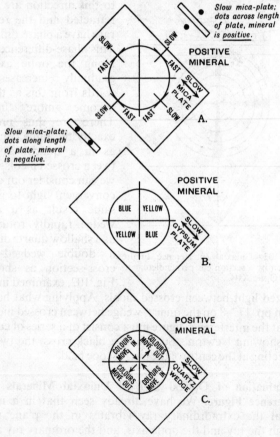

FIG. 103. Determination of Sign of Uniaxial Mineral

accessory plate. If the extraordinary ray is *slow, the mineral is optically positive,* if *fast, the mineral is optically negative* (cf. p. 170). The procedure is as follows:

(i) With the *Mica-plate.* Suppose that a slow-along mica-plate is placed over a plate in which the extraordinary ray is slow, as shown in Fig. 103A. The slow plate helps the extra-ordinary ray in the NW and SE quadrants and so the inter-ference colours rise; the slow-along plate opposes in the NE and SW quadrants, and for some retardation compensation is produced and two black spots appear, one in the NE and one in the SW quadrant. Since our mineral has the extraordinary ray slow, it is *positive,* so that the production of two black spots *across* the length of the slow plate can be used as the test. In optically *negative* minerals the two black spots appear *along* the length of the slow plate.

(ii) With the *Gypsum-plate.* This plate is used in the same way as the mica-plate. Reference to the description of the gypsum-plate on p. 178, shows that opposition gives a red or yellow tint and help gives a blue tint. If a slow-along gypsum-plate is used, yellow quadrants correspond to the dark spots produced by the mica-plate, as shown in Fig. 103B.

(iii) With the *Quartz-wedge.* With a slow-along quartz-wedge and a positive mineral, in the position shown in Fig. 103C, the colours of the wedge move in towards the centre in the NW and SE quadrants where wedge and mineral help one another, and move out from the centre in the NE and SW quadrants where wedge and mineral are opposed.

The production of a centred uniaxial interference figure is an important point in mineral diagnosis with the microscope, for it indicates that the mineral belongs to the tetragonal or hexa-gonal crystal systems. The determination of the optic sign further limits the possibilities, and, in some cases, the finding of the position of the *c*-axis, i.e. the optic axis, is of use.

Biaxial Minerals. In *biaxial* crystals there are *two optic axes,* or two directions along which there is no double refraction and along which light travels with a single velocity. Crystals belonging to the orthorhombic, monoclinic and triclinic crystal systems are biaxial, so that our final classification of minerals according to their optical characters is :—

ISOTROPIC. Cubic system.
UNIAXIAL Tetragonal and hexagonal systems.
BIAXIAL. Orthorhombic, monoclinic and triclinic systems.

We have already mentioned that in biaxial minerals there are three principal vibration-directions—X fastest, Y intermediate, Z slowest. The *principal optic planes* of biaxial crystals are the three planes at right angles in which the three principal vibration-directions interesect. The ray vibrating parallel to X is considered to have the refractive index α, that parallel to Y the refractive index β, and that parallel to Z the refractive index γ. We may construct a triaxial ellipsoid which has its three axes proportional to α, β and γ. This ellipsoid is called the *indicatrix* and is shown in Fig. 104. The properties of the indicatrix are beyond the scope

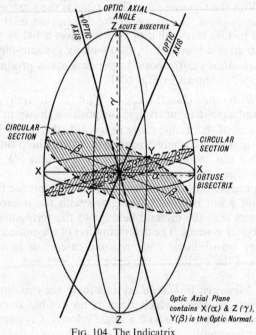

FIG. 104. The Indicatrix

of this book and the indicatrix is used here for making definitions. Examination of the indicatrix shows that there are only two *circular sections*, these being symmetrically placed, and their radius is the intermediate refractive index β. A cross-section of the indicatrix represents the refractive indices of the rays that vibrate in the plane of the section and advance along a line perpendicular to this plane. It is seen, therefore, that the circular sections are something like the section of a uniaxial mineral perpendicular to the optic axis, so that the perpendiculars to the circular sections are called the *optic axes*, the plane containing the optic axes (and therefore the vibration directions X and Z) is the *optic axial plane*, and Y, normal to this plane, is the *optic normal*.

The position of the optic axes in different crystals depends upon the relative values of α, β and γ. The angle between the optic axes is called the *optic axial angle*. The vibration-directions X and Z each bisect the acute or obtuse angle between the optic axes. The vibration-direction in the *acute* angle is called the *acute bisectrix*, that in the obtuse angle is the *obtuse bisectrix*. By analogy with uniaxial minerals (which, after all, may be considered as a special case of biaxial crystals in which two of the three main refractive indices of the latter are equal) the optical sign of biaxial minerals is defined as follows:

When Z, *slow* vibration-direction, is the *acute* bisectrix, mineral is optically *positive*.

When X, *fast* vibration-direction, is the *acute* bisectrix, mineral is optically *negative*.

Interference Figures in Biaxial Minerals. Interference figures given by two sections of biaxial minerals are briefly considered here. These sections are (1) perpendicular to the acute bisectrix and (2) perpendicular to an optic axis.

(i) *The Interference Figure given by a section perpendicular to the acute bisectrix.* In this section the two optic axes emerge. The interference figure shows a series of coloured ovals arranged about two centres or eyes, the points of emergence of the optic axes, and joining up into larger ovals with dimpled sides farther

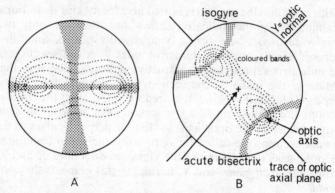

FIG. 105. Biaxial Interference Figure, perpendicular to Acute
Bisectrix: in A, the optic axial plane is parallel to a nicol plane,
and in B is at 45° to this plane.

from these centres. These coloured ovals correspond to the coloured rings seen in interference figures of uniaxial minerals and are due to the same causes. In addition to the coloured ovals, the biaxial interference figure shows dark brushes, or *isogyres*. The behaviour of these as the stage is rotated is important. When the line joining the eyes of the figure is parallel to one of the nicol planes the brushes form a cross, one arm joining the eyes and the other being perpendicular to this midway between the eyes. When the stage is rotated 45° so that the line joining the eyes lies say NW and SE, the black cross separates into two hyperbolae, one passing through each eye. The biaxial interference figure in the two positions is shown diagrammatically in Fig. 105.

The eyes mark the points of emergence of the *optic axes*; the nearer the eyes are together the smaller is the *optic axial angle*. The line joining the eyes is the trace of the *optic axial plane*. The acute bisectrix emerges midway between the two eyes and the *optic normal* is at right angles to the optic axial plane.

(ii) *The Interference Figure given by a section perpendicular to an Optic Axis*. This figure shows one isogyre or dark bar which crosses a series of nearly circular coloured bands. The isogyre is straight when parallel to a nicol plane but curved in an intermediate position, the convex side of the curve being towards the position of the acute bisectrix. This section is an important one in the determination of the optical sign; it is given by sections showing the lowest polarization colours.

Determination of the Optical Sign of Biaxial Minerals from the Interference Figure. Only two methods of performing this operation are given in this book; other methods are to be found in the larger textbooks.

(i) *In a section perpendicular to the acute bisectrix.* The interference figure is observed in the ordinary way, and the stage rotated till the optic axial plane is in the 45° position. Convergent light is removed and the mineral plate viewed in parallel polarized light between crossed nicols. The fast or slow character of the vibrations parallel either with the optic normal or the trace of the optic axial plane is determined in the usual way with an accessory plate. Suppose the vibrations parallel to the optic normal are found to be slow; those in the trace of the optic axial plane are therefore fast—but this is the direction of the obtuse bisectrix, so that the *acute bisectrix* is slow and the mineral is *positive*.

(ii) *In a section perpendicular to an optic axis.* It is sufficient for us to give the rule without explanation. The figure is placed in the 45° position, and a slow-along gypsum-plate inserted along the optic axial plane. For a positive mineral, a *yellow* colour appears on the *convex* side of the brush, and a *blue* colour on the *concave* side.

Thin Sections of Rocks and Minerals. For the examination of rocks and minerals under the microscope thin slices or sections are required. A chip of the rock is ground perfectly flat on one side, with emery or carborundum powder on a glass or metal plate. The grinding is begun with a coarse powder, and continued with finer powders until a perfectly smooth flat surface is obtained. In transferring the chip from one grinding plate to a finer, it is necessary to wash the chip free from emery, or otherwise the grades of grinding powder become mixed on the plates. A glass slip (3in. × 1in.) is taken, and a small amount of Canada balsam put in its centre and heated gently until sufficient turpentine or xylol has been driven off to cause the balsam to become hard and compact when cool. The correct moment to cease heating the balsam is judged by taking up a small quantity on forceps, and, by opening the forceps, causing a bridge or thread of balsam to be formed. If the balsam has been sufficiently heated, this bridge will be hard and brittle when cool. The chip of rock is then placed on the slip with the flattened side in contact with the balsam and glass. By pressing the chip air bubbles are removed from the film of balsam between the chip and glass. When cool, the chip will be firmly cemented to the glass slip. The next operation consists in grinding down the thick chip as in the first process, beginning with the coarse and ending with the finest powder. Great care is necessary during the final stages, or the rock will be completely rubbed away. The thickness of the slice can be judged by the polarization colours given by some recognizable mineral, such as quartz, and grinding must be continued until this mineral shows its usual polarization colours—for quartz, grey and yellow of the First Order. After the final grinding, the section is carefully washed, and all remaining balsam scraped from around the rock. The slice is then covered with fresh balsam and heated again to a slightly less extent than before. When the balsam is of the right consistency, a very thin sheet of glass—the cover-slip—is carefully placed over the rock and pressed down so that no air bubbles are present. Any balsam round the cover-slip is removed by methylated spirit. The result of these operations is what is known as a thin section or a thin slice of the rock or mineral.

The above process requires much patience and skill for good results to be obtained, but in the field some amount of information concerning rocks can be obtained by rubbing down the chip on a grindstone and using a whetstone, carborundum file, or similar device, for the final grinding. By these and other methods which will suggest themselves to the practical man there are obtained slices which, though thick and uneven, are yet capable of yielding information when examined by a pocket lens.

Synopsis of the Examination of Minerals Under the Petrological Microscope. The examination of minerals in thin section under the microscope consists of four groups of operations, namely, those carried out by using (1) ordinary light, (2) polarized light, (3) crossed nicols, (4) convergent light.

(1) ORDINARY LIGHT. In order to examine thin sections of

minerals in ordinary light, both polarizer and analyser are removed. The following properties can be observed:

(a) *Crystalline Form*. Accurate conclusions as to the crystalline form of a mineral can only be arrived at after the observation of a large number of sections of the mineral, but sometimes form observations are significant. Thus, nepheline is a hexagonal mineral, and in thin sections often appears as hexagonal cross-sections and rectangular longitudinal sections.

(b) *Cleavage* (p. 60). Cleavage appears in thin sections as one or more sets of parallel cracks. The number of cleavage-cracks, the number of sets of cleavages and their inclination one to another depend on the direction in which the section is cut. Thus the prismatic cleavage of hornblende appears in transverse sections as two sets of lines meeting at angles approximately 120°, whereas longitudinal sections show only one set of cleavage-cracks (Fig. 140).

(c) *Inclusions*. In some minerals included material is arranged on a definite plan, and serves as a useful diagnostic character. In leucite, inclusions occur in concentric and radial patterns (see Fig. 145, p. 434). The well-known cross-shaped inclusions in chiastolite are another example.

(d) *Transparency*. Most rock-forming minerals are transparent in thin section; some few, such as magnetite and pyrite, are opaque.

(e) *Colour*. In section a mineral may be colourless, as with quartz, or coloured as with biotite, and in most cases the colour is distinctive of a given mineral. Sections of one mineral may show several colours, or shades of the same colour, owing to its being a pleochroic mineral which has been cut in several different crystallographic directions in the same slice.

(f) *Refractive Index* (p. 164). A mineral of a refractive index markedly different from that of the medium in which it is mounted has well-marked borders, any cleavage-cracks it possesses are boldly shown, and its surface exhibits a pitted or shagreened appearance. The relative refractive indices of two minerals or of a mineral and the mount are judged by the Becke effect or the Shadow method.

(2) POLARIZED LIGHT. In order to examine thin sections in polarized light, the lower nicol or polarizer is put into place between the mirror and the stage. The following may be observed:

(a) *Pleochroism* (p. 183). Pleochroic minerals show a change in the quantity or quality of their colour, this change depending upon the direction followed by the polarized light as it traverses the mineral. Pleochroism is observed by rotating the polarizer. It is remembered that the light leaving the polarizer vibrates parallel to the short axis of this, so that it is possible to relate a colour with a vibration-direction in the crystal, as explained on p. 184.

(b) *Pleochroic Halos* (p. 184). The presence of pleochroic halos —small spots more strongly pleochroic than the main part of the mineral—may be observed by rotating the lower nicol. They are characteristic of certain minerals.

(c) *Twinkling*. The phenomenon of twinkling is well seen in calcite. In calcite the refractive index for the ordinary ray is 1·66, for the extraordinary ray 1·49. The refractive index of balsam is 1·54. If a granular mosaic of calcite is examined in polarized light, it is obvious that some grains will transmit the ordinary ray and some the extraordinary ray. Those grains transmitting the ordinary ray have a refractive index much greater than that of balsam, and therefore their borders will be well marked; those grains transmitting the extraordinary ray have a refractive index a little lower than that of balsam, and their borders will not be so strongly marked. Hence, when the polarizer is rotated beneath the slice of calcite, certain grains have alternately strongly marked and slightly marked borders, and a *twinkling* effect is noticed.

(3) CROSSED NICOLS. For the examination of minerals between crossed nicols both polarizer and analyser are inserted, and the nicols or polars are so arranged that their vibration-planes are at right angles to one another, or are *crossed*.

(a) *Isotropism and Anisotropism*. All sections of *isotropic* transparent minerals are black between crossed nicols. Such minerals belong to the Cubic System. Minerals of the other crystal systems are *anisotropic*, but it must be remembered that basal sections of uniaxial minerals are black between crossed nicols.

(b) *Extinction and Polarization Colours* (pp. 176–7). When a section, other than a basal section of a uniaxial mineral, is examined between crossed nicols it becomes black or is *extinguished* in four positions 90° apart during one rotation. Extinction may occur parallel to a crystal edge, or to a prominent cleavage, and then the crystal shows *straight extinction* with respect to that particular crystallographic direction. Extinction may occur when the particular crystallographic direction makes an angle with the cross-wires. This is *oblique extinction*, and the angle made between the crystallographic direction and the cross-wires—the *angle of extinction*—is of diagnostic value.

In positions intermediate between the positions of extinction the mineral shows *polarization colours*. The polarization colour shown by a section of a mineral depends upon the thickness of the slice, the direction of the slice in the crystal and the character of the mineral. In uniaxial minerals, the maximum polarization colours are shown in sections parallel to the optic axis or crystallographic *c*-axis.

(c) *Twinning* (p. 151). Twinning can be best seen between crossed nicols. The two halves of the twin, or the two sets of twin lamellae, give different polarization colours and extinguish at different angles. Plagioclase feldspar is usually twinned, and between crossed nicols an apparently homogeneous crystal shows two sets of twin lamellae, each set extinguishing in a different position from that of the other (Fig. 144, p. 428).

(d) *Alteration*. Alteration can be observed in ordinary light, but its nature is best seen between crossed nicols. An altered mineral is usually turbid or cloudy, and alteration products may be developed along cleavages, cracks or otherwise. Between crossed nicols, altered minerals usually show *aggregate polarization*, because the originally homogeneous crystal has been converted by alteration into a multitude of irregularly arranged crystals of the alteration product.

(e) *Elongation* (p. 179). Some minerals occur as elongated crystals. The sign of elongation can be determined by producing compensation with a quartz-wedge with the mineral in the 45° position between crossed nicols. If the vibrations along the elongation-direction are slow, the mineral is said to have positive elongation, if fast, negative elongation.

(4) CONVERGENT LIGHT. For the production of interference figures, the condenser is inserted below the stage, the nicols are crossed, and either the Bertrand lens put into position or the eye-piece is removed.

(a) *Interference Figures* (pp. 185, 189). Crystals belonging to to the Tetragonal and Hexagonal crystal systems give a *uniaxial interference figure*. The most suitable figure is given by basal sections perpendicular to the optic axis or the *c*-axis, and consists of a black cross and a series of coloured rings. Crystals of the Orthorhombic, Monoclinic and Triclinic systems give a *biaxial interference figure*. Sections perpendicular to the acute bisectrix show a figure consisting of black brushes and a series of coloured ovals arranged about two eyes, the points of emergence of the optic axes. When the optic axial plane is parallel to a nicol plane, the brushes form a cross. When the stage is rotated, this cross breaks up into two hyperbolae, one passing through each eye. Sections of biaxial minerals perpendicular to an optic axis show a series of coloured ovals traversed by a single brush.

(b) *Determination of the Optical Sign.* In *uniaxial* minerals the optical sign is determined from the centred interference figure by using an accessory plate. Remember that the extraordinary ray vibrates radially and the ordinary ray tangentially. If the extra-ordinary ray is determined by the quartz-wedge, etc., to be slow, the mineral is optically positive. In a positive mineral using a slow mica-plate, two black dots are formed across the length of the slow plate; with a slow gypsum-plate, yellow spots replace these black dots; with a slow quartz-wedge the colours move in in the quadrants along the length of the wedge. In *biaxial* minerals, when Z, slow vibration-direction, is the acute bisectrix, the mineral is optically positive. In a section perpendicular to the acute bisectrix, the optic axial plane joining the two eyes is placed in the 45° position, convergent light removed, and the slow or fast character of the vibrations in the trace of the optic axial plane determined by an accessory plate. If these vibrations are fast, then the acute bisectrix is slow and the mineral is optically positive. In sections perpendicular to an optic axis, the interference figure is placed in the 45° position and a slow gypsum-plate inserted along the trace of the optic axial plane. For a positive mineral, a yellow colour appears on the convex side of the brush and a blue colour on the concave side.

The Microscopic Investigation of Ore-Minerals

The investigation of opaque minerals, such as the ore-minerals, by *reflected light* has become an important branch of mineralogical study. The problems of ore-deposition, order of crystallization of ores in a mineral vein, the replacement of ore-minerals, and allied phenomena are of great importance in economic mineralogy. The discussion of this subject is beyond the scope of this book, and only a few remarks are given here, mainly to indicate the possibilities of this study.

The Ore Microscope. Opaque minerals are examined with a microscope fitted with a *vertical illuminator* placed at right angles to the tube of the microscope and above the objective. In the vertical illuminator there is a *reflector*, preferably a glass disc coated with material with a high refractive index, or a totally reflecting prism. By means of the reflector light is thrown down on the surface of the opaque mineral and is reflected therefrom back through the objective, through the glass disc or behind the prism of the vertical illuminator, and so reaches the eye-piece and is observed by the eye. A polarizer can be inserted between the reflector and the source of light.

The Polished Specimen. The ore to be examined is ground flat and smoothed in the ordinary way and then embedded in a plastic block, usually of a size and shape to fit into some kind of mechanical polishing machine. Either by rotation in this machine or by rubbing on a rotating lap sprinkled with a fine polishing powder such as magnesia, the final polish is put on the ore-specimen. Finally, the polished surface is made parallel to the stage of the microscope by various levelling devices.

Examination of the Polished Specimen. The polished specimen is examined in various ways. Many observations are similar to those made on transparent minerals—these deal with such properties as the shapes of the grains of the different minerals, their parting, cleavage, twinning, zoning and, of course, the mutual relations of the several minerals composing the ore. Minerals belonging to the

Cubic System are distinguished from those of the other systems by their isotropic character; their colour and brightness remain unchanged during the rotation of the specimen under polarized light. Between crossed nicols, isotropic minerals can again be distinguished from anisotropic minerals, the latter showing change in colour or brightness and extinguishing four times in a complete rotation of the stage.

Hardness tests may be applied to the minerals of the polished specimen by observing the results of scratching with a weighted needle, or of indentation with a punch under a known load. During polishing, the softer minerals are worn away slightly when compared with the harder minerals and so the different constituents of the ore may become revealed. These differences may be emphasized by etching of the polished surface by various acids and this operation may also bring up twinning and zoning in individual constituents.

Microchemical Tests. A series of microchemical tests may be used to identify, or to confirm the identification of, the various minerals discovered in the examination of the polished specimen by the means already outlined. A minute portion of each constituent of the ore sample is scraped off and placed upon an ordinary glass-slip. These tiny samples are then subjected to chemical tests. Their reactions to various reagents are studied under the microscope and from the crystalline form, colour and general character of the crystallization or precipitates produced during the reactions the nature of the ore-minerals present can be determined. Microchemical tests are also employed in the discrimination of certain minerals, such as calcite and dolomite, as explained on p. 278.

CHAPTER V

THE OCCURRENCE OF MINERALS

INTRODUCTION

In this chapter there is given an outline of the various modes of occurrence of minerals in the earth's crust, the chief purpose of this outline being admittedly to explain certain terms used in the description of minerals in the second part of the book.

The individual portions of the earth's crust, the *rocks*, are composed of one or more minerals mixed together, and we have to consider here the characters, classification and occurrence of rocks. A brief account of the common rock-types is given, and mention made of features of importance in economic mineralogy: this is followed by a summary of the characters of mineral deposits.

The Classification of Rocks

Rocks are usually divided into *three* great classes:

1. *Igneous Rocks*. Those consolidated from molten rock-material or *magma*.
2. *Sedimentary Rocks*. Those produced by the deposition of material arising from the physical and chemical decay of pre-existing rocks at the earth's surface.
3. *Metamorphic Rocks*. Those produced from pre-existing rocks by the action of high temperatures and strong pressures deep in the earth's crust.

1. Igneous Rocks

Magma. Within or below the earth's crust there can arise bodies of molten rock-material called *magma*. The fundamental cause of magma-formation is the development of abnormally high temperatures for a given depth, often connected with earth-movements of different kinds. This magma is forced higher into the crust and so penetrates and invades rocks of all kinds. It may stop before it reaches the surface, cool and solidify—the resulting rocks are *igneous rocks*, and since this type of igneous rock invades the surrounding rocks, or country-rocks as they are conveniently called, it is styled *intrusive*, and further, since it consolidates below the surface, it is often called *plutonic*. Some of the magma, however, may reach the surface of the earth and be poured out or blown out upon it—these igneous rocks are called *extrusive* or *volcanic*. The two great divisions of the igneous rocks are thus *intrusive* and *extrusive*.

Forms of Igneous Rocks. The extrusive rocks may form *lava-flows*, or the extrusive material may be blown to pieces and fall around the volcano to form *beds* of *pyroclastic rock*. Intrusive igneous rocks show a variety of more or less distinct forms. *Dykes* result from the forcing apart by the magma of more or less vertical fissures, so that the resultant igneous rock has a wall-like form, and appears at the surface as a straight band of rock often traceable over great distances. Dykes occupy cracks resulting from tensions acting tangentially to the surface of the crust, and often occur in a group or *swarm* of more or less parallel members. *Sills* or *intrusive sheets* result from the solidification of magma forced between more or less horizontal planes, and give evidence of radial pressure. A dyke may pass into a sill, or a sill into a dyke. A *laccolith* is formed by the consolidation of a low flattish dome of magma which has a flat floor and an arched roof, and may be considered as due to the swelling of a sheet. The floor cannot be depressed on the accession of more magma, and so relief is attained by forcing the roof upwards. *Batholiths* are igneous bodies usually of granitic rocks that cut across earlier structures in the country-rocks, and whose plan increases in area with depth. They are usually of great dimensions, the terms *stocks* and *bosses* being applied to those of smaller size. *Necks*, *plugs* and *pipes* are approximately cylindrical conduits filled with igneous or pyroclastic material.

Nature of Magma. The composition, both mineralogical and chemical, of the igneous rock obviously depends on the composition of the magma whose consolidation gave rise to it. Magmas consist of two parts—(1) non-volatile components, with fusion-points up to 1000°C or over, and (2) components of a highly volatile character. Many thousands of igneous rocks have been analysed, and the possible variation in their compositions may be roughly expressed as follows:

SiO_2	35–75%
Al_2O_3	0–25%
FeO and Fe_2O_3	0–20%
MgO	0–45%
CaO	0–20%
Na_2O	0–16%
K_2O	0–12%

The dominant component of the magma is silica, SiO_2; the other six oxides combine with the silica to form the very important group of minerals known as the *rock-forming silicates*. In natural magmas there are certain restrictions in the quantities of these oxides as constituents. It is generally true that high content of silica, alumina and alkalies is accompanied by low lime, magnesia and iron-oxide; and high lime, magnesia and iron-oxide is characteristic of magmas with low silica and low or moderate alumina and alkalies. The approximate composition of a magma may be expressed in terms of the oxides listed above, and these oxides are components of the non-volatile portion of the magma. The volatile constituents are present in magma only in small amounts but, as shown below, under certain circumstances they may become concentrated during the solidification and then give rise to phenomena of great importance in the formation of ore-deposits.

Crystallization of Magma. On cooling, the constituents of the magma unite to form the *rock-forming silicates*. Often there is an excess of silica over that required to form the silicates, and this appears as the mineral quartz in the igneous rock. Iron-oxides are formed in small amount. The most important and characteristic of the minerals of igneous rocks are:—

Quartz, SiO_2.

Feldspar Group.

Orthoclase, $KAlSi_3O_8$.

Plagioclase,

Albite, $NaAlSi_3O_8$: Anorthite, $CaAl_2Si_2O_8$.

Feldspathoid Group.

Nepheline, $NaAlSiO_4$.
Leucite, $KAlSi_2O_6$.

Mica Group.

Muscovite, $KAl_2[(AlSi_3)O_{10}](OH)_2$.
Biotite, $K(Mg,Fe)_3[(AlSi_3)O_{10}](OH)_2$.

Olivine, $(Mg,Fe)_2SiO_4$.

Pyroxene Group.

Enstatite-hypersthene Series, $(Mg,Fe)SiO_3$.
Augite, complex silicate of Mg,Fe,Ca,Na,Al.

Hornblende, complex silicate of Mg,Fe,Ca,Na,Al.

Iron-oxides.

Magnetite, Fe_3O_4.
Ilmenite, $(FeTi)_2O_3$.

The kinds of minerals of any given igneous rock depend on the bulk composition of the magma. If the magma is rich in alkalies, silica and alumina, the minerals feldspar, quartz and mica are likely; if the magma is low in silica but high in alkalies and alumina, the feldspathoids may be expected; in a magma rich in lime, magnesia and iron, *ferromagnesian minerals*, those rich in iron and magnesia, will crystallize out.

The Textures of Igneous Rocks. The relative sizes of the component mineral grains and the relations of these to each other give the *texture* of rocks. The texture of igneous rocks depends in a general way upon the manner of cooling of the magma. During *slow* cooling, the atoms have time to arrange themselves and so to produce *large* crystals; more *rapid* cooling gives rise to many centres of crystallization and *smaller* crystals result; still *more rapid* cooling produces a *dense* fine-grained rock, whilst with *very rapid* cooling where no time is allowed for the atoms to arrange themselves into the space-lattice of crystals, *glass* results. In this way, *coarse-grained*, *fine-grained*, *dense*, and *glassy igneous rocks* are formed.

Two other textures may be noted, first, *even-grained* texture in which the grains are approximately of the same size and, second, *porphyritic* texture, in which the components are of two sizes, large crystals called *phenocrysts* lying in a finer-grained base, the *groundmass*.

Texture thus depends on the cooling-history of the magma. Intrusive rocks have usually cooled slowly so that they are in general coarse-grained. Extrusive rocks, poured out on the surface, are rapidly chilled so that in general they are fine in grain or even glassy.

In *intrusive rocks*, the texture depends broadly upon the size of the igneous body, but in any one body the texture often varies from a fine-grained, more quickly cooled, marginal type to a coarse-grained, slowly cooled, interior type. The recognition of the fine-grained margins—the *chilled edges*—is important in determining the mutual age-relations of a series of igneous rocks. In the *extrusive rocks*, thick lava flows may have coarse centres. Further, in these rocks, the gases present in the magma expand on the release of pressure consequent on extrusion and often give almond-shaped gas cavities or *vesicles*, the resulting texture being called *vesicular*; often the vesicles are filled with later-formed minerals, so producing *amygdales* and *amygdaloidal* texture.

Porphyritic texture occupies an intermediate position; the formation of large crystals, the phenocrysts, under conditions of slow cooling was followed by more rapid cooling during which the finer-grained groundmass consolidated.

Sequence of Events in the Consolidation of Magma. We have seen that the magma is composed of non-volatile and volatile constituents. When with falling temperature the magma begins to crystallize, the non-volatile constituents come out to form the rock-forming minerals, silicates mostly, characteristic of the usual igneous rocks: this stage is called the *orthomagmatic*. The separation of the non-volatile constituents leads to a concentration of the volatile components, with the result that the residual portions of the magma are rich in fluxes, various gases and vapours. The liquid part of the residual magma forms the *pegmatitic stage*, and gaseous part the *pneumatolytic stage*. The final consolidation of the magma takes place from the last hot water-rich solutions and provides the *hydrothermal stage*. Various types of ore-deposits are associated with these various stages of consolidation.

Classification of Igneous Rocks. A broad classification of igneous rocks can be made by paying attention to two factors:

(1) The amounts and kinds of the constituent minerals, these depending upon the composition of the magma.
(2) The kind of texture, this depending upon the cooling-history of the magma.

On the basis of the *silica-percentage*, the igneous rocks can be divided into *acid* rocks with more than about 66% SiO_2, *intermediate* with between about 66% and 52% SiO_2, *basic* with between about 52% and 45% SiO_2, and *ultrabasic* with less than about 45% SiO_2.

On the basis of the actual minerals that make the igneous rocks we can classify them into the following groups:

A. *Those with Quartz and Feldspar.*
    A.1. *Quartz + dominant orthoclase ± mica:*
      Granite, coarse-grained, mostly deep-seated or plutonic.
      Quartz-porphyry, Quartz-felsite, porphyritic, mostly in dykes.
      Rhyolite, fine-grained, often with flow-structure, mostly extrusive.
      Obsidian, Pitchstone, glassy, mostly extrusive.
      [*Pegmatite,* coarse-grained, and *aplite,* fine-grained, quartz-feldspar rocks formed by the consolidation of the last portion of the magma.]

    A.2. *Quartz + dominant plagioclase + mica ± hornblende.*
      Granodiorite, coarse-grained, plutonic.
      Granodiorite-porphyry, porphyritic, mostly in dykes.
      Dacite, fine-grained, mostly extrusive.
      Obsidian, pitchstone, glassy.

B. *Those with dominant Feldspar, no quartz.*
    B.1. *Dominant alkali-feldspar (orthoclase or sodic plagioclase) ± mica ± hornblende.*
      Syenite, coarse-grained, plutonic.
      Syenite-porphyry, porphyritic, mostly in dykes.
      Trachyte, fine-grained, mostly extrusive.
      Pitchstone, glassy.

B.2. *Dominant sodic or intermediate plagioclase $\pm$ mica $\pm$ hornblende.*

Diorite, coarse-grained, plutonic.

Diorite-porphyrite, porphyritic, mostly in dykes.

Andesite, fine-grained, mostly extrusive.

Pitchstone, glassy.

B.3. *Dominant calcic plagioclase + pyroxene $\pm$ olivine.*

Gabbro, coarse-grained; *norite* (plagioclase + hypersthene), coarse-grained.

Dolerite, fine-grained, in dykes and sills.

Basalt, fine-grained, often partly glassy; dykes, sills, lava-flows.

Tachylyte, glassy, margins to small intrusions.

B.4. *Feldspar and Feldspathoids.*

Nepheline- and *leucite-syenites,* coarse-grained.

Phonolite, fine-grained, dykes, sills or flows: of syenite composition.

Nepheline- and *leucite-basalts*, fine-grained.

C. *Those with dominant ferromagnesian minerals.*

Rocks composed dominantly of olivine (*dunite, peridotite*), pyroxene (*pyroxenite*) or hornblende *(hornblendite)* or mixtures of these three minerals.

The Pyroclastic Rocks. As already noted, the violence of the volcanic eruption may blow the magma to pieces, and, in this process, disrupt also portions of the adjacent country-rocks. The resulting mixed fragments fall either around the volcanic orifice, or are dispersed over larger areas where they may take their place in other deposits being laid down at that time. The chief types of these *pyroclastic deposits* are coarsely fragmental rocks such as *agglomerate*, and finer-grained material such as *tuff* and *ash*.

2. Sedimentary Rocks

Origin. Sedimentary rocks are derived from pre-existing rocks exposed at the earth's surface. These earlier rocks are broken down by the agents of mechanical and chemical *weathering* such as frost, rain, ice, moving water and chemical decay. The material thus provided is in most cases moved from its place of origin and deposited elsewhere, the transport being performed as solid particles or in solution. Obdurate minerals such as quartz are transported unchanged. The less stable feldspars and ferromagnesian minerals are decomposed to give new products such as the clay-minerals—hydrous alumino-silicates—and iron and manganese hydroxides, as well as solutions containing the cations Ca, Mg, Na and K. Calcium and magnesium are deposited as the carbonates, calcite or dolomite, either by chemical action or by the activity of organisms. Under special circumstances of concentration and evaporation, the most soluble materials of weathering are precipitated as rock-salt, gypsum, anhydrite and salts of potassium and magnesium. By these various processes, the transported products of weathering become separated into distinct groups that can be used as the basis of the classification of the sedimentary rocks.

The sedimentary rocks are usually laid down in layers, one on top of the other, which differ to a more or less well-marked degree in composition, grain-size, colour or some other property. Such layers are called *beds* or *strata*. The separation-planes between the beds are *bedding-planes*, and the whole set of beds shows *stratification*. A group of rather similar beds is called a *formation*.

The consolidation of the original fragmentary material is brought about either by *welding* by pressure due to the weight of overlying beds, or by *cementation* whereby the constituent fragments are cemented together by the deposition between them of a binding material such as calcium carbonate, silica or iron-oxides.

Classification of Sedimentary Rocks. Two broad groups of sedimentary rocks can be recognized on the basis of origin: (i) Detrital, (ii) Chemical-organic.

(i) *Detrital Sedimentary Rocks.* This group is formed by the deposition of transported *particles* of minerals or rocks. The particles are either inherited unchanged from pre-existing rocks or derived as new minerals from them. They are classified by the grain-size of the constituent particles into (a) *pebbly* or psephitic, (b) *sandy* or psammitic, and (c) *clayey* or pelitic. In the first two groups the particles are largely inherited, in the last they are chiefly newly-formed minerals.

(a) *Pebbly* sedimentary rocks are consolidated gravels of various types. The constituent pebbles are large; if they are angular the rock is called a *breccia*, if they are rounded, a *conglomerate*. It must be noted that breccias may be formed in various ways, for instance by the fracture and re-cementing of a rock during faulting.

(b) *Sandy* rocks consist of small grains, mostly of quartz, cemented by a scanty bond of silica, iron-oxide, calcium carbonate, clayey material, etc. The chief type is *sandstone*. *Grit* is a psammitic rock made up of angular fragments.

(c) *Clayey* rocks consist of minute crystalline particles (*clay-minerals*) together with detrital grains. Examples are: *clay*, retaining enough moisture to be plastic; *mudstone*, containing little moisture and so not plastic, but still not fissile; *shale*, a non-plastic clay-rock splitting along its bedding planes. To this series may here be added *slate* which is a rock of clay-composition with a well-developed cleavage—*slaty cleavage*—not often coincident with the original bedding planes. Slate is really a metamorphic rock and its slaty cleavage results from the parallel orientation of flaky minerals formed when a shale has been subjected to considerable pressure.

(ii) *Chemical-organic Sedimentary Rocks*. This class is formed by the precipitation of material from solution or by organic processes. The group may be sub-divided by composition into the following:—carbonate, siliceous, ferruginous, aluminous, phosphatic, saline and carbonaceous sedimentary rocks.

(a) *Carbonate deposits*. Many *limestones* result from the accumulation of fragments, consisting of calcium carbonate, calcite, of shells, tests and hard parts of molluscs, corals, etc. Calcareous algae precipitate calcium carbonate and such deposits make thick beds in certain formations. The direct precipitation of calcium carbonate has given rise to certain types of *chemical limestones*, such as the stalactitic deposits in limestone-caves and *calc-tufa* deposits formed around calcareous springs. It is probable that some beds of *dolomite*, $CaMg(CO_3)_2$, have been formed as chemical precipitates.

(b) *Siliceous Deposits*. Some siliceous deposits arise by the collection of the remains of organisms which have a skeleton of silica, such as radiolaria and diatoms. The accumulation of the siliceous frustules of diatoms leads to deposits of *diatomite*, *tripoli* or *Kieselguhr*. Examples of the chemical precipitation of silica are given by the *siliceous sinter* formed around certain hot springs and by certain types of *flint* and *chert*.

(c) *Ferruginous Deposits*. Ferruginous deposits are considered to be due in some cases to the agency of iron-depositing bacteria. Precipitates of iron carbonate are represented by *bog iron-ore* and the *clayband* and *blackband ironstones*. The hydrated iron silicate, *chamosite*, $(Fe_3Al_2Si_2O_{15}Aq.)$ is an important constituent of many sedimentary iron-ores. *Laterite* is a residual clay enriched in ferric hydroxide as a result of chemical weathering in the humid tropics.

(d) *Aluminous Deposits*. Deposits of clay enriched in aluminium hydroxides and known as *bauxite* are formed in the same way as the laterites just mentioned, by extreme chemical decay. Bauxite is the important ore of aluminium.

(e) *Phosphatic Deposits*. Deposits of *phosphates* are produced by the accumulation of organic materials containing calcium phosphate such as the skeletons of fish, shells of crustacea, etc. or by the collection of phosphate-bearing bird-droppings, *guano*.

(f) *Saline Deposits*. The drying-up of enclosed bodies of salt water causes the precipitation of the dissolved salts with the formation of *saline residues*. An example is provided by the great German salt deposits of Stassfurt where the gradual evaporation of a body of salt water is shown by a succession of saline deposits beginning with dolomite and calcite, followed by a sequence of gypsum and anhydrite, then rock-salt, and the various bittern constituents such as polyhalite, kieserite and carnallite (see p. 237).

(g) *Carbonaceous Deposits* result from the accumulation of various types of plant material: examples are *peat*, *lignite* and the several classes of *coal*.

3. Metamorphic Rocks

Metamorphism. The third great group of rocks results from changes in pre-existent rocks brought about by alterations in the pressure and temperature to which the rocks are subjected, these alterations being accompanied in many cases by migration of material. Under the new conditions, certain of the original minerals are no longer stable and give place to minerals more fitted to the new environment. In addition to new minerals, new textures arise during this series of changes or *metamorphism*.

A type of pressure important in metamorphism is directed pressure or *stress*; the application of stress gives rise to shearing movements in the rock and is held to cause such of the new-formed minerals as have a platy habit to arrange themselves with their plates approximately parallel. A similar arrangement may result from the growth of platy minerals along pre-existing planes of slaty cleavage or bedding. This parallel texture is known as *schistosity*; a coarse banding produced by lenses of different schistose materials gives *foliation*. Temperature changes in metamorphism may be restricted to the immediate vicinity of an intrusive igneous body, and metamorphism under such conditions of high temperature and low stress is said to be *contact-metamorphism*. Metamorphism under low temperature and strong shearing stress, such as takes place along great fracture belts of the crust, is called *dislocation-metamorphism*. Metamorphism due to a rise in temperature, an increase in directed pressure and possibly an influx of energetic fluids, is called *regional metamorphism* as it affects the rocks of considerable portions of the crust.

Minerals of the Metamorphic Rocks. Quartz, feldspars, micas, pyroxenes and hornblendes, similar in most respects to those of the igneous rocks, occur under metamorphic conditions. Typical metamorphic minerals are garnet, andalusite, kyanite and sillimanite, cordierite, talc, chlorite, epidote and staurolite. The carbonates, calcite and dolomite, are the main constituents of metamorphosed limestones. The mineral assemblage produced depends upon the composition of the original rock and the conditions of its metamorphism, as shown in the following paragraphs.

Contact-Metamorphism. In thermal or contact-metamorphism a *metamorphic aureole* is formed around the igneous body responsible. The minerals produced by the rise in temperature depend upon the composition of the country-rocks affected. Sandstones are converted into *quartzite*. Clayey rocks pass into tough hard fine-grained rocks known as *hornfelses* and characterized by the occurrence of such aluminium silicates as andalusite and silli-manite, together with cordierite and biotite. Pure limestones are changed into *marbles*, impure limestones into *calc-silicate hornfelses* composed of various silicates such as wollastonite, diopside and tremolite. Basic igneous rocks are often recrystal-lized into finer-grained *basic granulites* composed of basic plagio-clase, pyroxene, hornblende and biotite. The effects of the contact-metamorphism become more marked as the igneous mass is approached.

Near the contact of intrusive and country-rock, the composi-tion of the latter may be markedly altered by addition of material derived from the igneous body, and important ore-deposits may arise by this process.

Dislocation-Metamorphism. In dislocation-metamorphism, the constituent minerals of a rock are distorted or broken-up, stream-lined fragments of resistant minerals often being set in a smeared-out groundmass of the softer minerals. The final product of this *cataclasis* is a fine-grained, often banded, rock called *mylonite* made up of minute angular grains enclosed in a still finer dust-like base. Mylonites are developed along zones of extensive movement at no great depth in the earth's crust; deeper down, flaky minerals such as chlorite and mica may be crystallized and by their parallel arrangements produce a fine schistosity in the deformed rock.

Regional Metamorphism. Regionally metamorphosed rocks vary in their *metamorphic grade* according to the physical controls, especially that of temperature, which governed their transforma-tion. Low-grade rocks were formed under conditions of low tem-perature and strong pressure, especially stress; medium-grade under higher temperatures and moderate stress and hydrostatic pressure; high-grade under high temperatures and high pressures, especially hydrostatic. The derivatives of the commonest rocks metamorphosed to these grades are now indicated.

At a *low grade*, clayey rocks are metamorphosed to give *slates* (p. 207), sandy rocks provide *quartzites* showing only a crude cleavage and carbonate-rocks are transformed into fine-grained *marbles*—both the last-named rocks show signs of cataclasis.

Under the same conditions, basic igneous rocks become *green-schists*, with abundant chlorite and amphibole, whilst ultrabasic rocks are converted into *serpentine* and *talc-schists*. *Medium-grade* rocks are typified by the *mica-schists*, derived from clayey parents, and showing micas and quartz and scarcer garnet and other aluminous silicates. *Quartzites* and *marbles* in this grade are coarse granular rocks. Basic igneous rocks are converted into *hornblende-schists* or *amphibolites* composed of plagioclose and hornblende. Rocks of *high grade* mostly show a coarse foliation resulting from increased chemical mobility during their transformation. Some *gneisses* are banded with quartzo-feldspathic streaks, in many cases of metasomatic (p. 212) origin. Mixtures of host-rock and granitic material are *migmatites* and intimate mixing results in *granitization*.

Mineral Deposits

We may describe a *mineral deposit* as a rock or mineral that is of economic value and repays working. Deposits worked essentially because they contain valuable elements are usually called *ore-deposits*. An *ore* is a mixture of the valuable mineral, the *ore-mineral*, and unwanted minerals, the *gangue*. The metal content of an ore is it *tenor*.

Syngenetic and Epigenetic Mineral Deposits:—Two groups of mineral deposits may be established, *syngenetic and epigenetic*, depending on their time-relations to the rocks enclosing or associated with them. *Syngenetic deposits* are formed at the *same time* as the associated rocks. One example is the type of ore-deposit called magmatic segregation which arises by the collection together of useful minerals during the orthomagmatic stage of consolidation of a magma, and is exemplified by the segregation of chromite in ultrabasic igneous rocks. Another example is provided by the sedimentary or bedded mineral deposits, such as the saline residues already noted, which are interbedded with other sedimentary rocks of closely similar age.

Epigenetic Deposits are formed *later than* the enclosing or associated rocks. Some of these deposits have filled or opened fissures in the country-rocks and such bodies of ore are called *veins* or *lodes*. Veins have great depth and length but no great thickness compared with these dimensions. In other cases the ore is deposited in the interstices of the country-rock and then forms *impregnations*. Again, certain ore-deposits *replace* the country-rock and have then an irregular form and shade off into the adjacent rocks. In ore-deposits associated with contact-metamorphism, the shape of the ore-body is irregular and related to the margin of the igneous rock. Mineral deposits may be subjected to the process of metamorphism and so partake of the characteristics of the associated country-rocks.

The complete or partial replacement of a pre-existing rock by an ore-body is an example of the geological process of *metasomatism* (*Greek*, change of body), which is the process in rock-transformation involving changes in bulk chemical composition. The term is applied not only to deposits formed through the action of circulating waters derived from higher levels in the crust but also to those brought about by solutions from igneous or deep-seated sources. Important ore-deposits result from this *pyrometasomatism* (p. 214) at igneous contacts; heated water from igneous sources may cause *propylitization* in the affected rocks, particularly in andesite lavas, marked by the production of such low-temperature minerals as chlorite and serpentine. In some deposits the valuable constituent has been leached from other rocks, transported in solution and precipitated with replacement in a new position– as examples of this latter type of metasomatic deposit may be given the iron-ore deposits of Cleveland where iron carbonate replaces limestone with the preservation, in many cases, of the original limestone structure and the pseudomorphism of fossils by siderite.

Types of Mineral Deposits

We have seen (p. 9) that most of the elements essential for modern industry are exceedingly scarce in the common rocks of the crust so that some kind of concentration is necessary before a workable mineral deposit can be formed. This concentration may occur in two environments to give *endogenetic* and *exogenetic* deposits.

Endogenetic deposits result from processes taking place or originating *within* the crust, such as the solidification of a magma or the operations of metamorphism. Exogenetic deposits are due to *surface-processes* such as deposition in a salt-lake or sorting by river-action.

Endogenetic Mineral Deposits

In the Table below is given a classification of endogenetic deposits with suggestions of the temperatures at which the various types may be formed.

Endogenetic Mineral Deposits

1. *Igneous or Magmatic*
    A. Orthomagmatic, 700-1500°C
    B. Pegmatitic-Pneumatolytic, ± 575°C
    C. Pyrometasomatic, 500-800°C
    D. Hydrothermal, 50-500°C
2. *Metamorphic*, c. 400°C

A. ORTHOMAGMATIC DEPOSITS. *Magmatic segregations* result from the concentration of minerals of economic value in particular parts of a cooling magma. The ore-body often shades off gradually into the igneous rock and is usually marginal in position. In this way have been formed important deposits of native metals such as platinum, metallic oxides such as magnetite and ilmenite, and sulphides such as pyrrhotite and chalcopyrite. The segregation of chromite has already been mentioned on p. 211.

B. PEGMATITIC-PNEUMATOLYTIC DEPOSITS. Pegmatites, as already mentioned, represent the residual portion of the magma and in them are often concentrated valuable minerals which occur only in accessory amounts in the main body of the igneous rock. Further, the richness in fluxes of the pegmatitic magma leads to the growth of minerals of large size and so makes exploitation profitable. Pegmatite magma is very fluid and is intruded as dykes, stringers, and veins mainly around the borders of granite masses. Thus are formed economic deposits of such minerals as feldspar, quartz, mica and gemstones.

During the later stages of consolidation of magmas, heated gases of great chemical activity stream into the adjacent country-rock, and in this way *pneumatolytic ore-deposits* are formed. These deposits usually accompany intrusions of granite and are associated with a special set of minerals which form a useful guide to the prospector—tourmaline, topaz, zinnwaldite, lepidolite, fluor, axinite. Examples of pneumatolytic ore-deposits are the tinstone deposits of Cornwall connected with the granites of that area, and the Norwegian apatite deposits associated with gabbro intrusions.

C. PYROMETASOMATIC DEPOSITS. The *pyrometasomatic or contact deposits* are found in the country-rock at its contact with an intrusive body. Heated solutions carrying ore-materials pass from the igneous rock into the country-rock and there produce metasomatism of suitable rocks with the deposition of the ore. The country-rock most favoured for pyrometasomatic ore-deposition is limestone. The chief ore-minerals formed in this way are sulphides such as chalcopyrite, pyrite, zinc blende, and oxides such as magnetite and hematite. With the ores are associated a characteristic set of minerals, the *skarn*-minerals, such as iron-garnets, iron-pyroxenes, wollastonite, epidote and idocrase.

D. HYDROTHERMAL DEPOSITS. This process of ore-deposition in which the ore-body has been deposited from aqueous solution represents the final stage in the consolidation of a magma. It consists of the evolution of heated waters of great chemical activity and capable of dissolving and transporting most metals of economic value. These solutions may find their way great distances from their parent source but eventually, by cooling or chemical reaction, they are compelled to deposit their load, and do so either in cavities or fissures (lodes or veins), or between the grains of sediments, etc. (impregnations). By consideration of the mineral assemblage, both ore and gangue, found in any particular deposit, the hydrothermal mineral deposits may be grouped into three types, hypothermal, mesothermal and epithermal, formed at high, medium and low temperatures respectively. Many famous deposits of gold, lead-zinc, copper and mercury belong to the hydrothermal group.

E. METAMORPHIC MINERAL DEPOSITS. Mineral deposits already in existence may be subjected to metamorphic changes of all types. During metamorphism impure or low-grade ores may become purified and form workable deposits. Thus, hydrated iron-ores give place to hematite and magnetite deposits. Many non-metallic mineral deposits, such as those of the aluminium silicates, graphite and asbestos, are of metamorphic origin.

Exogenetic Mineral Deposits

The exogenetic mineral deposits formed by surface-processes are of course sedimentary rocks. They can therefore be classified into three groups:

> **A.** Weathering Residual Deposits
>
> **B.** Detrital, Alluvial or Placer Deposits
>
> **C.** Chemical-organic Deposits

A. WEATHERING RESIDUAL DEPOSITS. Weathering may result in the concentration of an ore-mineral by the removal of unwanted constituents of the original rock, as in some gold and tin deposits. Profound chemical weathering results in the formation of residual clays enriched in aluminium, iron or manganese, as in the bauxites (p. 322).

B. DETRITAL, ALLUVIAL OR PLACER DEPOSITS. These deposits are formed by stream- or wave-action, and are found where the velocity and hence the carrying power of the currents have decreased. Such deposits consist, therefore, of gravelly or sandy material which becomes finer as the depositing stream, for instance, is followed from its mountain source. In these alluvial deposits, the minerals are concentrated into fractions according to specific gravity, and minerals of somewhat similar specific gravity are thus found together and form useful guides to the prospector. Chief of the alluvial deposits are the gold placers, with the gold being associated magnetite ('black sand'), chromite, zircon, etc. Alluvial gem deposits, platinum, tin and wolfram alluvials are other examples. Alluvials are thus the result of the breaking-up of the parent rock, and the subsequent transportation and concentration of minerals which may exist in very small quantity in the rocks at the head of the river-system; hence, by following the alluvial deposits upstream, it is often possible to locate the original home of the valuable minerals—gold is traced to a gold-bearing quartz-vein, platinum to ultrabasic igneous rocks, gems to metamorphic limestones, etc. Ancient alluvial deposits occur 'fossil' in the geological formations; among such deposits may possibly be placed the gold-bearing Banket of the Rand.

C. CHEMICAL-ORGANIC MINERAL DEPOSITS. This class of deposits has already been illustrated in the description of the chemically- and organically-formed sedimentary rocks given earlier at pp. 207-208. Examples of such bedded deposits are the limestones and dolomites, diatomite, many important iron-ores, some copper deposits, phosphates and the saline residues or evaporates. Coals and bitumens are of course also included in this group.

Supergene Enrichment of Ore Deposits

Like other rocks, ore deposits undergo weathering and decomposition at their outcrops, this weathered upper part of the deposit being known as the *gossan*. In this region of the ore-body the minerals of depth have been altered into oxy-salts. For example, a lode carrying galena and blende at depth might consist in the gossan of cerussite, anglesite, smithsonite, hemimorphite and other oxy-salts of lead and zinc. Native metals frequently occur in the gossan, and often concentration takes place by the removal of lighter or less stable material, thereby producing rich residual deposits capping a lower-grade ore.

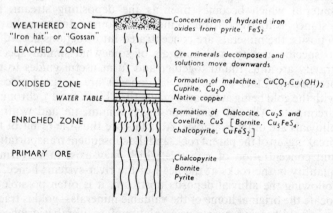

WEATHERED ZONE "Iron hat" or "Gossan" — Concentration of hydrated iron oxides from pyrite. FeS_2

LEACHED ZONE — Ore minerals decomposed and solutions move downwards

OXIDISED ZONE — Formation of malachite, $CuCO_3.Cu(OH)_2$ Cuprite. Cu_2O
WATER TABLE — Native copper

ENRICHED ZONE — Formation of Chalcocite, Cu_2S and Covellite, CuS [Bornite, Cu_5FeS_4, chalcopyrite, $CuFeS_2$]

PRIMARY ORE — Chalcopyrite Bornite Pyrite

FIG. 106. Supergene Enrichment in a Copper Lode

The process known as *supergene or secondary enrichment* is of great economic importance in silver, copper and other lodes. The oxy-salts produced in the gossan by weathering are carried deeper into the lode by descending waters. In the zone intermediate between the weathered outcrop and the unaltered sulphides—that is, in the *enriched zone*—chemical action takes place between the descending waters bearing oxy-salts and the unaltered sulphides, with the result that a new series of minerals is formed whose members are often very rich in the valuable metal of the lode. By this chemical concentration workable ore-bodies may result from rather low-grade ores.

In the diagram of a copper lode given in Fig. 106, the water-table is the level below which the rocks were saturated with water at the time of the enrichment of the lode. Very different operations take place above and below the water-table. The nature of these operations and their products are shown in the figure. The student will find it instructive to obtain the percentage of copper of each of the minerals from Part II of this book and insert the values against each zone.

GEOLOGICAL HISTORY

The sedimentary rocks of the crust that contain fossil-remains have been arranged in the order of their deposition, the complete succession being shown in the *stratigraphical column* of Fig. 107 with the oldest rocks at the bottom and the youngest at the top. In this column, each named portion indicates a geological *system* of rocks deposited in a *period* of time. The oldest system that contains fossils is called the *Cambrian*; rocks older than this are *Pre-Cambrian*. It will be noted from the age-data given in Fig. 107 that the Cambrian and later rocks were deposited in the last sixth of geological time.

Igneous rocks are dated in different ways. Lavas take their place in the succession of dated sedimentary rocks. Intrusives are later than the youngest rock that they invade and contact-metamorphose, and an upper age-limit can sometimes be fixed by the discovery of pebbles of the igneous rock in conglomerates of known age.

Ages in terms of millions of years of igneous and metamorphic rocks and of certain sedimentary rocks are obtained by methods

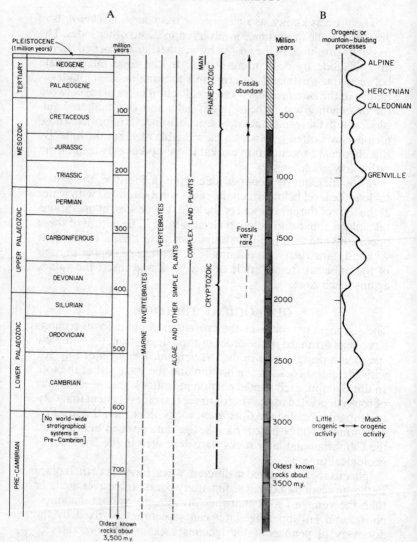

FIG. 107. Record of Earth-History

involving radioactive decay of certain elements in the rock-forming mineral, as explained on p. 71. This time-scale is included in the Stratigraphical Column of Fig. 107A and the complete time-scale for the whole 3,500+ million years of earth history is given in Fig. 107B.

The whole of geological time can be divided up into *geological cycles*, a method especially valuable for the immense span of Pre-Cambrian time. In its most complete form, a geological cycle is a record of three events; first comes the deposition in restricted belts of thick piles of sediments and lavas, especially basic types; then follows the deformation of this pile by *orogenesis* or mountain-building, a process usually marked also by regional metamorphism, granitization and, in late stages, the intrusion of granites; in the final phase, the whole folded belt is elevated into mountain ranges and subjected to erosion with the deposition of coarse detrital sediments. The whole cycle may take some hundreds of millions of years for its completion. Many very important ore-deposits are connected with the late granite episodes of the orogenic stage. The main orogenies recorded in the Northern Hemisphere are shown in Fig 107B.

CHAPTER VI

THE CLASSIFICATION OF MINERALS

The purpose of this short chapter is to introduce the arrangement of minerals adopted for their description in Part II of this book. Minerals may be classified in several ways, nearly all dependent either upon their chemical components or the structure or forms of their crystals.

The chemical classifications usually used begin with the elements and then follow with subdivisions based on the anion groups present. In the scheme employed by the great American mineralogist, J. D. Dana, the following subdivisions were made:

I Native Elements.

II Sulphides, selenides, tellurides, arsenides, antimonides.

III Sulpho-salts—sulpharsenites, sulphantimonites, sulphobismuthites.

IV Haloids—chlorides, bromides, iodides, fluorides.

V Oxides.

VI Oxygen salts—carbonates; silicates, titanates; niobates, tantalates; phosphates, arsenates, vanadates; antimonates; nitrates; borates, uranates; sulphates, chromates, tellurates; tungstates, molybdates.

VII Salts of organic acids—oxalates, mellates, etc.

VIII Hydrocarbon compounds.

Modern variants of this arrangement follow much the same lines. All chemical classifications have some definite relation to crystal-structure and more direct applications of the latter have been made either in terms of crystal symmetry or of structure.

In this book it is considered more advantageous for our purpose to employ a combined *economic and chemical classification*, directed to meeting the needs of the beginner in economic geology. It seems best to group minerals in the first place according to the useful element or group of elements contained in them; this is, in effect, a grouping of *ores* of each element. Thus, it is an advantage for our ends to place all minerals economically important as sources of lead together, and with these naturally are put other lead minerals. This method of grouping is best for the mineralogist determining minerals by blowpipe analysis or by chemical means. Difficulties arise in this classification in deciding which is the more important element in minerals containing two elements of economic value, or whether to place, for example, iron silicate with iron or silicon. Further, this classification scatters the large chemical groups, such as the carbonates, sulphates, etc., through many of the divisions based on elements. But most of these defects can be remedied by cross-references or other devices.

The first step in our classification, therefore, is to assemble the minerals, so far as is possible, into economic groups according to elements. The second step is to arrange these elements in reasonable associations and to deal with these associations in a reasonable order. For this purpose, we use the *Periodic Classification* of the elements already described and illustrated on pp. 26–28; as there explained this classification brings together elements of somewhat similar chemical properties. In the version of the Periodic Table given overleaf only those elements whose minerals are dealt with in this book are shown. Other elements represented in nature only by rare or unimportant minerals are not considered.

The groups and sub-groups obtained from the Table overleaf may be described in different sequences, depending on special aspects of the descriptions. In this book we deal with them in the order shown in the Table and now set out below:

Group I	(a)	Lithium, Sodium, Potassium.
	(b)	Copper, Silver, Gold.
Group II	(a)	Calcium, Strontium, Barium, (Radium).
	(b)	Beryllium, Magnesium, Zinc, Cadmium, Mercury.
Group III	(b)	Boron, Aluminium.
Group IV	(a)	Titanium, Zirconium, [Cerium], Thorium.
	(b)	Carbon, Silicon, Tin, Lead.

PERIODIC TABLE OF ELEMENTS OF COMMON MINERALS

GROUPS → Periods ↓	I a	I b	II a	II b	III a	III b	IV a	IV b	V a	V b	VI a	VI b	VII a	VII b	VIII
2	Li 3			Be 4		B 5		C 6		N 7				F 9	
3	Na 11			Mg 12		Al 13		Si 14		P 15		S 16		Cl 17	
4	K 19	Cu 29	Ca 20	Zn 30			Ti 22		V 23	As 33	Cr 24	Se 34	Mn 25	Br 35	Fe 26 (a) Co 27 Ni 28
5		Ag 47	Sr 38	Cd 48			Zr 40	Sn 50	Nb 41	Sb 51	Mo 42	Te 52		I 53	Ru 44 (b) Rh 45 Pd 46
6		Au 79	Ba 56	Hg 80		Ce 58		Pb 82	Ta 73	Bi 83	W 74				Os 76 Ir 77 Pt 78
7			Ra 88			Th 90					U 92				

Group V   (a) Vanadium, Niobium, Tantalum.

           (b) Nitrogen, Phosphorus, Arsenic, Antimony, Bismuth.

Group VI   (a) Chromium, Molybdenum, Tungsten, Uranium.

           (b) Sulphur, Selenium, Tellurium.

Group VII (a) Manganese.

           (b) Fluorine, Chlorine, Bromine, Iodine.

Group VIII (a) Iron, Cobalt, Nickel.

           (b) Ruthenium, Rhodium, Palladium, Osmium, Iridium, Platinum.

The great group of the *rock-forming silicates* is here dealt with under silicon. So far as is possible, the silicates are grouped according to the structural arrangements of their constituent atoms. For the needs of the geological student, the optical properties of the important rock-forming silicates are given in some detail, special attention being paid to diagnostic features. As occasion offers, some details of the crystal-structure are given for other minerals, such as metal-elements, carbonates and other compounds.

Before we enter on the formal description of minerals given in Part II it is worth summarizing the arrangement of industrial minerals used in the 1967 *Annual Review of Mining* issued by the *Mining Journal*. From this summary, the importance, complexity and variety of mineral products in modern industry will be apparent. The arrangement is:—

The Precious Metals: gold, silver, the platinum metals.

The Older Major Metals: copper, tin, lead, zinc.

The Light Metals: aluminium, titanium, magnesium.

The Steel Industry Metals: iron ore, steel, nickel, manganese, chromium, cobalt, tungsten, columbium, tantalum, molybdenum, vanadium.

The Fuel Minerals: coal, oil, natural gas.

The Nuclear Metals: uranium, caesium and rubidium, zirconium, hafnium, beryllium, thorium and the rare earths.

The Electronics Metals and Minerals: cadmium, mercury, mica, silica, germanium, tellurium, selenium, rhenium, indium, gallium.

Chemical Metals and Minerals: sulphur, phosphate rock, potash, boron, barytes, lithium, antimony, fluorspar, bismuth.

Insulants and Refractories: asbestos, magnesite, sillimanite, perlite, graphite, vermiculite.

Gemstones and Abrasives: diamonds, gemstones, natural abrasives.

PART II
DESCRIPTION OF MINERALS

Abbreviations Used

Composition	COMP.
Crystal System	CRYST. SYST.
Crystal Structure	CRYST. STRUCT.
Common Form	COM. FORM
Cleavage	CLEAV.
Fracture	FRACT.
Hardness	H.
Specific Gravity	SP. GR.
Optical Properties	OPT. PROPS.
Special Properties	SPEC. PROPS.

GROUP 1A
LITHIUM, SODIUM, POTASSIUM

LITHIUM MINERALS

Lithium (Li) does not occur in a free state in nature, nor are its compounds very abundant as minerals. The characteristic mode of occurrence of lithium minerals is in the pegmatites, where there has been a concentration of this somewhat rare element during the later stages of consolidation of granitic magma. The commonest lithium minerals are:

Lithium-bearing Micas: Lepidolite or *Lithium-mica*, and *Zinnwaldite* or *Lithium-iron-mica*. These two minerals are described with the *Micas* on pp. 402, 404.

Petalite, $Li(AlSi_4)O_{10}$.

Amblygonite, $Li(F,OH)AlPO_4$.

Spodumene, $LiAlSi_2O_6$.

Spodumene, petalite, amblygonite and lepidolite are sources of lithium salts. Spodumene is a member of the pyroxene family, but in view of its special importance as an ore of lithium it is here considered under that element. Lithium carbonate is produced in increasing amount from brines in western U.S.A. Lithium salts are used in batteries, in ceramics, as a flux and in medicine.

SPODUMENE

COMP. Lithium aluminium silicate, $LiAlSi_2O_6$ (see p. 380).

CRYST. SYST. Monoclinic. COM. FORM. Sometimes in large crystals made up of pinacoids, domes and the prism; also found massive and with broad cleavage-surfaces. CLEAV. Perfect parallel to the prism (110), a very good *parting* parallel to the front pinacoid (100). COLOUR. Greyish or greenish; two gem-varieties are *Hiddenite*, which is emerald-green, and *Kunzite*, lilac-coloured. LUSTRE. Pearly, but vitreous on fractured surfaces; translucent to subtranslucent. H. 6·5–7. SP. GR. 3·1–3·2.

TESTS. Heated before the blowpipe, swells up, becomes opaque and finally fuses to a colourless glass; colours the flame red, due to lithium, if not heated too much; not acted upon by acids.

VARIETIES. *Hiddenite*, the emerald-green, and *Kunzite*, the lilac, transparent varieties of gem quality.

OCCURRENCE. Occurs as large crystals in granitic pegmatites of many localities, accompanied by the other lithium minerals listed above. In pegmatite dykes in the Black Hills, South Dakota, spodumene occurs in enormous crystals, one being recorded measuring 42ft. by 6ft. by 3ft. and estimated to weigh 65 tons—these deposits are worked for lithium. Other pegmatite deposits are mined in King's Mountain, N. Carolina, and in Brazil, Rhodesia, S.W. Africa, South Africa and Western Australia.

USES. Spodumene is a chief source of the raw materials for the manufacture of lithium salts; the varieties hiddenite and kunzite are cut as gems.

SODIUM MINERALS

Sodium (Na) does not occur native. Metallic sodium is produced by the decomposition of a melt of sodium chloride by an electric current. Sodium is a soft silver-white metal, easily tarnishing in the air; it decomposes water, forming sodium hydroxide and hydrogen. It is used in heat-exchange systems.

Sodium forms 2·8 per cent of the earth's crust. It is a constituent of many very important *rock-forming silicates*, chief of which is *albite*, $NaAlSi_3O_8$, the sodic end-member of the plagioclase feldspars. A less common feldspar, *anorthoclase*, $(Na,K)AlSi_3O_8$, crystallizes from sodium-rich magmas, as do *sodium-pyroxenes*, *sodium-amphiboles*, and the feldspathoids, *nepheline*, *sodalite* and *hauyne*. *Scapolite*, and *paragonite* the sodium-mica, are other sodium-bearing silicates. Amongst the important group of silicates known as the *zeolites*, sodium-bearing types are represented by *heulandite*, *analcite*, *natrolite*, *stilbite* and others. All these and other sodium-bearing members are described at the appropriate pages in the account of the silicates.

The non-silicate sodium minerals mostly occur as *saline residues* deposited by the evaporation of enclosed bodies of salt water, as described on p. 208, and dealt with under rock-salt, below. Sodium chloride, NaCl, is the compound of the most frequent occurrence in nature, and is procured for industrial purposes either as rock-salt or by the evaporation of sea-water. Compared with these vast supplies, other sodium compounds are relatively insignificant, though there are important deposits of the carbonate, sulphate and nitrate.

TESTS. Sodium salts colour the blowpipe flame intense yellow. After heating before the blowpipe, sodium compounds give an alkaline reaction with litmus; this reaction is, however, also given by salts of the other alkalies and of the alkaline earths.

The Sodium minerals dealt with here are:

Chloride. Rock-salt, Halite, NaCl.

Nitrate. Soda-nitre, $NaNO_3$.

Sulphates.
- Thenardite, Na_2SO_4.
- Mirabilite, $Na_2SO_4.10H_2O$.
- Glauberite, $Na_2Ca(SO_4)_2$.

Carbonates.
- Thermonatrite, $Na_2CO_3.H_2O$.
- Natron, $Na_2CO_3.10H_2O$.
- Trona, $Na_3H(CO_3)_2. H_2O$.
- Gaylussite, $Na_2Ca(CO_3)_2.5H_2O$.

To these may be added:

Borates. Borax, $Na_2B_4O_7.10H_2O$, see p. 314.

Ulexite, $NaCaB_5O_9.8H_2O$, see p. 316.

Fluoride. Cryolite, Na_3AlF_6 see p. 324.

Borax and ulexite are worked for their boron content, and are dealt with here under Boron; similarly, cryolite is an ore of aluminium, and is considered with the other aluminium minerals.

ROCK-SALT, HALITE, Common Salt

COMP. Sodium chloride, NaCl; calcium sulphate, calcium chloride and magnesium chloride are usually present, and sometimes also magnesium sulphate; the presence of magnesium compounds causes the mineral to become wet and lumpy.

CRYST. SYST. Cubic; Galena Type or Hexoctahedral Class.

CRYST. STRUCT. Type of the rock-salt structure, described on

FIG. 108. Hopper Crystal of Rock-salt

pp. 76–77 and illustrated in Fig. 18, p. 77. COM. FORM. Common crystals in cubes, rarely octahedra; cubes often with hollow faces, giving *hopper* crystals, illustrated in Fig. 108: also occurs massive and granular, rarely fibrous. CLEAV. Perfect cubic. COLOUR. Colourless or white when pure; often yellow or red, sometimes blue, amethystine or purple tints. LUSTRE. Vitreous; transparent to translucent. TASTE. Saline; soluble in water. FRACT. Conchoidal, brittle. H. 2–2·5. SP. GR 2·2.

TESTS. Taste quite distinctive; colours the flame deep yellow, due to sodium; crackles and decrepitates when heated; gives the usual blue chlorine flame with copper oxide in the microcosmic salt bead; in solution gives a white precipitate of silver chloride on addition of silver nitrate solution.

OCCURRENCE. Deposits of rock-salt occur as extensive geological beds, and are the result of the evaporation of enclosed or partly enclosed bodies of sea-water. During this concentration and evaporation, the salts separate out in a definite order. In the great German deposits at Stassfurt the order is:
1. Dolomite and calcite—calcium and magnesium carbonates.
2. Gypsum and anhydrite—calcium sulphates.
3. Rock-salt—sodium chloride.
4. Polyhalite—calcium magnesium potassium sulphate.
5. Kieserite—magnesium sulphate.
6. Carnallite—potassium magnesium chloride.
  Hence there is a fairly regular sequence in the deposits from bottom to top in the order named, the least soluble minerals, the calcium and magnesium carbonates, being at the bottom, and the most soluble minerals, polyhalite, kieserite and carnallite—the *bitterns*—being at the top. In most cases of the drying-up of saline waters, however, the process has not been carried so far as at Stassfurt, and the bitterns are usually absent from the succession. Salt beds occur at various geological horizons—in the Silurian and Carboniferous of Michigan, New York State and Ontario: in the Permian of Stassfurt, Germany: in the Trias of Cheshire,

England; Lorraine, France; Wurtemberg, Germany; Salzburg, Austria: and in the Tertiary of Wieliczka, Poland. Countries producing over four million tons of salt annually are the U.S.A., China, U.S.S.R., Britain, Canada, West Germany and India. The salt is extracted by ordinary mining by shafts and galleries, or by pumping the brine from the salt bed to the surface, and there recovering the salt by evaporation. Salt is present in the waters of the ocean, and vast inland lakes of salt water exist, such as the Dead Sea, the Great Salt Lake of Utah, etc. Since sea-water contains only about 3·5 per cent of total dissolved material in it, the formation of beds of salt thousands of feet in thickness such as occur in nature requires an explanation. It is suggested that thick salt beds could be laid down by the continual replenishment of a lagoon, such as the Gulf of Karabugas in the Caspian, with supplies of salt water, and the consequent enrichment in salt of the waters of the lagoon by continued evaporation. Beds of salt differ from other rocks in their reaction to pressures; salt flows whilst other rocks fracture or fold when subjected to crustal movements. Salt glaciers have been described from Persia, and 'intrusive' plugs of salt—the *salt domes*—are of great economic importance in the Gulf States of the United States, since they have provided the proper conditions for the accumulation of vast reservoirs of petroleum.

USES. Rock-salt is used for culinary and preserving purposes, and especially in a great number of chemical manufacturing processes—such as the manufacture of sodium carbonate for glass-making, soap-making, etc.

SODA-NITRE, NITRATINE, Nitrate of Soda, Chile Saltpetre

COMP. Sodium nitrate, $NaNO_3$.

CRYST. SYST. Hexagonal, ditrigonal-scalenohedral. CRYST. STRUCT. Isostructural with calcite (p. 271). COM. FORM. Crystals not common, of Calcite Type; usually as efflorescences in crusts, or massive granular. CLEAV. Perfect rhombohedral.

COLOUR. White, grey, yellow, greenish, purple, and reddish-brown. LUSTRE. Vitreous; transparent. TASTE. Cooling; soluble in water. H. 1·5–2. SP. GR. 2·29.

TESTS. Deflagrates less violently than nitre when heated, and colours the flame yellow, by which and its deliquescence it may be distinguished from that mineral.

OCCURRENCE. So soluble a mineral as soda-nitre can occur in workable quantity only in regions of very low rainfall. Economically important deposits are found in the Atacama Desert of northern Chile; the production from these deposits exceeded three million tons in 1929, but is now considerably less. The soda-nitre occurs, mixed with sodium chloride, sulphate and borate, and with clayey and sandy material, in beds up to six feet thick. The sodium nitrate forms 14–25 per cent of this *caliche*, as the material is termed, and is accompanied by 2–3 per cent of potassium nitrate, and up to 1 per cent of sodium iodate, the last being an important source of iodine. These remarkable deposits have most likely been leached from surrounding volcanic rocks, and owe their preservation to the very dry climate of the area.

USE. As a source of nitrates used in explosives and fertilizers.

The Sodium Sulphate Minerals

OCCURRENCE. Sodium sulphate occurs in nature as thenardite (Na_2SO_4) and mirabilite ($Na_2SO_4.10H_2O$), and as glauberite $Na_2Ca(SO_4)_2$ and other double salts of which bloedite $Na_2Mg(SO_4)_2.4H_2O$ is perhaps the chief. These minerals, associated with other related salts such as rock-salt (NaCl), natron ($Na_2CO_3.10H_2O$) and epsomite ($MgSO_4.7H_2O$), are deposited by the concentration of the waters of 'alkali' lakes in desert regions of the western states of the United States, and elsewhere. These materials have been leached from the rocks, usually of Mesozoic age, of the drainage areas of the inland basins. In the case of the deposits of sodium sulphate of western Canada, it has been suggested that the sodium salts have been leached from the surrounding drift deposits by percolating waters.

PRODUCTION. Soviet Russia, Canada and the United States appear to be the chief producers of sodium sulphate; it is estimated that over 170,000 tons were produced in the Gulf of Kara-bugas area (see p. 231) of the Caspian in 1946; Canada during 1966 produced over 300,000 tons; the United States in the same year produced about twice this amount.

USES. Sodium sulphate, or salt cake, is of great importance in chemical industry, in pulp and paper making, glass-making, the refining of nickel, in dyeing and tanning, and in the manufacture of paint.

THENARDITE

COMP. Anhydrous sodium sulphate, Na_2SO_4.

CRYST. SYST. Orthorhombic. COM. FORM. Prismatic or tabular crystals, often twinned; often as crusts or as a powder. COLOUR. Whitish. LUSTRE. Vitreous; takes up moisture from the air and becomes opaque. TASTE. Faintly saline; very soluble in water. H. 2·5 SP. GR. 2·68.

OCCURRENCE. In saline residues of the alkali lakes of the western United States and Canada.

MIRABILITE, Glauber Salt

COMP. Hydrous sodium sulphate, $Na_2SO_4.10H_2O$.

CRYST. SYST. Monoclinic. COM. FORM. In crystals like those of augite in shape, or in long needle-like forms; also in efflorescent crusts and in solution in mineral waters. CLEAV. Perfect parallel to the orthopinacoid (100). COLOUR. White or yellow. LUSTRE. Vitreous; translucent to opaque. TASTE. Cooling, saline, and bitter; soluble in water. H. 1·5–2. SP. GR. 1·48.

TESTS. Gives water on heating in closed tube; yellow flame of sodium; black stain on moistening with water on a silver coin the residue obtained by heating on charcoal; when exposed to dry air, mirabilite loses water and goes to powder.

OCCURRENCE. In the residues of alkali lakes, as at the Great Salt Lake of Utah, the alkaline lakes of Wyoming and other western states, in Saskatchewan, Canada, and elsewhere; sulphur is produced from mirabilite obtained from the Gulf of Karabugas (p. 231).

GLAUBERITE

COMP. Anhydrous sulphate of sodium and calcium, $Na_2Ca(SO_4)_2$.

CRYST. SYST. Monoclinic. COM. FORM. Crystals tabular parallel to the basal plane. CLEAV. Perfect basal. COLOUR. Yellow, red grey. STREAK. White. LUSTRE. Vitreous. FRACT. Conchoidal. H. 2·5–3. SP. GR. 2·8.

OCCURRENCE. As a saline residue associated with the other sodium minerals of this type at the localities already cited. Deposits of glauberite that were of great economic importance up to a few years ago occur in the valley of the Ebro, Spain, and bear the characters of a saline residue.

The Sodium Carbonate Minerals

The chief naturally occurring sodium carbonates are natron ($Na_2CO_3.10H_2O$), trona ($Na_3H(CO_3)_2.2H_2O$), thermonatrite ($Na_2CO_3.H_2O$) and gaylussite ($Na_2Ca(CO_3)_2.5H_2O$). These minerals, together with other sodium salts occurring as saline residues, are deposited from the waters of alkaline lakes. At Owen's Lake, California, the order of deposition is trona: sodium sulphate: sodium chloride: natron.

PRODUCTION. The naturally occurring sodium carbonates are not produced in great amount, since the artificially produced material is cheaper. In 1966 the United States produced $1\frac{3}{4}$ million tons and in 1959 over 150,000 tons came from Kenya.

USES. Sodium carbonate is extensively employed in the manufacture of chemicals, glass, soap, detergents and paper, and in the bleaching, dyeing and printing of various fabrics.

THERMONATRITE

COMP. Hydrous sodium carbonate, $Na_2CO_3.H_2O$.

CRYST. SYST. Orthorhombic. COM. FORM. Crystals and efflorescent crusts.

OCCURRENCE. As a saline residue.

NATRON

COMP. Hydrous sodium carbonate, $Na_2CO_3.10H_2O$.

CRYST. SYST. Monoclinic. COM. FORM. Usually in solution, but also as efflorescent crusts. COLOUR. White, grey or yellowish. LUSTRE. Vitreous or earthy. TASTE. Alkaline; very soluble in water. H. 1–1·5. SP. GR. 1·46.

TESTS. Effervesces with acid; gives water on heating in closed tube; yellow sodium flame.

OCCURRENCE. Found in solution in the soda lakes of Egypt. western United States, and elsewhere: occurs in saline residues, as in British Columbia.

TRONA, Urao

COMP. Hydrous acid sodium carbonate, $Na_3H(CO_3)_2.2H_2O$.
CRYST. SYST. Monoclinic. COM. FORM. Fibrous or columnar layers and masses. CLEAV. Perfect parallel to the orthopinacoid. COLOUR. Grey or yellowish-white. LUSTRE. Vitreous; translucent. TASTE. Alkaline; soluble in water. H. 2·5–3. SP. GR. 2·13.

TESTS. Effervesces with acids; gives water on heating in closed tube: yellow flame of sodium.

OCCURRENCE. In saline residues, with other minerals formed in this way, in California, Mexico, Fezzan, and Egypt.

GAYLUSSITE

COMP. Hydrous carbonate of sodium and calcium, $Na_2Ca(CO_3)_2.5H_2O$.
CRYST. SYST. Monoclinic. COM. FORM. Flattened wedge-shaped crystals. CLEAV. Perfect prismatic. COLOUR. White. H. 2–3. SP. GR. 1·99.

TESTS. Partly soluble in water; effervesces with acids; colours flame yellow.

OCCURRENCE. As a saline residue.

POTASSIUM MINERALS

Potassium (K) does not occur native. The metal is prepared in a similar fashion to sodium, and its chemical and physical properties are very like those of that metal. Formerly potassium salts were procured for the most part from vegetable matter, this being burnt and the soluble portion of the ashes dissolved in water. The plants, however, have in the first instance procured their potassium from soils which have resulted more or less from the decomposition of igneous rocks containing *orthoclase feldspar*, $KAlSi_3O_8$. Examples of other potash-bearing silicates are *leucite*, $KAlSi_2O_6$, and *muscovite*, *potash-mica*, $KAl_2(AlSi_3)O_{10}(OH)_2$. Potash-bearing zeolites are represented by *apophyllite* and *harmotome*. The potash-bearing silicates are described with the other rock-forming silicates.

The extraction of potash from such silicates is a complex and costly process, and more readily accessible supplies of potassium compounds are available in the *saline residues* (see p. 208). The most interesting deposit of this kind is that of Stassfurt, Germany (p. 230); in this, the most important potassium compound is *carnallite*, $KMgCl_3.6H_2O$—other potassium minerals occurring there being *kainite*, $KCl.MgSO_4.3H_2O$, *polyhalite*, $K_2Ca_2Mg(SO_4)_4.2H_2O$, and *sylvine*, KCl. A third natural source of potash is found in the mineral *alunite*, $KAl_3(SO_4)_2(OH)_6$, and in this case also the extraction of potash is not a complicated process. Some deposits of *nitre*, KNO_3, are of organic origin; nitre occurs also in small amount in the sodium nitrate deposits of Chile, described on p. 232. Sea-water contains about 0·04 per cent of potassium salts, and the recovery of such salts supplies a small proportion of the potash production; similarly, the waters of salt lakes, such as the Dead Sea, contain potash salts, and are exploited.

The following gives a summary of the potash production in normal years, considered under the heads of the minerals exploited. Italy produces a small quantity of *leucite*. The overwhelming proportion of the production comes from the *saline residues*, the chief producers being West Germany, France, East Germany, U.S.S.R., U.S.A. and Canada, the two last being likely to become the most important. Large deposits have been discovered in Yorkshire. Alunite is produced by Korea, Italy, U.S.S.R., Japan, Australia and Spain. Nitre production is largely from India, where it is of organic origin, and Chile, where it is of inorganic origin in the sodium nitrate deposits. Evaporation of sea-water, and especially of the waters of saline lakes such as the Dead Sea, produces a small but increasing amount of potash salts.

The most important use of potash salts is as fertilizers; other uses are in the manufacture of explosives, and in chemical and metallurgical processes. The average annual demand for potash salts in a normal year approaches fourteen million metric tons of potassa, K_2O.

TESTS. Potassium compounds give a lilac flame coloration, which is, however, masked by sodium and other elements; the flame should be viewed through blue glass or an indigo prism, whereby elements other than potassium are eliminated. Fused potassium compounds give an alkaline reaction with litmus. For the detection of small quantities of potassium compounds in solution a few drops of platinic chloride produce in such a solution after prolonged stirring a precipitate of minute yellow crystals of potassium platinochloride, K_2PtCl_6.

The potassium minerals, other than silicates, dealt with here are:—

$$\textit{Chlorides} \begin{cases} \text{Sylvine, KCl.} \\ \text{Carnallite, } KMgCl_3.6H_2O. \\ \text{Kainite, } KCl.MgSO_4.3H_2O. \end{cases}$$

$$\textit{Sulphates} \begin{cases} \text{Polyhalite, } K_2Ca_2Mg(SO_4)_4.2H_2O. \\ \text{Alunite, } KAl_3(SO_4)_2(OH)_6. \end{cases}$$

$\textit{Nitrate} \quad \text{Nitre, } KNO_3.$

SYLVINE, Sylvite

COMP. Potassium chloride, KCl.

CRYST. SYST. Cubic, hexoctahedral. CRYST. STRUCT. Isostructural with rock-salt (pp. 76–7). COM. FORM. Cube modified by octahedron; also crystalline, massive and granular. CLEAV. Perfect cubic. COLOUR. Colourless or white. LUSTRE. Vitreous. TASTE. Saline, more bitter than that of rock-salt. H. 2. SP. GR. 2.

TESTS. Soluble in water; lilac flame of potassium; gives blue flame in the copper oxide-microcosmic salt bead test.

OCCURRENCE. Occurs as a saline residue in the Stassfurt and other salt deposits, associated with rock-salt and carnallite; also around the fumaroles of Vesuvius.

CARNALLITE

COMP. Hydrous chloride of potassium and magnesium, $KMgCl_3.6H_2O$; chlorides of sodium and potassium often present.

CRYST. SYST. Orthorhombic. COM. FORM. Crystals rare; occurs massive and granular. CLEAV. None. FRACT. Conchoidal. COLOUR. White, but usually pink or reddish, from admixture with iron oxide. LUSTRE. Shining and greasy; transparent to translucent. TASTE. Bitter; soluble in water. H. Probably 3. SP. GR. 1·60.

TESTS. Gives water on heating in the closed tube; lilac flame of potassium; blue flame of chloride in copper oxide-microcosmic salt bead test; pink mass due to magnesium by heating with cobalt nitrate the residue from roasting on charcoal.

OCCURRENCE. Occurs as a saline residue at Stassfurt, and represents the final stage in the drying-up of the salt-lake.

USES. In the natural state as a fertilizer; also as a source of potassium salts.

KAINITE

COMP. Hydrous magnesium potassium chloride and sulphate, $KCl.MgSO_4.3H_2O$.

CRYST. SYST. Monoclinic. COM. FORM. Granular or in crusts. COLOUR. White. H. 3. SP. GR. 2·1.

OCCURRENCE. In the upper parts of saline residues, such as Stassfurt, where it is in part due to the leaching-out of magnesium chloride from the carnallite zone.

POLYHALITE

COMP. Hydrous triple sulphate of potassium, magnesium and calcium, $K_2Ca_2Mg(SO_4)_4.2H_2O$.

CRYST. SYST. Triclinic. COM. FORM. In compact lamellar masses. COLOUR. Pinkish or reddish.

OCCURRENCE. In saline residues as at Stassfurt, where it forms a layer, about 50 metres thick, above the rock-salt layer and below the carnallite and kieserite ($MgSO_4.H_2O$) layer.

ALUNITE, Alumstone

COMP. Basic sulphate of potassium and aluminium, $KAl_3(SO_4)_2(OH)_6$.

CRYST. SYST. Hexagonal, trigonal. COM. FORM. Crystals uncommon—small rhombohedra with basal plane; usually found massive, granular, fibrous and sometimes earthy. CLEAV. Good parallel to basal plane. COLOUR. White, greyish or reddish. LUSTRE. Of crystals, vitreous; of massive kinds, frequently dull. FRACT. Of crystals, flat conchoidal or uneven; of massive varieties, splintery, and sometimes earthy; brittle. H. 3·5-4. SP. GR. 2·6.

TESTS. Heated in closed tube, gives water, on intense heating gives sulphurous fumes; heated with cobalt nitrate gives blue colour; sulphur given by the silver coin test; insoluble, therefore no taste.

OCCURRENCE. As an alteration or replacement of trachytes and rhyolites, in which it forms seams and pockets, as in Italy, Spain, Korea, New South Wales, U.S.S.R. and in Nevada, Utah and Colorado.

USES. As a source of potassium and aluminium salts.

NITRE, Saltpetre, Nitrate of Potash

COMP. Potassium nitrate, KNO_3.

CRYST. SYST. Orthorhombic. COM. FORM. Acicular crystals; also in silky tufts and thin crusts. COLOUR. White. LUSTRE. Vitreous; subtransparent. TASTE. Saline and cooling. H. 2. SP. GR. 2·1

TESTS. Lilac flame of potassium; soluble in water; deflagrates on heating on charcoal; brittle.

OCCURRENCE. Occurs in considerable quantities in the soil of certain countries—India, Egypt, Algeria, Iran and Spain; also occurs in the loose earth forming the floors of natural caves, as in Kentucky, Tennessee and the Mississippi valley; as noted on p. 232, the sodium nitrate deposits of Chile contain some 2–3 per cent of potassium nitrate, and Chile now supplies considerable quantities of the salt; nitre is artificially manufactured from refuse animal and vegetable matter, which is mixed with calcareous soil—calcium nitrate is thus formed, and this when treated with potassium carbonate yields nitre.

USES. Nitre is used in the manufacture of explosives, in metallurgical and chemical processes, and as a fertilizer.

GROUP 1B
COPPER, SILVER, GOLD

COPPER MINERALS

Copper (Cu) is a widely distributed and abundant element in combination, and is also found in the native state. The metal copper has a specific gravity of about 8·9, and melts at about 1,100°C. It is a comparatively soft but extremely tough metal, very ductile and malleable when pure, and, next to silver, the best conductor of electricity. These properties are a consequence of its cubic close-packed or face-centred cubic structure as described and figured on pp. 21–22. Figs. 19C and 7.

Copper is obtained from its ores (usually sulphide) by an elaborate series of metallurgical operations, commonly consisting of roasting, to expel part of the combined sulphur, fusion in blast or reverberatory furnaces for the production of a concentrated double sulphide of copper and iron called *matte*, and the conversion of the matte to crude metallic copper in a reverberatory or Bessemer converter. Blast-furnaces are sometimes used for ores rich in sulphur for the production of copper matte, or for the further concentration of matte from the first reverberatory furnace fusion; the former operation, the smelting of sulphur-rich ores, or pyritic smelting, utilises the heat produced by the oxidation of the sulphides, with the aid of little or no fuel. Oxidized ores may be reduced in a blast-furnace with coal or coke, but are best smelted in admixture with sulphide ores. In the case of native copper, the ore is crushed and the metal, separated from its gangue by dressing, melted in some form of reverberatory furnace. Copper of the necessary purity for use as conducting wires is obtained from the crude metal by electrolysis. From poor ores and residues from the pyrites used in the manufacture of sulphuric acid, copper is obtained by roasting with common salt, leaching out the soluble copper chloride with water, and deposition of the metal on scrap iron or by electrolysis.

The most important use of copper is in the electrical industry, both as a conductor and for electrical machinery. It is also of great importance in the construction of machinery generally, in the motor-car industry, and in chemical engineering. Copper is extensively used in the manufacture of alloys, such as bronze, gunmetal and bell-metal (copper and tin), brass (copper, zinc and sometimes tin), nickel silver (copper, zinc and nickel), and some others of great technical importance, such as phosphor and manganese bronzes (copper and tin, with small percentages of phosphorus or ferro-manganese respectively), silicon bronze, monel metal, etc. Copper salts are employed in various industrial processes—the chloride is used as a disinfectant and in chemical operations, the sulphate is employed in the printing and dyeing of textiles, for preventing rot in timber, and as a fungicide.

The chief sources of copper are native copper, chalcopyrite, chalcocite, erubescite, and malachite together with cupriferous pyrite—i.e. pyrite containing a few per cent of chalcopyrite; associated with copper ores are frequently found the following metals, recoverable at one or other stage of their treatment: gold, silver, platinum, palladium, bismuth, etc. Copper ores usually carry a very small percentage of copper; thus, some ores worked for copper with profit have only 0·5 per cent of the metal, and perhaps the average copper content of all copper ore production is not more than 2 per cent. Copper ores occur in a variety of ways: magmatic segregations, in veins and lodes, in contact-metamorphic deposits, in bedded deposits, etc. Examples of these types are given under the descriptions of the copper minerals. The sulphides of copper which occur in depth in copper lodes are converted by oxidation and other chemical actions in the surface portion to the native metal, oxides and oxy-salts (see also pp. 244–5, 247, 251–5.

The world mine-production of copper in 1967 was over five million tons. The main producers were the Katanga area of the Congo and Zambia, U.S.A., Chile, U.S.S.R. and Canada. Other countries producing important amounts are Australia, Southwest and South Africa, Mexico, Philippines, Peru, Yugoslavia, Cyprus, China, Finland, Austria, Turkey, Japan and Spain.

TESTS. Copper oxides colour the flame emerald-green when moistened with nitric acid, and copper chloride colours the flame an intense sky-blue. The microcosmic salt bead is blue in the oxidizing flame, and opaque red in the reducing flame. The borax bead is somewhat similar. When heated with sodium carbonate and carbon on charcoal, copper compounds give a reddish mass, which speedily blackens. Dilute solutions of copper minerals in acids become deep blue on addition of ammonia.

The copper minerals considered here are:

Element	Native Copper, Cu.
Oxides	Cuprite, Cu_2O.
	Tenorite, CuO.
Sulphides	Chalcopyrite, Copper Pyrites, $CuFeS_2$.
	Chalcocite, Copper Glance, Cu_2S.
	Covelline, CuS.
	Bornite, Erubescite, Cu_5FeS_4.
'Grey Coppers'	Tetrahedrite, $(Cu,Fe)_{12}Sb_4S_{13}$.
	Tennantite, $(Cu,Fe)_{12}As_4S_{13}$.
	Famatinite, Cu_3SbS_4.
	Enargite, Cu_3AsS_4.
	(Bournonite, $CuPbSbS_3$).
Sulphate	Chalcanthite, $CuSO_4.5H_2O$.
Carbonates	Malachite, $Cu_2CO_3(OH)_2$.
	Azurite, $Cu_3(CO_3)_2(OH)_2$.
Silicates	Chrysocolla, $CuSiO_3.2H_2O$.
	Dioptase, $CuSiO_3.H_2O$.
Chloride	Atacamite, $Cu_2(OH)_3Cl$.

NATIVE COPPER

COMP. Pure copper, sometimes containing a little silver and bismuth.

CRYST. SYST. Cubic. COM. FORM. In frequently-twinned crystals; often massive; sometimes in thin sheets or plates, filling narrow fissures; arborescent forms are also of frequent occurrence, and at times native copper is found in confused threads. CLEAV. None. COLOUR. Copper-red. STREAK. Metallic and shining FRACT. Hackly; ductile and malleable. H. 2·5–3. SP. GR. 8·9.

Tests. Before the blowpipe, copper fuses easily and becomes coated with black oxide of copper; dissolves in nitric acid, and affords a blue solution on addition of ammonia.

Occurrence. Native copper occurs as a hydrothermal and metasomatic deposit filling cracks, or amygdales and forming partial replacements in basic lava-flows, or building the cement of associated conglomerates, or as a cement in other rocks such as sandstones. The most important deposits of native copper are those of the Lake Superior region, where the copper occurs in a series of ancient lava-flows and conglomerates; the native copper is found in the amygdaloids—the volcanic rocks—in the Calumet conglomerate, and in deposits of vein type of metasomatic origin. Its origin is a matter of discussion, but possibly it is derived from the basic volcanic rocks. Native copper occurs associated with a basic intrusive rock at Monte Catini, Italy. Native copper also occurs in the upper part of copper lodes—the zone of weathering, and rich but mostly small deposits of this type have been worked.

Use. As an ore of copper.

CUPRITE, Red Oxide of Copper

Comp. Copper oxide, Cu_2O; copper 88·8 per cent.

Cryst. Syst. Cubic. Com. Form. Crystals showing the octahedron and rhombdodecahedron; sometimes massive or earthy, and occasionally capillary. Cleav. Not good, parallel to faces of the octahedron. Colour. Different shades of red, especially cochineal-red. Streak. Brownish-red and shining.

Lustre. Adamantine, or submetallic to earthy; subtransparent to nearly opaque. Fract. Conchoidal, uneven; brittle. H. 3·5–4. Sp. Gr. 5·8–6·15.

Tests. Before the blowpipe, colours the flame emerald-green; on charcoal fuses to a globule of metallic copper; with the fluxes gives the usual copper reactions; soluble in acids.

Varieties. *Ruby Copper*, crystallized cuprite; *Tile Ore,* a red or reddish-brown earthy variety, generally containing oxide of iron; *Chalcotrichite* (Greek, *chalkos*, copper and *thrix*, hair), a variety consisting of delicate, straight, interlacing fibrous crystals of a beautiful cochineal-red colour.

OCCURRENCE. Occurs in the upper oxidized zone of copper lodes (see p. 216), as in Cornwall, Chessy (France), Linares (Spain), Bisbee (Arizona), Chile, Peru, Burra Burra (Australia).

USE. As an ore of copper.

TENORITE, Melaconite

COMP. Copper oxide, CuO; copper, 79·85 per cent.

COM. FORM. Occurs mostly in a black powder; also in dull black masses, and in botryoidal concretions; sometimes in shining and flexible scales. H. 3–4. SP. GR. 6·25, when massive.

TESTS. Gives the usual copper reactions with the fluxes, but is infusible alone in the oxidizing flame; soluble in acids.

OCCURRENCE. In the oxidized zone of weathered copper lodes. Abundant in the copper mines of the Mississippi Valley and in Tennessee, U.S.A., where it is worked.

CHALCOPYRITE, COPPER PYRITES

COMP. Sulphide of copper and iron, $CuFeS_2$; copper 34·5 per cent.

CRYST. SYST. Tetragonal. COM. FORM. Crystals often resemble those of tetrahedral type; wedge-shaped forms are common, and the crystals are frequently twinned; generally found massive. COLOUR. Brass-yellow, frequently with a tarnish, which is sometimes iridescent. STREAK. Greenish-black, very slightly shining. LUSTRE. Metallic; opaque. FRACT. Conchoidal, uneven. H. 3·5–4. SP. GR. 4·1–4·3.

TESTS. Decrepitates when heated in the closed tube and gives a sublimate of sulphur; before the blowpipe on charcoal, fuses to a metallic magnetic globule, and gives off sulphurous fumes; with fluxes, affords reactions for both copper and iron; chalcopyrite may be distinguished from pyrite by its inferior hardness, chalcopyrite crumbling when cut with a knife and pyrite resisting the attempt to cut it—pyrite emits sparks when struck with steel, chalcopyrite does not; the powder of pyrite is black, of chalcopyrite greenish-black; chalcopyrite may be distinguished from gold by its brittle nature and its non-malleability, gold being soft, malleable and easily cut with a knife—chalcopyrite is soluble in nitric acid, gold is not.

OCCURRENCE. Chalcopyrite is the principal commercial source of copper. It occurs in a number of ways. Magmatic segregations of chalcopyrite are known, but are not important. Pneumatolytic veins with chalcopyrite occur in Cornwall, Norway, Oregon, South Australia, Chile. An important mode of occurrence is as hydro-thermal or metasomatic veins, as in California, Montana, Arizona, Alaska and Canada. Pyrometasomatic deposits are important; in these the chalcopyrite occurs with other sulphides and skarn minerals at or near the contact between bodies of intrusive granodioritic rock and limestone; examples are Clifton-Morenci, and Bisbee, Arizona; Bingham, Utah; Alaska, Canada, Australia, Japan and Korea. Chalcopyrite, associated with the dominant bornite, occurs in the famous Kupferschiefer of Permian age at Mansfeld, Germany, where the copper minerals occur as grains in a shale; the copper minerals are possibly of sedimentary origin and were deposited at the same time as the shale; some authorities, however, hold that the ore-minerals are epigenetic and derived from an igneous source.

CHALCOCITE, Copper Glance, Redruthite, Vitreous Copper Ore

COMP. Copper sulphide, Cu_2S; copper 79·8 per cent; traces of iron are usually present.

CRYST. SYST. Orthorhombic. COM. FORM. Crystals combinations of prism and pinacoids, twinning frequently giving a stellate grouping of three individuals; usually massive and with a granular or compact structure. CLEAV. Poor prismatic. COLOUR. Blackish lead-grey; often with a bluish or greenish tarnish. STREAK. Same as the colour; sometimes shining. LUSTRE. Metallic. FRACT. Conchoidal. H. 2·5–3. SP. GR. 5·5–5·8.

TESTS. Heated on charcoal, boils and ultimately fuses to a globule of copper; soluble in hot nitric acid, but leaving a precipitate of sulphur; other reactions for copper are given.

OCCURRENCE. Chalcocite is a very valuable copper ore. It is formed by the alteration of primary copper sulphides in the zone of secondary enrichment, often through the agency of meteoric waters. It occurs in veins or beds associated with other copper minerals, as in Cornwall, Siberia, Kongsberg (Norway), Monte Catini (Italy), Mexico, Peru, Chile, Butte (Montana), and Zambia. Chalcocite 'blankets' are of secondary formation due to the alteration of original chalcopyrite deposits by descending waters; the very important 'Porphyry Coppers' belong to this type of deposit.

COVELLINE, COVELLITE

COMP. Copper sulphide, CuS; copper 66·4 per cent.

CRYST. SYST. Hexagonal. COM. FORM. Crystals hexagonal plates; usually massive. CLEAV. Perfect basal. COLOUR. Indigo-blue. H. 1·5–2. SP. GR. 4·6.

OCCURRENCE. Occurs at many localities in the zone of secondary enrichment of copper lodes, etc. This zone is situated between the gossan and the unaltered zone, and in it occur products of the chemical reactions between the original vein-stuff and descending solutions, and it often forms the richest part of the lode (see p. 217 and Fig. 106).

BORNITE, ERUBESCITE, Variegated Copper Ore

COMP. Sulphide of copper and iron; the composition varies, the ratio of iron and copper being expressed for the average by the formula, Cu_5FeS_4.

CRYST. SYST. Cubic. COM. FORM. Crystals showing cube and octahedron known, but most commonly massive. COLOUR. Coppery red or pinchbeck brown; tarnishes speedily on exposure and becomes iridescent (*Peacock Ore*). STREAK. Pale greyish-black, and slightly shining. LUSTRE. Metallic. FRACT. Conchoidal, uneven; brittle. H. 3. SP. GR. 4·9–5·4.

TESTS. Heated in open tube, yields sulphur dioxide fumes, but gives no sublimate; fuses in the reducing flame to a brittle magnetic globule; soluble in nitric acid, leaving a deposit of sulphur; characterized by its tarnish and by the red colour of fresh surfaces.

OCCURRENCE. Bornite is a valuable ore of copper. It occurs as a primary deposit in many copper lodes, and as a constituent of the zone of secondary enrichment of these. It is found in some of the Cornish mines, where it is known as *horse-flesh ore*. In many occurrences, bornite is closely associated with igneous magmas, appearing either as magmatic segregations (e.g. Namaqualand) or as rather later deposits connected with pegmatites or end-stage consolidation. Bornite forms veins with quartz, or with quartz and chalcopyrite. Finally, it is the dominant constituent of the bedded copper deposit of the Kupferschiefer of Mansfeld, Germany, referred to on p. 246.

Sulphides of Copper, Arsenic and Antimony—The 'Grey Coppers'

These sulphides form a closely related series of minerals, the members of which usually occur together and are connected by transitional types of intermediate composition. The chief members are:

    Tetrahedrite, $(Cu,Fe)_{12}Sb_4S_{13}$.
    Tennantite, $(Cu,Fe)_{12}As_4S_{13}$.
    Famatinite, Cu_3SbS_4.
    Enargite, Cu_3AsS_4.
    Bournonite, $CuPbSbS_3$.

These minerals occur associated with other ores of copper in veins and replacement-deposits; economically important deposits are those of South America (Chile, Peru and Argentina) and United States (Butte, Montana). These ores are often important as sources of metals other than copper; for instance, the Butte deposits produce copper, silver, gold, lead and zinc.

TETRAHEDRITE, Grey Copper, Fahlerz

COMP. Sulphide of copper and antimony, $(Cu,Fe)_{12}Sb_4S_{13}$; part of the copper is often replaced by iron, zinc, silver, or mercury; part of the antimony is often replaced by arsenic, seldom by bismuth. Tetrahedrite sometimes contains 30 per cent silver in place of part of the copper, and is then called *argentiferous grey copper ore*, *silver fahlerz* or *freibergite;* this variety is more of the nature of stephanite (p. 259).

CRYST. SYST. Cubic, Tetrahedrite Type (see p. 106). COM. FORM. Tetrahedral crystals, usually modified and frequently twinned;

also massive with a compact, granular, or crypto-crystalline structure. COLOUR. Between steel-grey and iron-black. STREAK. Nearly the same as the colour. LUSTRE. Metallic; opaque, but very thin splinters are subtranslucent and appear cherry-red by transmitted light. FRACT. Subconchoidal or uneven; rather brittle. H. 3–4·5. SP. GR. 4·5–5·1.

TESTS. This mineral varies in its chemical behaviour according to the different substances which the varieties contain; in the closed tube they all fuse, affording a deep red sublimate of antimony sulphide; in the open tube, tetrahedrite fuses and gives off sulphurous fumes, and forms a white sublimate inside the tube; the mercurial varieties give minute globules of quicksilver; on charcoal, tetrahedrite fuses and, according to the constituents present, yields white encrustations of antimony oxide, arsenic oxide or zinc oxide, or a yellow encrustation of lead oxide—the zinc encrustation becoming green when moistened with cobalt nitrate and reheated; heated before the blowpipe with sodium carbonate, tetrahedrite yields scales of metallic copper; tetrahedrite is decomposed by nitric acid—arsenic oxide, antimony oxide and sulphur remaining.

OCCURRENCE. Occurs associated with other ores of copper, and also with siderite, galena and blende (Idaho and British Columbia). Localities for tetrahedrite are Levant (Cornwall), Andreasberg (Harz), Freiberg (Saxony), Pribram (Czechoslovakia), Chile, Bolivia, and Montana and Colorado (U.S.A.).

TENNANTITE

COMP. Sulphide of copper and arsenic, $(Cu,Fe)_{12}As_4S_{13}$; often contains antimony, causing it to grade towards tetrahedrite.

CRYST. SYST. Cubic, Tetrahedrite Type. COM. FORM. Tetrahedral crystals; usually massive and compact. COLOUR. Blackish lead-grey to iron-black. H. 3–4. SP. GR. 4·37–4·49.

OCCURRENCE. Occurs associated with other copper ores in Cornwall, Freiberg (Saxony), Colorado, and Butte, Montana.

FAMATINITE

COMP. Sulphide of copper and antimony, Cu_3SbS_4.

FORM. Isomorphous with enargite. COLOUR. Greyish to copper-red.

OCCURRENCE. With enargite in the veins of the Sierra Famatina, Argentina, and other South American deposits.

ENARGITE

COMP. Sulphide of copper and arsenic, Cu_3AsS_4; antimony is often present, when the mineral grades into famatinite.

CRYST. SYST. Orthorhombic. COM. FORM. Usually occurs massive and granular; small crystals are known and are often repeatedly twinned. CLEAV. Good prismatic. COLOUR. Greyish-black to iron-black. LUSTRE. Metallic. H. 3. SP. GR. 4·44.

TESTS. Heated on charcoal fuses and yields encrustations of oxides of arsenic, and usually of antimony and maybe zinc; in open tube gives sulphur and arsenic fumes, and forms a white sublimate of arsenic oxide; with fluxes yields a copper residue.

OCCURRENCE. An important ore of copper in copper veins of the Sierra Famatina in Argentina, and in Chile and Peru; occurs in abundance with other ores of copper in veins in monzonite at Butte, Montana.

BOURNONITE, Wheel Ore, Endellionite

COMP. Sulphide of copper, lead and antimony, $CuPbSbS_3$.

FIG. 109. Bournonite, Wheel Ore

CRYST. SYST. Orthorhombic. COM. FORM. Modified prisms, often twinned, producing a cruciform or cogwheel like arrangement, whence the name *wheel ore*, or *Radelerz*, given to it by the German miners (see Fig. 109); also occurs massive. COLOUR. Steel-grey or lead-grey, and sometimes blackish. STREAK. Same as colour. LUSTRE. Metallic; opaque. FRACT. Conchoidal or uneven; brittle. H. 2·5–3. SP. GR. 5·7–5·9.

TESTS. On charcoal, fuses easily, giving at first a white encrustation of antimony oxide, and afterwards a yellow one of lead oxide;

the residue heated with sodium carbonate on charcoal yields reddish flakes of metallic copper.

OCCURRENCE. Occurs with other ores of copper; first found at St. Endellion in Cornwall; also occurs at Kapnic (Rumania), Clausthal, Andreasberg and Neudorf (in the Harz), Chile, Bolivia, etc.

CHALCANTHITE, BLUE VITRIOL, Cyanosite, Copper Vitriol

COMP. Hydrous sulphate of copper, $CuSO_4.5H_2O$; copper 25.4 per cent.

CRYST. SYST. Triclinic. COM. FORM. In flattened crystals; also compact massive, stalactitic, and encrusting. COLOUR. Sky-blue, sometimes greenish. LUSTRE. Vitreous; subtransparent to translucent. FRACT. Rather brittle. TASTE. Nauseous and metallic. H. 2·5. SP. GR. 2·12–2·3.

TESTS. Heated in the closed tube, gives water; soluble in water, the solution coating a clean strip of iron with metallic copper—by this means much copper may be procured from the water pumped from copper mines; heated on charcoal with sodium carbonate and carbon, chalcanthite yields metallic copper.

OCCURRENCE. Chalcanthite results from the alteration of chalcopyrite and other copper sulphides, and occurs therefore in the zone of weathering of copper lodes, as in Cornwall, in the Rammelsberg mine in the Harz and abundantly in Chile.

Brochanthite or *Waringtonite*, and *Langite* are other hydrated sulphates of copper, of an emerald green colour, and occurring in copper gossans.

FIG. 110. Malachite, Stalactitic

MALACHITE

COMP. Basic carbonate of copper, $Cu_2CO_3(OH)_2$; copper 57·3 per cent.

CRYST. SYST. Monoclinic. COM. FORM. Commonly occurs massive, encrusting, stalactitic or stalagmitic, and with a smooth mammillated or botryoidal surface (see Fig. 110); internal structure often divergently fibrous and compact; also occurs granular and earthy. COLOUR. Bright green; different shades of the colour often follow a concentrically banded arrangement. STREAK. Of the uncrystallized variety, paler than the colour. LUSTRE. Slightly silky on surfaces broken parallel with the fibrous structure; sometimes earthy or dull; the crystals have an adamantine lustre inclining to vitreous, and they are translucent to subtranslucent; in the massive, mammillated, stalactitic or stalagmitic conditions, the mineral is opaque. H. 3·5–4. SP. GR. 3·9–4.

TESTS. Heated in closed tube, it gives off water and blackens; dissolves with effervescence in acids; before the blowpipe, alone, fuses and colours the flame emerald-green; on charcoal reduced to metallic copper; colours borax bead green.

OCCURRENCE. Malachite is found in the zone of weathering or oxidation of copper deposits, lodes or other types. Some localities are:—Redruth (Cornwall), Chessy (France), Nishni Tagilsk (Siberia), Burra Burra mine (South Australia), Chile, Pennsylvania, Tennessee. The colour-banding marks the successive deposits of the mineral which has in many cases resulted from the percolation of water through copper-bearing rocks and the subsequent deposition of the dissolved carbonate in fissures or cavities, the solution having apparently dripped in slowly, and the water evaporated, thus forming a series of layers in the same way that stalactites and stalagmites are formed by the percolation of water through limestone; very large masses of malachite have been procured from Siberia and Australia. The most remarkable deposit of malachite is probably that of the Katanga region of the Congo and the adjacent part of Zambia; this field produced over 800,000 tons of copper in 1966; the ores are malachite, other ores of the oxidized zone, such as azurite, chrysocolla, melaconite, chalcocite, together with, at some mines, chalcopyrite and bornite; the ores occur disseminated through sedimentary rocks such as dolomites, dolomitic sandstones, and feldspathic sandstones, and probably represent the weathered upper part of an enormous disseminated deposit, since chalcopyrite and other sulphides have been encountered at depth; opinion on the origin of the ores is divided—one view is that they are due to emanations from a granite mass, whilst another view regards them as of sedimentary origin.

USES. Malachite is a valuable ore of copper; it is also cut and polished and used for ornamental purposes.

AZURITE, CHESSYLITE, Blue Carbonate of Copper

COMP. Basic carbonate of copper, $Cu_3(CO_3)_2(OH)_2$; copper 55·1 per cent; compare the composition of malachite.

CRYST. SYST. Monoclinic. COM. FORM. Modified prisms; usually massive or earthy. COLOUR. Deep azure blue, whence the name. STREAK. Blue, but lighter than the colour. LUSTRE. Of crystals, vitreous verging on adamantine; crystals are transparent to nearly opaque, massive and earthy varieties are opaque. FRACT. Conchoidal; brittle. H. 3·5–4. SP. GR. 3·7–3·8.

TESTS. As for malachite (see p. 252); distinguished from malachite by its azure-blue colour.

OCCURRENCE. Azurite is found associated with other oxidized copper minerals—malachite, etc.—in the zone of weathering of copper lodes and deposits; when occurring in sufficient quantity, it is a valuable ore of copper; some localities are:—Redruth (Cornwall), Chessy (France), Katanga, etc.

CHRYSOCOLLA

COMP. Very variable from the presence of impurities, but it is essentially a hydrous silicate of copper, $CuSiO_3.2H_2O$. It is sometimes considered to contain a variable amount of free silica, and some specimens are sufficiently hard to scratch glass and to be cut and polished for jewellery.

COM. FORM. Amorphous; usually occurs in encrustations or in thin seams; botryoidal and massive, with a compact and enamel-like texture. COLOUR. Bluish-green, sky-blue, or turquoise-blue. STREAK. White when pure. LUSTRE. Vitreous to earthy; translucent to opaque. FRACT. Conchoidal; rather sectile; brittle when translucent. H. 2–4, see under composition, above. SP. GR. 2–2·2.

TESTS. Heated in closed tube, blackens and gives off water; before blowpipe, infusible, but decrepitates and colours the flame emerald-green; with sodium carbonate on charcoal, yields metallic copper; is decomposed by acids without effervescence, which serves to distinguish it from malachite—it is not, however, completely soluble; with the fluxes, it gives the usual copper reactions.

OCCURRENCE. Occurs in the zone of weathering of copper lodes and deposits, and when found in sufficient quantity, it constitutes a copper ore of some value, and is easily reduced when mixed with limestone; it seldom, however, yields more than 10 per cent copper; localities are:—Lizard (Cornwall), Schneeberg (Saxony), Kupferberg (Bavaria), Nishni Tagilsk (Siberia), Adelaide (South Australia), Chile, United States, and Katanga.

DIOPTASE, Emerald Copper

COMP. Hydrous silicate of copper, $CuSiO_3.H_2O$.

CRYST. SYST. Hexagonal-rhombohedral. COM. FORM. Crystals combinations of prism and rhombohedron; sometimes found massive. CLEAV. Perfect rhombohedral. COLOUR. Emerald-green. STREAK. Green. LUSTRE. Vitreous; transparent to subtranslucent. FRACT. Conchoidal or uneven; brittle. H. 5. SP. GR. 3·2–3·3.

TESTS. Resembles chrysocolla in its reactions, but differs from it in yielding gelatinous silica when dissolved in hydrochloric acid.

OCCURRENCE. In the zone of weathering of copper lodes, as in Chile, Siberia and Arizona; dioptase, from its rarity, is of no industrial importance.

ATACAMITE, Remolinite

COMP. Basic chloride of copper, $Cu_2(OH)_3Cl$; copper 59·4 per cent.

CRYST. SYST. Orthorhombic. COM. FORM. Prismatic, crystals not common; frequently massive and lamellar. COLOUR. Bright deep green to blackish-green. STREAK. Apple-green. LUSTRE. Adamantine to vitreous; translucent to subtranslucent. H. 3–3·5 SP. GR. 3·76.

TESTS. Heated in closed tube, gives off water and forms a grey sublimate; easily soluble in acids; on charcoal fuses in time to metallic copper, colouring the flame azure-blue (chloride), and forming a brownish and greyish-white deposit on the charcoal, which voltilizes in the reducing flame, again giving an azure-blue coloration.

OCCURRENCE. Occurs in the zone of weathering of copper lodes, especially when this weathering has been effected under desert-conditions; occurs at Botallack Mine, St. Just, Cornwall, Los Remolinos and the Atacama Desert, South America, and at Linares, Spain, and Burra Mine, South Australia.

Some Other Minerals of the Oxidized Zone of Copper Lodes

Mention may be made of a few of the many hydrated oxysalts of copper that accompany the more abundant minerals of this type already described. *Libethenite*. A hydrated phosphate of copper, $4CuO.P_2O_5.H_2O$, of a dark olive-green colour, occurring both crystallized and massive in Cornwall, Hungary, the Urals, etc. *Phosphochalcite, or Pseudomalachite*. A hydrated phosphate of copper, $6CuO.P_2O_5.3H_2O$, of an emerald or blackish-green colour, sometimes crystallized, but mostly encrusting and massive. H. 4·5–5. It is found near Bonn on the Rhine, in Hungary and Cornwall. *Liroconite*. A hydrated arsenate of aluminium and copper, variable in composition; monoclinic; colour, sky-blue or verdigris-green. H. 2·5; occurs in Cornwall, Hungary, etc. *Clinoclase* and *Olivenite* are also arsenates of copper.

SILVER MINERALS

Silver (Ag) occurs in nature in the free state, occasionally 99 per cent pure, but generally containing copper, gold and other metals. It is a white metal which, next to gold, is the most malleable and ductile of all metals and the best conductor of heat and electricity. These properties of silver depend on its atomic structure, that of a face-centred cube (see p. 21, Figs. 7 and 19C) in which it is like copper and gold. The specific gravity of silver is 10·5, and its melting point is 960·5°C. It is unaltered by dry or moist air. Silver occurs also as sulphide, sulpho-salts, arsenide, antimonide, and chloride, and also associated with ores of lead, zinc, copper and other metals.

There are two main classes of silver ores. The first is the *dry* or *siliceous* ores which are mined primarily for their silver content—the silver ores proper. But the greater part of the world production of silver is derived from the smelting of metalliferous ores, such as those of lead, copper and zinc, which contain a small percentage of silver—these are the *argentiferous* lead, copper, or zinc ores.

Silver is recovered from silver ores proper by cyanidation or by cupellation, and from argentiferous lead ores by melting with a small percentage of zinc, the silver being more soluble in the molten zinc than is the lead. Silver and gold may be recovered together in the form of an alloy, which is afterwards refined or 'parted'. Silver containing gold is called 'doré silver'. Refined silver usually contains from 997·5 to 999·0 parts of silver per 1,000 —pure silver being 1,000 fine. On the London market the price is quoted for 999 fine; the standard alloy employed in Britain for coin, plate and jewellery was 925 fine and is generally called standard silver. The addition of a small amount of copper produces an alloy having a lower melting point, a greater hardness and affording a sharper casting than pure silver. Formerly the chief uses of silver were in coinage, plate and jewellery but now its industrial uses far exceed these; in 1966 nearly four times as much was used in industry as in coinage. Industrial uses are in the manufacture of electronic components, electrical machinery, mirrors, electroplate and batteries, in medicine, photography, glass-making, etc.

The average production of silver in recent years has approached 230 million ounces troy. The main production comes from seven countries, in order of importance, Mexico, U.S.A., Peru, Canada, U.S.S.R., Australia and Japan; a dozen other countries each produce over a million ounces annually. It is interesting to note that of the total production of the United States in 1959, 45 per cent was derived from dry or siliceous ores and 55 per cent from base-metal ores, the latter figure having fallen during recent years. The production from this latter category was distributed as follows:—22·1 per cent from copper ores, 5·8 from lead ores, 5·1 from zinc ores and 22·2 from mixed base-metal ores.

Silver ores occur as veins, replacement-deposits, contact-metamorphic deposits or as alluvials. The most important primary ore is argentite, Ag_2S. The upper parts of silver deposits or lodes are weathered with the production of cerargyrite, $AgCl$, which is often accompanied by bromyrite, AgBr, and iodyrite, AgI; in several cases of such gossans, cerargyrite occurs above bromyrite and below this latter comes iodyrite. Below this halide zone there is in many silver lodes a zone of secondary enrichment in which native silver and rich secondary sulphides are developed; below this zone come the primary deposits in which the ore is usually much poorer.

TESTS. Silver compounds, heated with sodium carbonate and charcoal on charcoal, give a silver-white bead; this bead is malleable, and does not tarnish. Hydrochloric acid, added to a solution of silver in nitric acid, produces a dense white precipitate of silver chloride which is soluble in ammonia—silver beads obtained by fusion with sodium carbonate may be tested in this way.

The silver minerals considered here are:

Element	Native Silver, Ag.
Sulphide	Argentite, Silver Glance, Ag_2S.
Complex Sulphides	Stephanite, Ag_5SbS_4.
	Pyrargyrite, Ag_3SbS_3.
	Proustite, Ag_3AsS_3.
	Freieslebenite, $(Pb,Ag)_8Sb_5S_{12}$.
	Polybasite, $(Ag,Cu)_{16}(Sb,As)_2S_{11}$.
Telluride	Hessite, Ag_2Te.
Chloride	Cerargyrite, Horn Silver, AgCl.

NATIVE SILVER

COMP. Silver, Ag, but usually associated with it are small amounts of other metals, such as copper, gold, mercury, platinum, bismuth, etc.

CRYST. SYST. Cubic. COM. FORM. Distorted crystals; mostly found filiform, arborescent or massive. COLOUR. Silver-white, tarnishing readily. STREAK. Silver-white, and shining. LUSTRE. Metallic. FRACT. Hackly; sectile, malleable and ductile. H. 2·5–3. SP. GR. 10·1–11·1.

TESTS. Soluble in nitric acid—a clean piece of copper immersed in the solution becomes coated with silver, and a pinch of common salt or a drop of hydrochloric acid when added to the solution throws down a white precipitate of silver chloride, soluble in ammonia; before the blowpipe on charcoal, silver fuses readily to a silver globule which crystallizes on cooling.

OCCURRENCE. Native silver occurs in the upper parts of silver sulphide lodes below the chloride capping, and is often concentrated to form rich deposits, as in Mexico, the Comstock Lode (Nevada), Broken Hill (N.S.W.), Peru, and many other silver-mining districts. Primary native silver occurs in strings and veins with silver sulphides—one group of such deposits is exemplified by that of Kongsberg (Norway), where large masses of silver have been found; in this and allied deposits (e.g. Andreasberg in the Harz) the formation of the silver is connected with zeolitization, that is, the deposition of the zeolite silicates by heated waters passing through the rocks. In another group of primary native silver veins the metal occurs in cobalt-nickel veins, as at Annaberg (Saxony) and Cobalt (Ontario). Native silver is assocated with native copper in the Lake Superior region (see p. 244).

ARGENTITE, SILVER GLANCE

COMP. Silver sulphide, Ag_2S; silver 87·1 per cent.

CRYST. SYST. Cubic. COM. FORM. Crystals show cube and octahedron, usually distorted; also occurs reticulated, arborescent and, most commonly, massive. COLOUR. Blackish lead-grey. STREAK. Same as colour, and shining. LUSTRE. Metallic; opaque. FRACT. Small, subconchoidal, or uneven; sectile. H. 2–2·5. SP. GR. 7·19–7·36.

TESTS. Heated in open tube, gives off sulphurous fumes; on charcoal, in oxidizing flame, fuses with intumescence, gives off sulphurous fumes, and yields metallic silver; soluble in dilute nitric acid.

OCCURRENCE. Argentite is the most common primary ore of silver. It occurs in small quantity in the sedimentary Kupfer-schiefer of Mansfeld, Germany (see p. 246). Its main occurrence is in various types of veins; argentite-veins, in which it is accompanied by stephanite and polybasite, in propylitized volcanic rocks are important for Mexican silver production; argentite, associated with galena and blende, occurs in veins in the San Juan mining district of Colorado; gold-bearing argentite-quartz veins are common, localities being Tonapah (Nevada) and the famous Comstock Lode of Nevada; replacement-veins carrying argentite are typified by that of Portland Canal, British Columbia. Argentite accompanies native silver in the cobalt-nickel veins of Cobalt (Ontario) and Annaberg (Saxony), and in the silver stringers of Kongsberg (Norway), and in many German veins.

STEPHANITE, Brittle Silver Ore

COMP. Sulphide of silver and antimony, Ag_5SbS_4; silver 68·5 per cent.

CRYST. SYST. Orthorhombic. COM. FORM. Crystals commonly flat tabular prisms, often twinned; also massive and disseminated. COLOUR. Iron-black. STREAK. Iron-black. FRACT. Uneven; brittle. H. 2–2·5. SP. GR. 6·26.

TESTS. Heated in closed tube, fuses with decrepitation and gives a slight sublimate of antimony sulphide after long heating; on charcoal, it fuses to a dark metallic globule and encrusts the support with antimony oxide—the globule heated in the reducing flame with sodium carbonate yields metallic silver; stephanite is decomposed by dilute nitric acid, leaving a residue of sulphur and antimony oxide—a clean strip of copper placed in the solution becomes coated with silver, and hydrochloric acid added to the solution gives a white precipitate of silver chloride.

OCCURRENCE. Occurs with other primary silver ores in veins at many of the localities given for argentite—Mexico, Harz, Comstock Lode (Nevada), Freiberg (Saxony), Pribram (Czechoslovakia), Cornwall, etc.

Red Silver Ores, Ruby Silver

There are two species of Red Silver Ores, closely allied in structure, composition and mode of occurrence; they are:

Pyrargyrite, Ag_3SbS_3—Dark Red Silver Ore.
Proustite, Ag_3AsS_3—Light Red Silver Ore.

OCCURRENCE. The Ruby Silvers occur in veins and replacement-veins of various types accompanying other primary ores of silver; they often occur just below the enriched zone, as at Potosi (Bolivia); typical localities are Andreasberg (Harz), Freiberg (Saxony), Pribram (Czechoslovakia), Mexico, Potosi, Comstock Lode (Nevada), Cobalt (Ontario).

PYRARGYRITE, Dark Red Silver Ore

COMP. Sulphide of silver and antimony, Ag_3SbS_3; silver, 59·9 per cent.

CRYST. SYST. Hexagonal, Tourmaline Type. COM. FORM. Crystals, hexagonal prism terminated by rhombohedra, or variously modified, and sometimes twinned; commonly occurs massive. CLEAV. Fairly good rhombohedral. COLOUR. Black to cochineal-red. STREAK. Cochineal-red. LUSTRE. Metallic, adamantine; translucent to opaque. FRACT. Conchoidal. H. 2–3. SP. GR. 5·7–5·9.

TESTS. Heated in the open tube gives sulphurous fumes and a white sublimate of antimony oxide; heated on charcoal spirts and fuses easily to a globule of silver sulphide, and coats the support white—the globule heated with sodium carbonate and charcoal on charcoal yields metallic silver; pyrargyrite is decomposed by nitric acid, leaving a residue of sulphur and antimony oxide.

OCCURRENCE. See above.

PROUSTITE, Light Red Silver Ore

COMP. Sulphide of silver and arsenic, Ag_3AsS_3; silver, 65·4 per cent.

CRYST. SYST. Hexagonal, Tourmaline Type. COM. FORM. Pointed crystals; commonly granular and massive. COLOUR. Cochineal-red. STREAK. Cochineal-red. LUSTRE. Adamantine; subtransparent to subtranslucent. FRACT. Conchoidal or uneven. H. 2–2·5. SP. GR. 5·55–5·64.

TESTS. Heated in open tube, gives off sulphurous fumes and yields a white sublimate of arsenic oxide; heated on charcoal with sodium carbonate gives metallic silver; decomposed by nitric acid.

OCCURRENCE. See above, p. 259.

FREIESLEBENITE

COMP. Sulphide of silver, lead and antimony, $(Pb,Ag)_8Sb_5S_{12}$; silver about 22–23 per cent.

CRYST. SYST. Monoclinic. COM. FORM. In prismatic crystals and massive. COLOUR. Light steel-grey to dark lead-grey. LUSTRE. Metallic. FRACT. Subconchoidal or uneven; brittle; sectile. H. 2–2·5. SP. GR. 6–6·4.

TESTS. Heated on charcoal gives white sublimate of antimony oxide near assay; gives yellow encrustation when roasted with potassium iodide and sulphur, indicating lead; heated on charcoal in oxidizing flame gives metallic silver; in reducing flame gives lead.

OCCURRENCE. Found associated with other silver ores, galena, etc., in Spain, Saxony, Rumania, etc.

POLYBASITE

COMP. Sulphide of silver, antimony, copper and arsenic, $(Ag,Cu)_{16}(Sb,As)_2S_{11}$; silver about 70 per cent.

CRYST. SYST. Monoclinic. COM. FORM. Tabular prismatic crystals; usually massive. COLOUR. Iron-black. STREAK. Black. LUSTRE. Metallic. H. 2–3. SP. GR. 6–6·2.

TESTS. Heated in open tube, gives sulphurous fumes and sublimate of antimony and arsenic oxides; copper residue and silver bead given by lengthy heating with fluxes on charcoal.

VARIETY. *Pearcite*, an arsenical variety.

OCCURRENCE. In silver veins associated with other primary silver ores; in the argentite veins of Mexico, the argentite-gold-quartz veins of Tonopah (Nevada) and the Comstock Lode, in the replacement-veins of Portland Canal (British Columbia) and in the silver deposits of San Juan (Colorado), Pribram (Czechoslovakia), Freiberg (Saxony), Andreasberg (Harz), Chile, etc.

HESSITE

COMP. Silver telluride, Ag_2Te.

CRYST. SYST. Cubic. COM. FORM. Massive. COLOUR. Lead-grey. LUSTRE. Metallic. H. 2·5; sectile. SP. GR. 8·4.

TESTS. Powdered mineral heated with strong sulphuric acid gives a reddish-violet solution, indicating tellurium; heated on charcoal gives silver bead.

VARIETIES. Hessite often contains some gold and so grades into petzite $(Ag,Au)_2Te$.

OCCURRENCE. Occurs with other tellurides in the Kalgoorlie goldfield, Western Australia, in the gold-veins of the Porcupine mining area of Ontario, and in various veins in Chile, Mexico, California, etc.

CERARGYRITE, KERARGYRITE, HORN SILVER

COMP. Silver chloride, $AgCl$; silver, 75·3 per cent.

CRYST. SYST. Cubic. COM. FORM. Cube; usually massive and waxlike, and frequently in encrustations. COLOUR. Pale shades of grey, sometimes greenish or bluish, and, when pure, colourless; assumes a brownish tint on exposure. STREAK. Shining. LUSTRE. Resinous, passing to adamantine; transparent to subtranslucent, or nearly opaque. FRACT. Somewhat conchoidal; sectile, cutting like wax. H. 2–3. SP. GR. 5·8.

TESTS. Soluble in ammonia, but not in nitric acid; fuses in the candle flame; on charcoal yields a globule of metallic silver; when placed in the microcosmic salt bead, to which copper oxide has been added, it gives when heated in the oxidizing flame an intense azure-blue colour to the flame—indicating chlorine; a plate of iron rubbed with the mineral becomes silvered.

OCCURRENCE. Occurs in the gossan or upper parts of silver veins, associated with other silver halides—bromyrite, $AgBr$, iodyrite AgI, and embolite, $Ag(Cl,Br)$, and arising by the action of descending waters containing chlorides, etc., on the oxidized primary ores; cerargyrite often forms extremely rich but small silver deposits; some localities are Freiberg (Saxony), Andreasberg (Harz), Broken Hill (N.S.W.), Atacama (Chile), Comstock Lode, Tonopah (Nevada), etc.

GOLD MINERALS

Gold (Au) occurs very widely diffused in nature, chiefly in the free state, but invariably alloyed with some proportion of silver or copper, and occasionally with bismuth, mercury and other metals. Native gold has been known to contain as much as 99·8 per cent gold, but as a rule ranges from 85 to 95 per cent, the balance being usually silver for the most part. Gold, when pure, is the most malleable and ductile of all metals, properties depending on its atomic structure, that of a face-centred cube (see p. 21, Figs. 7 and 19C). It becomes brittle, however, when it contains small amounts of bismuth, lead, arsenic, etc. It has a specific gravity of 19·3 and melts at about 1,060°C.

Native gold is recovered from alluvial deposits by some form of water concentration, followed by amalgamation with mercury. That occurring in veins is milled and ground previous to amalgamation in the case of 'free-milling' ores. When the ore is of a partly 'refractory' nature, or when the gold is very finely divided, cyanidation, i.e., solution of the gold in sodium cyanide, is employed. Often, treatment by mercury is followed by cyanidation of the tailings for the recovery of the unamalgamated fine and combined gold. Before the fine grinding of ores and treatment by cyanide that are the usual practice, it is sometimes necessary to roast ores to eliminate arsenic and antimony compounds which decompose the cyanide and cause excessive consumption. Gold is also recovered by chlorination, and by smelting with lead ores.

In addition to the native metal gold occurs in combination as tellurides, and possibly as selenides; large quantities of gold are obtained from sulphides with which it is probably mechanically mixed. Gold ores can thus be classed into two groups:

(1) *Free-milling ores*, from which native gold is recoverable by crushing and amalgamation, and (2) *refractory ores*, tellurides and auriferous sulphides, which yield their gold by complex smelting processes.

Gold has a remarkable position in world economy. Apart from its use for jewellery, it is employed for coinage, and for these purposes it is usually alloyed with silver or copper to withstand wear better. The purity or 'fineness' of gold is expressed in parts per 1000, the standard for coin in the British Commonwealth being 916·6 parts of gold to 83·4 of copper. In England the legal standard for jewellery is the carat of 22, 18, 15, 12 or 9 parts per 24. The fineness of gold in alloys therefore can be expressed either in carats or thousandths; thus pure or fine gold was 22 carats or 916·6 thousandths fine. For purposes of plate, jewellery, watch-cases, etc., the standard of 18 carats or 750 fine is legal, but the lower standards of 16 and 14 carats are also general. Less gold coin is minted at the present time, but gold is still required as a medium of exchange, measure of value and cover for paper currency.

The total world production, excluding the U.S.S.R., of gold in 1966 is estimated at nearly 42 million ounces, of which South Africa was responsible for nearly 31 million, Canada over 3 million, U.S.A. nearly 2 million, and Australia approaching 1 million. It is likely that the output from the U.S.S.R. exceeds that of South Africa. Other notable producers were Ghana, Rhodesia, Philippines, Colombia, Congo, Mexico, Japan, Nicaragua, Brazil and India. Of the gold produced in the U.S.A. in 1959, it is estimated that half came from gold ores proper, 28 per cent from base-metal ores and 22 per cent from alluvial deposits. During the last thirty years or so, the production from gold ores has fallen from about 70 per cent and that from base-metal ores has risen from about 7 per cent, the placer production keeping more or less constant.

Native gold occurs in veins of various types, and in alluvial deposits both modern and ancient; tellurides and auriferous sulphides also occur in veins. In the weathered parts of these vein deposits, gold may be concentrated mainly by removal of the useless associates. In gold-bearing quartz-pyrite veins, for example, the weathered portion may be made up of rusty quartz in which are gold nuggets; in the gossans of telluride veins, gold appears as mustard-gold—spongy, filmy and finely-divided free gold.

TESTS. The physical properties of gold serve to distinguish the native metal; its yellow colour, malleability, fusibility, high specific gravity, and insolubility in any one acid are distinctive. All gold compounds yield a gold bead when heated on charcoal with sodium carbonate. Tellurides, as such, are detected by tests given on p. 44.

The gold minerals dealt with here are:

Element	Native gold, Au.
Element with other metallic elements	Gold Amalgam, Au with Hg, Ag.

Tellurides
$\begin{cases} \text{Sylvanite, } (Au,Ag)Te_2. \\ \text{Calaverite, } (Au,Ag)Te_2. \\ \text{Petzite, } (Ag,Au)_2Te. \\ \text{Nagyagite, sulpho-telluride of lead} \\ \qquad \text{and gold.} \end{cases}$

NATIVE GOLD

COMP. Pure gold, Au, or gold alloyed with silver; the latter metal has been known to amount to over 26 per cent in *argentiferous gold* or *electrum*; copper, iron, palladium and rhodium have also been found in gold—43 per cent of rhodium has been reported in a variety called *rhodium gold* from Mexico; *maldonite*, a black variety containing bismuth and found in Australia, has a composition represented by Au_2Bi.

CRYST. SYST. Cubic. COM. FORM. Crystals rare, cube, octahedron and rhombdodecahedron; usually found in grains or scales in alluvial deposits associated with heavy obdurate minerals such as garnet, zircon, etc.; also rounded masses, called nuggets, in alluvial deposits or embedded in quartz veins—the Welcome Stranger nugget found in Victoria contained £10,000 worth of gold; occasionally the metal occurs in strings, threads, etc., and the following names indicate the varied forms assumed—*grain gold, thread gold, wire gold, foil gold, moss gold, tree gold, mustard gold, sponge gold*. COLOUR. The yellow colour of native gold from different localities varies in shade; some specimens from Kashmir possess a coppery or bronze yellow colour; when much silver is present the metal may appear almost silver-white; in thin leaves by transmitted light gold is green. LUSTRE. Metallic; opaque. FRACT. Hackly; very ductile, malleable, and sectile, being easily cut with a knife. H. 2·5–3. SP. GR. 12–20—the variation being due to the metals with which the gold specimen may happen to be alloyed.

TESTS. The colour, combined with the malleability, weight, and sectility of gold, distinguish it from other minerals; iron pyrites, which has sometimes been mistaken for gold, cannot be cut with a knife, whilst chalcopyrite crumbles beneath the blade and gives a greenish-black streak; heated on charcoal, all gold compounds give a yellow malleable globule of gold.

OCCURRENCE. Exceedingly important sources of gold are the deposits of placer or alluvial type; these deposits are derived from the weathering and disintegration of the primary gold-bearing rock; gold is found in the residual or lateritic deposits formed on the outcrop of weathered gold-bearing rock—*eluvial placers*; the great placer deposits, however, occur in the valleys and may be of recent or ancient date, and may be at the surface or at a great depth (*deep lead*); concentration of gold may arise by many processes of sedimentation, such as wind action, wave action (*beach placers*) and especially by river action; examples of shallow placers are those of the Urals and Siberia, India, China, Gold Coast, Alaska (Klondyke), British Columbia, Western United States (California, Montana, etc), South America (Equador, Chile, Bolivia, etc.), Australia (Ballarat, River Torrens, etc.); alluvial gold is found in small quantities in the deposits of several British rivers, in Cornwall, North Wales, Leadhills (Scotland) and Sutherlandshire. Various ancient placers of little importance in gold production have been described, such as in the Permo-Carboniferous rocks of New South Wales, and in the basal Cambrian of the Black Hills, South Dakota. Possibly the greatest of all placers, whether ancient or modern, is that of the *Banket* or gold-bearing conglomerates of the Rand, South Africa, which supply a dominant part of the world's gold. These conglomerates occur in the Witwatersrand system of Pre-Cambrian age; conglomerates, composed of small quartz pebbles in a metamorphosed sandy matrix, are found at several horizons and carry gold. The origin of this gold is a subject of discussion—some hold that the Banket is an ancient deltaic placer deposit in which the detrital gold has undergone solution, redeposition and recrystallization—others believe that the gold was introduced long after the formation of the conglomerate. Alluvial or placer gold has been derived from primary gold deposits, usually true veins, the *reefs* of the gold miner. The veins usually consist of gold, quartz and pyrite; many of them occur in Tertiary volcanic rocks, chiefly andesites and rhyolites as in certain goldfields of New Zealand, Mexico, Transylvania and the Western United States. Deeper-seated gold-quartz veins are some in California and the Western United States, Ballarat and Bendigo in Victoria, New South Wales, Queensland, Nova Scotia, Alaska, Brazil, Austria, the Urals, etc.

In the great Porcupine gold-field of Ontario free gold is accompanied by tellurides. Argentite is associated with quartz and gold in the Comstock Lode and at Tonopah (Nevada) and elsewhere. Small gold deposits occur as replacements in limestone where gold-bearing jaspery rocks are found, and another minor type of deposit is of pyrometasomatic origin. Small but exceedingly rich deposits are found associated with alunite at Goldfield, Nevada. In Great Britain gold was worked in North Wales, in quartz veins in the Menevian slates of Middle Cambrian age.

GOLD AMALGAM

An amalgam composed of gold, mercury and silver, the gold averaging about 40 per cent. It is sometimes found crystallized, but usually in small white or yellowish-white grains which crumble easily. It is usually associated with platinum and has been reported from California, Colombia, Urals and elsewhere.

Gold Tellurides

OCCURRENCE. The gold tellurides occur mainly in veins and replacement-deposits in which they are associated with pyrite and other sulphides and often free gold. The chief localities at which gold telluride ores are worked are Kalgoorlie (Western Australia), Cripple Creek and Boulder Country (Colorado), Nagyag, etc. (Transylvania); telluride ores occur in the Porcupine (Ontario) gold veins, and in the gold-alunite deposits of Goldfield, Nevada.

GOLD TELLURIDES. The chief gold tellurides are:
Sylvanite, $(Au,Ag)Te_2$.
Calaverite, $(Au,Ag)Te_2$.
Petzite, $(Ag,Au)_2Te$.
Nagyagite, sulpho-telluride of Pb and Au.

OXIDATION OF TELLURIDE VEINS. In the upper parts of gold telluride veins the tellurides are decomposed, most of the tellurium is removed in solution and the gold set free as a fine powder, *mustard gold*.

SYLVANITE, Graphic Tellurium

COMP. Telluride of gold and silver $(Au,Ag)Te_2$, with gold 24·5 per cent, silver 13·4 per cent, if $Au = Ag$; antimony and lead are sometimes present.

CRYST. SYST. Monoclinic. COM. FORM. Crystals often arranged in more or less regular lines, bearing a fanciful resemblance to writing, whence the name *graphic tellurium*; also occurs massive and granular. COLOUR. Steel-grey to silver-white; sometimes yellowish. STREAK. Same as colour. LUSTRE. Metallic. FRACT. Uneven; brittle. CLEAV. Perfect orthopinacoidal. H. 1·5–2. SP. GR. 8–8·2.

TESTS. In the open tube, behaves like native tellurium (see p. 501); before the reducing flame, gives on charcoal a yellow malleable metallic globule after long heating—an encrustation of telluric oxide being formed on the charcoal; the powdered mineral heated with strong sulphuric acid gives a reddish-violet colour to the acid; decomposed by nitric acid leaving gold powder—hydrochloric acid added to the solution gives a dense white precipitate of silver chloride.

OCCURRENCE. See above, under Gold Tellurides.

CALAVERITE

COMP. Telluride of gold and silver, $(Au,Ag)Te_2$, with gold predominant.

CRYST. SYST. Monoclinic. COM. FORM. Crystals small and elongated; massive granular. COLOUR. Pale yellow. H. 2·5. SP. GR. 9.

TESTS. As for sylvanite, above; has a higher percentage of gold than sylvanite.

OCCURRENCE. See above, under Gold Tellurides.

PETZITE

COMP. Telluride of gold and silver, $(Ag,Au)_2Te$.

COM. FORM. Massive granular. COLOUR. Steel-grey to iron-black. FRACT. Uneven; brittle; slightly sectile. H. 2·5–3. SP. GR. 8·7–9·0.

TESTS. As for sylvanite, above.

OCCURRENCE. See above, under Gold Tellurides.

NAGYAGITE, Black Tellurium

COMP. Sulpho-telluride of lead and gold; antimony is usually present; gold varies from less than 6 to more than 12 per cent.

CRYST. SYST. Orthorhombic. COM. FORM. Usually massive or foliaceous. COLOUR. Dark lead-grey. STREAK. Dark lead-grey. LUSTRE. Metallic; opaque. FRACT. Sectile; thin laminae flexible. CLEAV. Perfect pinacoidal. H. 1–1·5. SP. GR. 6·8–7·2.

TESTS. Heated in open tube, gives a sublimate of antimonate and tellurate of lead, and of antimony and tellurium oxides in the higher parts of the tube; the antimony oxide volatilizes when reheated, and the tellurium oxide, at a high temperature, fuses to colourless transparent drops.

OCCURRENCE. See above.

GROUP 2A
CALCIUM, STRONTIUM, BARIUM, RADIUM

CALCIUM MINERALS

Calcium (Ca) does not occur in the free state in nature but its compounds are extremely abundant. It may be produced by the electrolysis of fused calcium chloride, but modern production is by the reduction of lime by aluminium in retorts under a low pressure. The metal calcium is being increasingly used in alloys, and in many metallurgical and chemical operations.

Although not occurring native, calcium nevertheless enters into the composition of a very considerable portion of the earth's crust, of which it forms about $3\frac{1}{2}$ per cent. Whole formations, such as the Chalk and the Carboniferous Limestone, consist almost entirely of calcium carbonate, while thick and thin beds of limestone are more or less common throughout the entire series of stratified rocks. Calcium enters also into the composition of many rock-forming silicates; chief of these are *anorthite feldspar*, $CaAl_2Si_2O_8$, *pyroxenes* and *amphiboles*, *garnets*, *scapolite*, *epidotes*, many *zeolites*, and *wollastonite*, $CaSiO_3$; these are described with the other rock-forming silicates in later pages.

The non-slicate calcium minerals are of great economic value, and their various uses are given in their descriptions below.

TESTS. Some calcium minerals give a brick-red flame coloration which is enhanced by moistening the substance with hydrochloric acid; fused calcium compounds give an alkaline reaction with litmus; on the addition of sulphuric acid to solutions containing calcium salts, a white precipitate of calcium sulphate is formed.

The following are the more important non-silicate calcium minerals:

Carbonates
- Calcite, $CaCO_3$ (hexagonal-trigonal).
- Aragonite, $CaCO_3$ (orthorhombic).
- Dolomite, $CaMg(CO_3)_2$.
- Gaylussite, $Na_2Ca(CO_3)_2.5H_2O$ (described with the *Sodium* minerals on p. 235).
- Barytocalcite, $BaCa(CO_3)_2$ (described with the *Barium* minerals on p. 290).

Sulphates
- Anhydrite, $CaSO_4$.
- Gypsum, $CaSO_4.2H_2O$.
- Glauberite, $Na_2Ca(SO_4)_2$ (described with the *Sodium* minerals on p. 233).
- Polyhalite, $K_2Ca_2Mg(SO_4)_4.2H_2O$ (described with the *Potassium* minerals on p. 238).

Phosphate   Apatite, $Ca_5(F,Cl)(PO_4)_3$.

Fluoride   Fluorspar, CaF_2.

Borates
- Ulexite, $NaCaB_5O_9.8H_2O$. (See p. 316.)
- Colemanite, $Ca_2B_6O_{11}.5H_2O$. (See p. 315.)

Colemanite and ulexite are worked for their boron content, and are described under Boron.

CALCITE, Calc Spar, Carbonate of Lime

COMP. Calcium carbonate, $CaCO_3$.

CRYST. SYST. Hexagonal, ditrigonal-scalenohedral, Calcite Type (p. 124). CRYST. STRUCT. The Ca-ions are situated at the corners of a rhombohedron and the CO_3-ions, represented by triangles in Fig. 111, are arranged with their centres midway along the rhombohedral edges and lie in horizontal planes. Physical and optical properties parallel and perpendicular to these planes differ greatly as, for example, the values of refractive index. Other common carbonates with Calcite structure are magnesite (Mg), siderite (Fe), smithsonite (Zn) and rhodochrosite (Mn). COM. FORM. Good crystals common—of three main habits: (1) *nail-head*, combination of flat rhombohedron (10$\bar{1}$2), and prism, (2) *dog-tooth*, combination of scalenohedron (21$\bar{3}$1) and prism, and (3) *prismatic*—two of these types are shown in Figs. 112 and 113; twinning is common in calcite crystals, the twin laws being on the

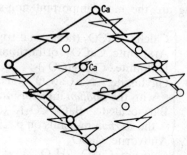

Fig. 111 The structure of Calcite, CaCO₃. Triangles represent the CO₃-groups.

basal pinacoid (0001) (see Fig. 86), and the rhombohedra (01$\bar{1}$2), and (10$\bar{1}$1), as described on p. 154 ; calcite also occurs fibrous, lamellar, stalactitic, nodular, granular, compact and earthy. CLEAV. Perfect parallel to the unit rhombohedron (10$\bar{1}$1) (see Fig. 63), powdered calcite consists of minute cleavage-rhombohedra (see p. 62).

COLOUR. Colourless or white, sometimes with grey, yellow, blue, red, brown or black tints. STREAK. White. LUSTRE. Vitreous to earthy; transparent to opaque. FRACT. Conchoidal, but difficult to observe owing to the perfect cleavage. H. 3; scratched by knife. SP. GR. 2·71.

OPT. PROPS. In ordinary light, calcite appears usually as shapeless grains traversed by excellent rhombohedral cleavages, giving one, two or three sets of intersecting lines; in polarized light twinkles when the nicol is rotated, due to the refractive index (1·658) for the ordinary ray being much higher than that of balsam, whilst that for the extraordinary ray (1.486) is lower; between crossed nicols polarizes in very high colours, giving a grey interspersed with points of pink, blue, etc.; twinning is very common; optically negative, uniaxial; calcite rarely occurs as a primary constituent of igneous rocks, but is common as an alteration or infiltration product, and is the dominant component of the limestones and of their metamorphic derivatives, the marbles.

TESTS. Infusible, but becomes highly luminous when heated; effervesces with evolution of carbon dioxide in cold dilute acid;

FIG. 112. Nail-head Spar

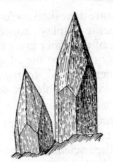

FIG. 113. Dog-tooth Spar

brick-red calcium flame; cleavage very distinctive; for the methods of distinguishing between calcite and aragonite see p. 277, and between calcite and dolomite see p. 278.

VARIETIES.—*Nail-head Spar*, crystals showing combination of flat rhombohedron and prism (see Fig. 112).

Dog-tooth Spar, crystals showing combination of scaleno-hedron and prism (see Fig. 113).

Iceland Spar, a very pure transparent form of calcite first brought from Iceland; it cleaves into perfect rhombohedra, and on account of its transparent character and high double refraction it is employed in the construction of the Nicol prism, as described on p. 170.

Satin Spar, a compact finely-fibrous variety with a satin-like lustre, which it displays to great advantage when polished; it has mostly been formed in veins or crevices in rocks, the fibres stretching across the crevices; the term 'satin-spar' is more commonly applied to the fibrous form of gypsum, described on p. 281; '*beef*' is a quarryman's term for fibrous calcite similar in habit to satin-spar.

Aphrite and *Argentine* are unimportant lamellar varieties of calcite.

Stalactites are pendant columns formed by the dripping of water charged with calcium carbonate from the roofs of caverns in limestone rocks and other favourable situations; successive layers of calcite are deposited one over another, so that a cross-section of the stalactite displays concentric rings of growth; the surplus dripping of the water gives rise to a similar deposit which forms in crusts one above the other on the floors of the caverns, this deposit being called *stalagmite*; beneath these stalagmitic crusts the remains of pre-historic cave-haunting men and animals have been found.

Oriental Alabaster, Algerian Onyx. These are stalagmitic varieties of calcite characterized by well-marked banding, and were used by the ancients for making ointment-jars; both names, however, are bad, since true alabaster is calcium sulphate, and onyx is a cryptocrystalline banded variety of silica.

Calcareous Tufa, Travertine, Calc Tufa are more or less cellular deposits of calcium carbonate derived from waters charged with calcareous matter in solution; at Matlock, Knaresborough and many other places where the springs are thus highly charged, twigs, bird's nests and other objects when immersed in the spring become encrusted with a hard coating of tufa; calcareous tufa sometimes forms thick beds, as in Italy, and is then used as a building-stone.

Agaric Mineral, Rock Milk, Rock Meal. These are white earthy varieties of calcite, softer than chalk, and deposited from solution in caverns, etc.

Chalk, a soft, white earthy carbonate of lime, forming thick and extensive beds in various parts of the world; it has been deposited from the waters of an ancient sea, as shown by the marine character of the fossils which it contains; it sometimes consists to some extent of the remains of microscopic organisms, foraminifera; it is suggested that the sea in which the English Chalk was laid down was margined by a desert area, so that no clayey or sandy material was contributed to the deposit forming in that sea.

Limestone, Marble. Limestone is a general term for carbonate of lime when occurring in extensive beds; it may be crystalline, oolitic (see below), or earthy, and, when impure, either argillaceous, siliceous, bituminous, ferruginous or dolomitic; all true marbles are limestones which have been crystallized by heat or pressure during metamorphic processes, but the name marble is often applied to some special type of non-metamorphic limestone;

the different names of limestones and marbles are derived from the locality where they are found, the formation in which they occur, the fossils which make up their substance, or from some peculiarity of structure, colour, etc. Examples are shell marble, ruin marble, crinoidal limestone, Carboniferous Limestone, etc.

Lithographic Stone, a very fine-grained variety of limestone used in printing.

Pisolite and *Oolite.* These varieties of limestones are formed of granules produced by the deposition of calcium carbonate in successive layers around small nuclei; pisolite differs from oolite in the larger size of the granules; oolitic structure is considered to be the result of a purely inorganic process in which the granules are washed backwards and forwards on beaches in sea-water saturated with calcium carbonate.

Anthraconite or *Stinkstone,* a dark-coloured limestone containing bituminous matter, and emitting a fetid odour when struck.

Fontainbleau Sandstone, a name given to calcite which contains a large admixture of sand, sometimes when concretionary containing 80 per cent, and even when crystallized in rhombohedra containing as much as 65 per cent; it was formerly found at Fontainbleau, in France.

Thinolite.—Interlacing crystals of yellow or brown calcite, occurring as tufa deposits in Nevada, Australia, etc.; the crystals are often of skeleton form.

OCCURRENCE. The occurrence of calcite has been indicated in the description of the varieties given above. Calcite may be either of an organic or of an inorganic, chemical origin, and both forms may be metamorphosed into marble.

USES. Carbonate of lime finds many different uses according to its purity and character: the varieties containing some clayey matter are burnt for cement, the purer varieties providing lime used in many industrial processes, e.g., the manufacture of bleaching powder, calcium carbide, glass, soap, paper, paints, etc.; enormous quantities of limestone of various kinds are used with clay in cement manufacture; limestone is an important road-metal especially for use with tar; marbles and crystalline limestones, and the more resistant calcareous rocks generally, are important building and ornamental stones; calcium carbonate is used as a flux in smelting; certain varieties of limestone are used in printing processes; chalk and lime are applied to the soil as a dressing; the clear transparent form of calcite, Iceland spar, is used in the construction of optical apparatus.

ARAGONITE

COMP. Calcium carbonate, $CaCO_3$, similar to calcite; sometimes contains 1–2 per cent of strontium carbonate, or other impurity. CRYST. SYST. Orthorhombic, Barytes Type. CRYST. STRUCT. The Aragonite structure differs from that of Calcite in that the CO_3-ions are not midway between the calcium layers and the co-ordinations are different; the symmetry is lowered to orthorhombic. Common carbonates with Aragonite structure are strontianite (Sr), witherite (Ba) and cerussite (Pb). COM. FORM. Prismatic crystals, often terminated by acute domes, giving sharp pointed crystals. Twinning is common, the twin plane being the prism (110); this twinning is often repeated, and since the angle between the prism faces is nearly 64°, pseudo-hexagonal twin crystals result, as illustrated in Fig. 114. These pseudo-hexagonal twins are distinguished from true hexagonal crystals by re-entrant angles; individual crystals are often many times twinned, and with alternately reversed striation on faces of prism and cleavage planes. Also occurs in groups of acicular crystals, often radiating columnar, as shown in Fig. 115; also occurs in globular, stalactitic, coralloidal or encrusting forms. CLEAV. Poor, parallel to the pinacoid (010). COLOUR. White, grey, yellowish, sometimes green or violet. LUSTRE. Vitreous: transparent to translucent. FRACT. Subconchoidal; brittle. H. 3·5–4. SP. GR. 2·94.

TESTS. Heated before the blowpipe, aragonite whitens and

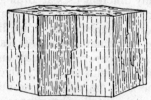

FIG. 114. Pseudo-hexagonal Twin of Aragonite

FIG. 115. Aragonite.

crumbles, changing to calcite; with cold dilute hydrochloric acid, gives off carbon dioxide; flame test, brick-red; aragonite is distinguished from calcite by the following tests: (1) The shape of the crystals, (2) the different cleavage and cleavage-fragments, (3) aragonite is harder than calcite, (4) aragonite has a higher specific gravity, (5) *Meigen's Test*, aragonite is stained with a solution of cobalt nitrate, whereas calcite is not; the mineral under observation is boiled with cobalt nitrate solution for a quarter of an hour, and then washed, a pink staining indicates aragonite, (6) *Leitmeier and Feigl's Test*, a solution of manganese sulphate of 11·8 gr. $MnSO_4.7H_2O$ in 100 c.c. water is prepared, some solid silver sulphate introduced, the whole heated, cooled and filtered, then one or two drops of dilute caustic soda solution are added, and after 1–2 hours the precipitate is filtered off; this solution is kept in an opaque bottle; for distinguishing between aragonite and calcite, the powder or slice is covered by the solution—aragonite at once turns grey and finally black, whilst calcite becomes only greyish after more than an hour; this is a good test for fine intergrowths of the two minerals.

VARIETIES. Aragonite occurs in crystallized, crystalline, massive or stalactitic varieties; *Flos Ferri* is a stalactitic coralloidal variety, which consists of beautiful snow-white divergent and ramifying branches, in many cases encrusting hematite; *pisolites* deposited at some hot springs are of aragonite.

OCCURRENCE. Aragonite occurs with beds of gypsum, or associated with iron-ore in the form of flos-ferri, or as a deposit from the waters of hot springs in oolitic or pisolitic forms. It is less stable than calcite, into which it passes on the application of heat or pressure. The tests of reef-building corals, some algae, etc., are composed of aragonite; this aragonite is changed into calcite by pressure, so that the upper parts of a coral island may be aragonite and the lower parts calcite.

DOLOMITE, Pearl Spar
COMP. Calcium magnesium carbonate, $CaMg(CO_3)_2$, CaO 30·4, MgO 21·7, CO_2 47·9 per cent. Ferrous iron may replace magnesium and when Fe > Mg, the mineral is called *ankerite*.
CRYST. SYST. Hexagonal, trigonal, rhombohedral (compare calcite). CRYST. STRUCT. Alternate atoms of Ca in the Calcite

structure are replaced by Mg, resulting in the lower symmetry of

dolomite. COM. FORM. Rhombohedral crystals, the unit rhombohedron (10$\bar{1}$1) being common (compare calcite); the faces of crystals are often curved, as shown in Fig. 116; twinning is common, on the basal plane and rhombohedron; also occurs massive and granular, forming extensive geological beds, in which

FIG. 116. Dolomite with Curved Faces

state it has a saccharoidal texture. CLEAV. Perfect parallel to the rhombohedron. COLOUR. White, often tinged with yellow and brown, and sometimes with red, green or black. LUSTRE. Of crystals, vitreous inclining to pearly; of massive varieties, dull and opaque. FRACT. Conchoidal or uneven; brittle. H. 3·5–4. SP. GR. 2·8–2·9.

OPT. PROPS. Under the microscope, dolomite is much like calcite, but has a slightly higher refractive index and birefringence, $\varepsilon = 1·500$, $\omega = 1·681$; a useful method of determining a rhombohedral carbonate in crushed material is to employ the immersion method (p. 166) for the determination of refractive indices; the refractive indices for the ray vibrating parallel with the short diagonal of cleavage-rhombs of various rhombohedral carbonates are: calcite, 1·566; dolomite, 1·588; magnesite, 1·599; siderite, 1·747.

TESTS. Before the blowpipe, dolomite behaves like calcite; cold acid acts very slightly on fragments, but in warm acid the mineral is readily dissolved with effervescence. *Lemberg's Test*. If calcite is boiled for fifteen minutes with a solution of aluminium chloride and logwood, it is stained pink, dolomite undergoing no such staining; ferric chloride may be used with equal advantage.

VARIETIES. *Pearl Spar* is a white, grey, pale yellowish or brownish variety, with a pearly lustre, occurring in small rhombohedra with curved faces, and frequently found associated with blende and galena; *Brown Spar, Rhomb Spar, Bitter Spar,* comprise the iron-bearing varieties which turn brown on exposure; *Miemite* is a yellowish-brown fibrous variety found at Miemo in Tuscany; *Ankerite* contains more ferrous iron than magnesium; it resembles Brown Spar, but when heated before the blowpipe on charcoal it becomes black and magnetic; it is found in the Styrian iron mines, and elsewhere; *Magnesian Limestone* is crystalline granular dolomite, occurring in massive beds of considerable extent in, for example, the Permian rocks of England.

OCCURRENCE. Dolomite occurs in extensive beds at many geological horizons; dolomite may be deposited directly from sea-water, but most dolomite beds have been formed by the alteration of limestones, the calcite of which is replaced by dolomite; dolomitization is often related to joints and fissures through which the solutions penetrated, and thick beds, as in the Dolomite Alps of Tyrol, may be completely changed to dolomite; as a result of this change, a shrinkage takes place and useful minerals may afterwards be deposited in the cracks so caused; the solutions giving rise to dolomitization are mainly derived from the sea, and an example of the change is seen in the conversion of the aragonite and calcite of coral reefs into dolomite by reaction with the magnesium salts contained in the sea-water; dolomite is also a common veinstone of metalliferous veins.

USES. Dolomite is an extremely important building material; it is also used for making refractory furnace linings, and as a source of carbon dioxide.

ANHYDRITE

COMP. Anhydrous calcium sulphate, $CaSO_4$.

CRYST. SYST. Orthorhombic. COM. FORM. Crystals, prismatic or tabular, combinations of prisms of the second and third orders and the three pinacoids, or of the two domes; sometimes occurs in cubes pseudomorphous after rock-salt; also commonly fibrous, lamellar, granular and compact, lamellar varieties sometimes contorted. CLEAV. Perfect parallel to the three pinacoids, thus giving rectangular fragments; front pinacoidal cleavage not so good as the other two. COLOUR. White, often with a grey, bluish or reddish tint. LUSTRE. On cleavage-planes, pearly; on the basal plane vitreous; transparent to subtranslucent. FRACT. Uneven; splintery in lamellar and fibrous varieties. H. 3–3·5. SP. GR. 2·93.

TESTS. Before the blowpipe, turns white, but does not exfoliate like gypsum, after a time yields an enamel-like bead; fused with sodium carbonate and charcoal, anhydrite gives a mass which blackens silver when moistened; anhydrite is soluble in boiling hydrochloric acid, a white precipitate being given on the addition of barium chloride; anhydrite is harder than gypsum, has three cleavages whilst gypsum has one, has a greater specific gravity, and does not yield water when heated in the closed tube.

VARIETIES. *Vulpinite* is a scaly granular variety, found at Vulpino in Lombardy; it is sometimes harder than common anhydrite, owing to the presence of silica, and is occasionally cut and polished for ornaments; *Tripestone* is a contorted concretionary form of anhydrite; *Muriacite* is a name sometimes applied to some of the crystallized varieties.

OCCURRENCE. Anhydrite occurs as a saline residue associated with gypsum and rock-salt, as in the Stassfurt, Germany, and in many similar deposits; it has been shown that anhydrite forms from gypsum in sea-water at 25°C, and so possibly the alternating bands of gypsum and anhydrite found for example in the German deposits may be annual layers: the question as to whether gypsum or anhydrite was the original mineral in many deposits has been much discussed; it seems most likely than anhydrite is the original mineral and its irregular conversion into gypsum accounts for the patchy association of the two minerals; anhydrite is associated with gypsum in the 'cap-rock' overlying salt-domes.

USES. Anhydrite is becoming of importance as a fertilizer, in the manufacture of plasters and cements, and of sulphates and sulphuric acid.

GYPSUM

COMP. Hydrated calcium sulphate, $CaSO_4.2H_2O$.

FIG. 117.
Gypsum

CRYST. SYST. Monoclinic, Gypsum Type. COM. FORM. Crystals common, combinations of prism, clinopinacoid and negative hemipyramid, flattened parallel to the clinopinacoid (see Fig. 117); twins are of two types, first, the *swallow-tail type* in which the twin-plane is the ortho-pinacoid (100), and, second, the *arrow-head type* with a hemi-orthodome (101) as the twin-plane; it must be remarked, however, that these popular names are not used consistently; crystals are also often in stellate interpenetrated groups; gypsum also occurs in laminated, granular, or compact masses, and in fibrous forms. CLEAV. Perfect parallel to the clinopinacoid (010), giving very thin, flexible, non-elastic plates; sectile. COLOUR. Crystals colourless; massive varieties, colourless or white, sometimes grey, yellowish or red. LUSTRE. Of clinopinacoidal faces and cleavage-planes, shining and pearly; of other faces, subvitreous; massive varieties generally glistening, but sometimes dull and earthy; fibrous forms, silky; gypsum is pellucidly transparent like glass, to translucent and even opaque from the admixture with impurities. H. 1·5–2; may be scratched easily with the finger nail. Sp. Gr. 2·3.

TESTS. Heated in the closed tube, gypsum gives water; in the flame test, gives the calcium flame, but not readily; fused with sodium carbonate, yields a mass which blackens silver when moistened; readily soluble in dilute hydrochloric acid.

VARIETIES. Selenite includes the crystallized forms of gypsum, a typical selenite crystal being shown in Fig. 117; *Alabaster* is a very fine-grained and compact snow-white or light-coloured massive variety; *Satin Spar* is the fibrous variety and has a silky lustre; *Gypsite* is gypsum mixed with sand and dirt.

OCCURRENCE. Gypsum is formed in three chief ways, (1) as a saline residue (see p. 208) arising by the evaporation of enclosed basins of sea-water, as at Stassfurt in Germany, in the United States, and elsewhere, such deposits being of great commercial value, (2) accompanying the dolomitization of limestone in the sea, and (3) by the formation of a calcium sulphate by the action of sulphuric acid, generated by the decomposition of pyrite, on the calcium carbonate of shells, etc., in clays—the good crystals of selenite found in many clay formations, such as the London Clay, Oxford Clay, etc., arising in this way.

USES. Gypsum is an important industrial mineral, over 42 million tons (including some anhydrite) being produced in 1959. The chief producers are U.S.A., Canada, Britain, France, U.S.S.R., Spain, Italy and Germany. Britain's annual production is well over 4 million tons. Gypsum is used as a retarder in cement, as a fertilizer, as a filler in various materials, such as paper, crayons, paint, rubber, etc., and in the manufacture of *Plaster of*

Paris, for which purpose the mineral is heated to expel some of its water of crystallization and then ground up. Calcined gypsum is extensively employed in the building trade, for the production of various types of plasters, sheets and boards, and for stucco work; it is also used as polishing-beds in the manufacture of plate-glass, and as an adulterant of foods.

APATITE

COMP. Fluo-phosphate or chloro-phosphate of calcium, the first being *fluor-apatite*, $Ca_5F(PO_4)_3$ and the second *chlor-apatite*, $Ca_5Cl(PO_4)_3$; usually chlorine and fluorine are both present. Hydroxyl can substitute for F or Cl.

CRYST. SYST. Hexagonal, Apatite symmetry type. COM. FORM. Crystals common, consisting of combinations of prism and pyramid with or without the basal plane, as shown in Fig. 118; also mammillated, concretionary and massive. CLEAV. Very poor, parallel to the basal plane. COLOUR. Usually pale sea-green or bluish-green, yellowish-green or yellow; sometimes different shades of blue, grey, red and brown; also white, and at times colourless and transparent. STREAK. White. LUSTRE. Sub-resinous, or vitreous; transparent to opaque.

FIG. 118. Apatite

FRACT. Conchoidal and uneven; brittle. H. 5. SP. GR. 3·17–3·23.

OPT. PROPS. Apatite occurs in igneous rocks as small prismatic crystals, giving lath-shaped longitudinal sections and hexagonal basal sections; refractive indices high, ($\omega = 1·633$ to $1·667$, $\varepsilon = 1·630$ to $1·664$, depending on the composition), and polarization colours low; basal sections are isotropic between crossed nicols and in convergent light when thick enough give a negative uniaxial interference figure; longitudinal sections give low grey polarization colours, extinguish straight, and show negative elongation, the fast ray vibrating parallel with the length of the crystal.

TESTS. Lustre rather distinctive; reactions to blowpipe and other tests depend on the composition; soluble in hydrochloric acid, the solution giving a precipitate of calcium sulphate on addition of sulphuric acid; red flame of calcium sometimes given, and, when moistened with sulphuric acid, apatite may show the blue-green flame of phosphorus; heated with magnesium and moistened, gives off phosphoretted hydrogen; heated with sulphuric acid, sometimes gives greasy bubbles of hydrofluoric acid; presence of chloride sometimes given by the copper oxide-microcosmic salt bead test.

VARIETIES. The two major varieties of natural phosphates are, (1) *Apatite*, which has a definite chemical composition, and (2) *Rock phosphates*, such as phosphorite, phosphatic limestone, guano, bone beds, etc., which have no definite chemical composition. *Phosphorite* is a variety of natural phosphate resulting, in some important occurrences, from the accumulation of organic remains and droppings upon desert islands, the calcite of the island rock being replaced by phosphates to form a mixture of calcium phosphate and unaltered calcite; phosphorite may show traces of the original structure of the parent rock, or may be concretionary or mammillated, in which case it is known as *staffelite*. *Coprolite* is a term applied more particularly to those masses of phosphate found in sedimentary rocks which exhibit a corrugated or convoluted form corresponding with what is supposed to have been the form of the internal casts of the intestines of certain saurians, fishes, etc.—coprolites being consequently regarded as the fossil excrement of those animals; the name coprolite has been loosely applied to *phosphatic concretions* which have formed round fossil shells or bones, and which have been worked at several geological horizons—for examples, in the Greensand, Gault and Crags of England. *Asparagus Stone* is the translucent greenish-yellow crystallized variety of apatite. *Osteolite* is a massive impure altered phosphate, usually having the appearance of lithographic stone.

OCCURRENCE. *Apatite* occurs as a primary constituent of igneous rocks, but only in accessory amount; it is also present in small quantity in most metamorphic rocks, and especially in crystalline limestones; workable deposits of apatite occur in pegmatitic and pneumatolytic veins, as in Ontario and Norway—in the case of the Canadian deposits it is possible that a part of the apatite rock represents a thermally-altered limestone; large apatite deposits are found in the alkaline syenites of the Kola peninsula in North Russia; dyke-like masses of apatite, rutile and ilmenite—called *nelsonite*—have been worked in Virginia. Rock phosphates, *phosphorite*, etc., occur in marine bedded deposits where the phosphate forms thick and extensive beds, layers of nodules, or the material of bone-beds, in residual deposits, and in replacement-deposits in which the original phosphate has been leached from overlying guano beds and has replaced the underlying calcareous rocks. Phosphate deposits are worked on a large scale on many of the Pacific Islands, in North Africa, in Florida, Tennessee and the Western States. The normal annual production is well over 70 million tons, the greatest part of this coming from the United States and North Africa, especially Morocco and Tunisia. Other important producers are the U.S.S.R. and Pacific islands such as Nauru.

USES. The most important use of apatite and phosphate rock is as fertilizers, only minor amounts being employed for the production of phosphorus chemicals.

FLUORSPAR, FLUORITE, Blue John, Derbyshire Spar

COMP. Calcium fluoride, CaF_2.

CRYST. SYST. Cubic. COM. FORM. Crystals of cubes very common Fig. 119), more rarely octahedra or tetrahexahedra; fluorspar also occurs compact and coarsely or fine granular. CLEAV. Perfect parallel to the octahedron. COLOUR. Colourless, white, green, purple, amethyst, yellow or blue. STREAK. White. LUSTRE. Vitreous; transparent to translucent. FRACT. Conchoidal to uneven; brittle. H. 4. SP. GR. 3–3·25.

OPT. PROPS. In thin section, fluorspar is colourless; refractive index (1·434) much lower than that of Canada balsam; isotropic between crossed nicols.

TESTS. Gives the reddish flame of calcium; heated with sulphuric acid, gives greasy bubbles of hydrofluoric acid gas, which cause a white film of silica to be deposited on a drop of water held on a glass rod at the mouth of the tube; heated with potassium bisulphate in a closed tube, gives hydrofluoric acid which attacks the glass to form silicon fluoride which is decomposed in the presence of water to give a white ring of silica on the tube.

VARIETY. *Blue John* is a purple or blue variety from Derbyshire, and is used for vases, etc.

OCCURRENCE. Fluorspar occurs in hydrothermal veins and replacement-deposits associated with galena, blende, quartz and barytes, such deposits being worked; it is found also in tin veins of pneumatolytic origin and as a minor constituent of some granites; fluorspar forms the cementing material in some sandstones, as in the Elgin Trias of Scotland. The chief countries producing fluorspar are Mexico, United States, U.S.S.R., China, Spain, Britain and France. The world production is approaching 3 million tons.

FIG. 119. Fluorspar

USES. The finest grade of fluorspar is used in enamelling iron for baths, etc., in the manufacture of opaque and opalescent glasses, and for the production of hydrofluoric acid; the inferior grades are used as a flux in steel-making and for foundry work; transparent fluorspar is being used in increasing quantities for the construction of lenses.

STRONTIUM MINERALS

Strontium (Sr) does not occur in a free state in nature, but may be prepared by the electrolysis of fused potassium and strontium chlorides. It much resembles calcium in its properties.

The chief minerals of strontium are the sulphate and the carbonate, and these are the source of the strontium compounds used in industry. These strontium minerals occur as nodular deposits in sedimentary rocks or as veins possibly of hydrothermal origin. Strontium compounds are used in the manufacture of pyrotechnics, such as flares and fuses, and fireworks in which red-coloured flames are required. Their former use in the purification of molasses has now dwindled but they are employed in the manufacture of ceramics, plastics, paints, etc., and in some electrolytic refining processes.

TESTS. Strontium compounds colour the blowpipe flame crimson. Fused strontium compounds give an alkaline reaction with litmus. With dilute sulphuric acid, solutions of strontium salts give a white precipitate of strontium sulphate.

The chief minerals of strontium are:

| Sulphate | Celestine, $SrSO_4$. |
| Carbonate | Strontianite, $SrCO_3$. |

CELESTINE, Celestite

COMP. Strontium sulphate, $SrSO_4$.

CRYST. SYST. Orthorhombic. COM. FORM. Tabular crystals resembling those of barytes—combinations of prism and basal pinacoid, together with the two domes; also fibrous, granular or massive. CLEAV. Perfect parallel to the basal pinacoid, and good parallel to the (210) prism. COLOUR. White, sometimes with a pale blue tint. LUSTRE. Vitreous, inclined to pearly at times; transparent to subtranslucent. FRACT. Imperfectly conchoidal; very brittle. H. 3–3·5. SP. GR. 3·96.

TESTS. Gives the crimson flame of strontium; insoluble in acids; fuses to a milk-white globule which gives an alkaline reaction; fused with sodium carbonate, gives a mass which blackens silver when moistened; celestine is distinguished from barytes by its coloration and by its granular character.

OCCURRENCE. Occurs as a sedimentary deposit associated with rock-salt, gypsum and clay—deposits of this type are worked near Bristol (Yate) and in Mexico, Morocco and the United States; these deposits are usually of nodular or lenticular form, and they are in many cases the source of the celestine found replacing limestones and other rocks. Celestine occurs also in the sulphur deposits of Sicily and in the 'cap-rock' of the salt-domes of the Gulf states of the United States.

USES As a source of strontium salts, which are used as described above.

STRONTIANITE

COMP. Strontium carbonate, $SrCO_3$; a small proportion of calcium carbonate is usually present.

CRYST. SYST. Orthorhombic. COM. FORM. Prismatic crystals, often acicular and divergent; frequently twinned like aragonite;

also fibrous or granular. CLEAV. Nearly perfect parallel to the unit prism. COLOUR. Pale green, yellow, grey and white. LUSTRE. Vitreous to resinous on fracture-surfaces; transparent to translucent. FRACT. Uneven, brittle. H. 3·5–4. SP. GR. 3·6–3·7.

TESTS. Gives the crimson flame of strontium; effervesces with hydrochloric acid and dissolves, the dilute solution giving a precipitate of strontium sulphate on addition of sulphuric acid.

OCCURRENCE. The chief commercial source of strontianite is from veins traversing Cretaceous marls and limestones in Westphalia, Germany; a vein of strontianite, 4 feet wide, cuts the Carboniferous Limestone at Green Laws Mine, Weardale, Durham; it occurs in veins with galena and barytes, etc., as in the original locality, Strontian in Argyllshire, and elsewhere; it also occurs as nodules, nests and geodes in limestones, where it may be an original deposit, a replacement-deposit, or formed by alteration of celestine.

USES. As a source of strontium salts which are used as described above.

BARIUM MINERALS

The metal barium (Ba) is procured by the electrolysis of fused barium chloride or by the reduction of barium oxides by aluminium in a vacuum furnace. It resembles calcium in its properties and is being used on an increasing scale in the production of certain alloys, for vacuum tube work, and as a hardener for lead.

Barium occurs in small amounts in many of the rock-forming silicates. The rare feldspar, *celsian*, is a barium aluminium silicate, $BaAl_2Si_2O_8$, corresponding to anorthite, $CaAl_2Si_2O_8$. Economically, the most important barium minerals are the sulphate (*barytes*) and carbonate (*witherite*), which are mined from vein deposits or from residual deposits resulting from the decay of rocks containing veins. The chief producers are U.S.A., West Germany, Canada, Mexico, Greece, U.S.S.R., Yugoslavia, Italy, Peru and China. Britain, France, Spain and Brazil produce significant amounts.

The sulphate is used in the manufacture of white pigment, lithopone, and in various processes listed below; the carbonate is chiefly of value as a source of barium salts.

TESTS. Barium compounds colour the blowpipe flame yellowish-green. Fused barium salts give an alkaline reaction with litmus. With dilute sulphuric acid, solutions of barium salts give a white precipitate of barium sulphate, $BaSO_4$. Barium minerals are usually whitish and have a high specific gravity.

The chief minerals of barium are:

Sulphate	Barytes, $BaSO_4$.
Carbonates	Witherite, $BaCO_3$.
	Bromlite, $(Ba,Ca)CO_3$.
	Barytocalcite, $BaCa(CO_3)_2$.

BARYTES, BARITE, Heavy Spar

COMP.—Barium sulphate, $BaSO_4$; strontium and calcium sulphates are often present as impurities.

FIG. 120. Barytes, showing Prism, Basal Plane, and Dome.

CRYST. SYST. Orthorhombic, Barytes Type. COM. FORM. Crystals common—combination of prism, basal pinacoid, and a dome, as shown in Fig: 120; also occurs massive, coarsely lamellar, granular, compact, columnar, and rarely stalactitic, with a radiating fibrous structure resembling that of wood. CLEAV. Perfect parallel to the basal plane, also perfect parallel to the prism (210). COLOUR. Colourless or white; often tinged with yellow, red, and brown; sometimes bluish. STREAK. White. LUSTRE. Vitreous, approaching resinous, and sometimes pearly; transparent to opaque. FRACT. Uneven; brittle. H. 3–3·5. SP. GR. 4·5.

TESTS. Heated before the blowpipe, barytes decrepitates and fuses with difficulty, colouring the flame yellowish-green; barytes is absorbed by the charcoal when fused with sodium carbonate, and the saturated charcoal when placed on a silver coin and moistened leaves a black stain; the high specific gravity of barytes is distinctive.

VARIETIES. *Cockscomb Barytes* shows tabular crystals arranged nearly parallel to one another. *Caulk* and *Boulder* are terms used in the Derbyshire mines, caulk being the white massive variety and boulder the crystallized type. *Bologna Stone* is a nodular and concretionary form of barytes.

OCCURRENCE. Barytes is a very common veinstone in lead and zinc veins, where it is associated with galena, blende, fluorspar, and quartz, deposits of this nature being worked in the North of England, the United States, etc. It also occurs as residual nodules, resulting from the decay of limestones, etc., containing barytes veins as in Virginia and Derbyshire. Barytes veins appear to be formed in various ways, some being of hydrothermal origin and others arising by leaching of barium compounds from rocks containing these. The cement of some sandstones, as for example the Triassic sandstone of Elgin and the Hemlock Stone of Nottingham, is barytes, and it seems clear that the mineral may be deposited by sedimentary processes.

USES. Barytes is used in the manufacture of white paint, especially to give weight to paper, for dressing poor-quality calico, etc., and in the production of wallpaper and asbestos goods and as a drilling mud, this last use accounting for three-quarters of the annual world output of nearly four million tons.

WITHERITE

COMP. Barium carbonate, $BaCO_3$.

CRYST. SYST. Orthorhombic. COM. FORM. Crystals are always repeatedly twinned (twin-plane the prism), this giving rise to six-sided prisms and bipyramids, which much resemble those of quartz, but re-entrant angles are sometimes observed—a group of witherite twins is shown in Fig. 121; witherite also occurs massive, often with a columnar or granular structure, or tuberose or bottryoidal. CLEAV. Poor parallel to the front pinacoid. COLOUR. White, yellowish or greyish. STREAK. White. LUSTRE. Vitreous; resinous on fracture-surfaces; subtransparent to translucent.

FRACT. Uneven; brittle. H. 3·5. SP. GR. 4·3.

TESTS. Colours the flame yellowish-green; effervesces with hydrochloric acid, the solution giving a dense white precipitate on the addition of sulphuric acid; the weight is noticeable.

FIG. 121. Witherite Twins

OCCURRENCE. Witherite occurs as a gangue mineral, associated with galena and barytes, in many of the veins of the North of England—important localities are Settlingstones and Fallowfield in Northumberland, New Brancepeth Colliery in Durham, many of the Alston veins of Cumberland, and near St. Asaph in North Wales.

USES. Witherite is the source of barium salts, and the finely divided sulphate required in certain processes is produced from it; small quantities are employed in the pottery industry.

BROMLITE, Alstonite

Bromlite is a carbonate of barium and calcium, $(Ba,Ca)CO_3$, in which calcium replaces barium in varying proportions. Bromlite is similar to witherite in properties and mode of occurrence; it is found in small quantity in several of the barytes-witherite veins of the North of England.

BARYTOCALCITE

COMP. Carbonate of barium and calcium, $BaCa(CO_3)_2$.

CRYST. SYST. Monoclinic. COM. FORM. In prismatic crystals; also massive. CLEAV. Perfect prismatic. COLOUR. White, greyish, or yellowish. LUSTRE. Vitreous or slightly resinous; transparent to translucent. FRACT. Uneven. H. 4. SP. GR. 3·6.

TESTS. Gives the yellowish-green flame of barium, rarely the brick-red flame of calcium; effervesces with hydrochloric acid; when heated on charcoal, the barium carbonate fuses and sinks into the block, leaving the calcium carbonate as an infusible mass.

OCCURRENCE. Occurs in barytes and lead veins at Alston Moor, Fallowfield, and elsewhere in the North of England.

RADIUM MINERALS

The salts of radium resemble those of barium in their chemical and physical properties, and radium minerals would be appropriately described here. The sources of radium are, however, uranium minerals in which the radium occurs in minute quantity, and so discussion of the properties, economics and uses of radium is placed with the description of the uranium minerals on pp. 493–6.

GROUP 2B
BERYLLIUM, MAGNESIUM, ZINC, CADMIUM, MERCURY

BERYLLIUM MINERALS

The metal beryllium (Be) does not occur native, but can be obtained by the electrolysis of its fused compounds or by the reduction of the oxide or fluoride. Beryllium is a white metal with a specific gravity of about 1·85, and is thus much lighter than aluminium. The metal is employed in the production of special alloys, mainly with copper, but also with iron and nickel, and in nuclear reactors as a source of neutrons. The ore from which beryllium is obtained is *beryl*, $Be_3Al_2(Si_6O_{18})$ which is available as a by-product in the mining of mica and feldspar deposits in pegmatites. The only other beryllium-bearing mineral described here is *chrysoberyl*, $BeAl_2O_4$. Both beryl and chrysoberyl are used as gemstones.

The beryllium minerals described here are:

Silicate	Beryl, $Be_3Al_2(Si_6O_{18})$.
Aluminate	Chrysoberyl, $BeAl_2O_4$.

The blowpipe tests for beryllium are not good, and the element is detected only by rather lengthy chemical tests.

BERYL

COMP. Beryllium aluminium silicate, $Be_3Al_2(Si_6O_{18})$. The structure of beryl is described on p. 351.

CRYST. SYST. Hexagonal, Beryl Type, described and figured on p. 121. COM. FORM. Crystals common, often of large size, one from Albany, Maine, being 18 feet in length and weighing 18 tons;

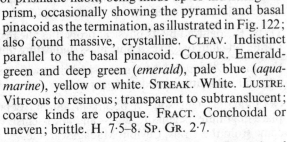

crystals are of prismatic habit, being made up of the hexagonal prism, occasionally showing the pyramid and basal pinacoid as the termination, as illustrated in Fig. 122; also found massive, crystalline. CLEAV. Indistinct parallel to the basal pinacoid. COLOUR. Emerald-green and deep green (*emerald*), pale blue (*aquamarine*), yellow or white. STREAK. White. LUSTRE. Vitreous to resinous; transparent to subtranslucent; coarse kinds are opaque. FRACT. Conchoidal or uneven; brittle. H. 7·5–8. SP. GR. 2·7.

TESTS. Heated before the blowpipe alone, beryl and its varieties become clouded, but otherwise un-altered, except that after protracted heating the edges of splinters become rounded; the mean refractive index of beryl is 1·58.

FIG. 122.
Beryl

VARIETIES. *Emerald* is an emerald-green or pale green variety used as a gemstone—the colour is due to a small content of chromium; *Aquamarine* is a pale blue variety; the term *Beryl* is applied to the coarse kinds which on account of their opacity are unfit for jewellery.

OCCURRENCE. Beryl occurs as an accessory mineral in acid igneous rocks, such as granites and pegmatites, in the druses or cavities of which large crystals often project, as in the Mourne Mountains granite; it is also found in metamorphic rocks of various types; the gem varieties come chiefly from Colombia where the beryl is found in veins of calcite cutting black Cretaceous shales, from Brazil in altered limestones and in pegmatites, from the Urals in mica-schists within a metamorphic aureole, and from a few less important localities. The chief producers for industry are India, Brazil, U.S.S.R. and Ruanda.

CHRYSOBERYL, Alexandrite

COMP. Beryllium aluminate or beryllium aluminium oxide, $BeAl_2O_4$.

CRYST. SYST. Orthorhombic. COM. FORM. Prismatic crystals, often twinned on a prism to produce stellate and six-sided forms. CLEAV. Distinct parallel to the prism (110). COLOUR. Shades of green. LUSTRE. Vitreous; translucent to transparent. FRACT. Conchoidal and uneven. H. 8·5. SP. GR. 3·6–3·8.

TESTS. Recognized by its physical properties; heated with cobalt nitrate on charcoal gives a blue mass, indicating aluminium.

VARIETIES. *Alexandrite* is a greenish variety which is reddish by artificial light, and is used as a gem.

OCCURRENCE. Chrysoberyl occurs in alluvial deposits, and in place in granite, pegmatite, gneiss and mica-schist. Gem varieties come from the Urals, Ceylon, Madagascar, etc.

MAGNESIUM MINERALS

Magnesium (Mg) is not found free native, but is prepared by electrolysis of a mixture of anhydrous magnesium chloride and potassium or sodium chloride. It is a silver-white metal, easily tarnishing to the oxide magnesia, MgO. The metal is employed in the manufacture of light alloys and castings, especially for air-craft, and in various metallurgical processes. It is manufactured from magnesium chloride recovered from sea-water, saline residues, or from the same compound produced from the carbonate, magnesite or from dolomite. The annual production is well over 176,000 tons, the chief producers being U.S.A., U.S.S.R., Norway, Canada, Italy, Britain, France and Japan.

Magnesium is estimated to constitute about 2·7 of the earth's crust. It enters into the composition of a large number of rock-forming silicates—one great group of these being the *ferro-magnesian silicates*, which includes such common rock-forming minerals as biotite, pyroxene, amphibole, olivine, etc. The magnesium-bearing silicates are described with the other silicates on pp. 355–438.

The chief magnesium mineral of economic importance is *magnesite*, $MgCO_3$, which is used for furnace-linings and other purposes mentioned below; *dolomite*, $MgCa(CO_3)_2$, also of great industrial importance, is described with the calcium minerals on p. 277. The sulphates, *epsomite*, $MgSO_4$. $7H_2O$, and *kieserite*, $MgSO_4.H_2O$, are used in chemical manufacture, tanning, etc.; the complex salts of magnesium, potassium and sometimes calcium, such as *polyhalite*, $K_2Ca_2Mg(SO_4)_4.2H_2O$, *kainite*,

$MgSO_4.KCl.3H_2O$, and *carnallite*, $KMgCl_3.6H_2O$, which occur as saline residues, are described with the potassium minerals on pp. 236–238. *Boracite*, $Mg_3B_7O_{13}Cl$, is a source of boron compounds and is accordingly dealt with under Boron on p. 316. *Spinel*, $MgO.Al_2O_3$, used as a gemstone, is described with aluminium on p. 320.

TESTS. The tests for magnesium are very unsatisfactory. Some magnesium compounds, heated on charcoal, moistened with cobalt nitrate and strongly reheated, give a pink residue. On the addition of sodium phosphate to an alkaline solution of a magnesium salt, a white precipitate of ammonium magnesium phosphate is thrown down.

The chief non-silicate magnesium minerals are:

Oxides	Periclase, MgO. Brucite, $Mg(OH)_2$.
Carbonates	Magnesite, $MgCO_3$. Dolomite, $MgCa(CO_3)_2$. (see p. 277)
Sulphates	Epsomite, $MgSO_4.7H_2O$. Kieserite, $MgSO_4.H_2O$. Polyhalite, $K_2Ca_2Mg(SO_4)_3.2H_2O$. (see p. 238). Kainite, $MgSO_4.KCl.3H_2O$ (see p. 238).
Chloride	Carnallite, $KMgCl_2.6H_2O$ (see p. 238).
Chloride and Borate	Boracite, $Mg_3B_7O_{13}Cl$ (see p. 316).
Aluminate	Spinel, $MgAl_2O_4$ (see p. 320).

PERICLASE, Native Magnesia

COMP. Magnesium oxide, MgO.

CRYST. SYST. Cubic.

CHARACTERS AND OCCURRENCE. Periclase occurs as dark green grains and octahedra, showing a perfect cubic cleavage, disseminated in masses of limestone caught up and contact-metamorphosed by the lavas of Monte Somma, Vesuvius, and elsewhere; the original limestone contained dolomite, and the magnesium carbonate of this dissociated on being subjected to heat into periclase and carbon-dioxide, the resulting rock being periclase-marble; periclase, however, is easily converted into brucite, $MgO.H_2O$, by hydration, so that brucite-marbles result.

BRUCITE

COMP. Magnesium hydroxide, $Mg(OH)_2$.

CRYST. SYST. Hexagonal-trigonal, Calcite type. COM. FORM. Prismatic and broad tabular crystals; usually found massive and foliaceous, sometimes fibrous—the laminae being easily separable and flexible, but not elastic, and the fibres being separable, and elastic. CLEAV. Perfect parallel to the basal pinacoid. COLOUR. White, often bluish, greyish and greenish. LUSTRE. On cleavage-planes, pearly, elsewhere between waxy and vitreous; fibrous kinds silky; translucent to subtranslucent. H. 2·5. SP. GR. 2·39.

OPT. PROPS. Colourless in section, with a moderate refractive index and strong double refraction, giving bright colours between crossed nicols; uniaxial positive—a character distinguishing it from talc, gypsum and muscovite.

TESTS. Heated in the closed tube, gives off water, and becomes opaque and friable; before the blowpipe, becomes brilliantly incandescent, and yields a pink mass when moistened with cobalt nitrate and strongly reheated; soluble in hydrochloric acid, which distinguishes it from talc and gypsum; brucite is distinguished from heulandite and stilbite by its infusibility.

OCCURRENCE. As stated in the account of periclase above, brucite is found in contact-metamorphosed impure limestones, called *pencatites* or *predazzites*—typical examples coming from the Tyrol, Skye, Assynt in Sutherland; it occurs also in veins traversing serpentine (a magnesium-rich rock) as in Unst, Shetland. Used for production of magnesium and refractories.

MAGNESITE

COMP. Magnesium carbonate, $MgCO_3$.

CRYST. SYST. Hexagonal-trigonal, Calcite type. COM. FORM. Crystals very rare, and resemble those of dolomite; magnesite is commonly massive and fibrous, sometimes very compact, and sometimes granular. CLEAV. Perfect rhombohedral in crystals. COLOUR. White, greyish-white, yellowish, or brown; commonly chalk-like. LUSTRE. Vitreous; fibrous varieties, earthy and dull; transparent to opaque. FRACT. Flat, conchoidal. H. 3·5–4·5. SP. GR. Of crystals, 3; of earthy varieties, from 2·8 to over 3.

TESTS. The compact chalk-like variety is quite distinctive; magnesite effervesces with hot acids; heated on charcoal, gives an incandescent mass, which, when moistened with cobalt nitrate and strongly reheated, turns pink.

OCCURRENCE. Economically important deposits of magnesite occur in two chief ways: the first is as irregular veins and fracture-zones in serpentine masses from which it has presumably been derived by the action of waters containing carbon-dioxide—such deposits are worked in Greece, India and elsewhere: the second type of deposit is found replacing dolomite and limestone, and is most likely due to the alteration of these rocks by solutions coming from an igneous magma—the Austrian deposits are of this type and supply an important part of the world production, and other similar deposits are worked in Manchuria, Washington and Quebec. Certain bedded deposits of magnesite are interpreted as saline residues.

USES. Magnesite is used in the production of carbon dioxide, magnesium and magnesium salts; its most important use is for refractory bricks, furnace-linings and crucibles; it is employed in the manufacture of special cements, and in the paper and sugar industries.

EPSOMITE, Epsom Salts

COMP. Hydrous magnesium sulphate, $MgSO_4.7H_2O$.

CRYST. SYST. Orthorhombic. COM. FORM. Crystals rare, prismatic; commonly in fibrous crusts, or botryoidal. COLOUR. White. LUSTRE. Vitreous; transparent to translucent. TASTE. Bitter and saline. H. 2–2·5. SP. GR. 1·68.

TESTS. Soluble; heated in the closed tube, gives water; heated with cobalt nitrate on charcoal, gives a pink residue; the mass obtained by heating with sodium carbonate on charcoal when moistened and placed on a silver coin, produces a black stain.

OCCURRENCE. Epsomite occurs in solution in sea-water and in mineral waters; it is deposited from the waters of saline lakes, as in British Columbia and Saskatchewan; it occurs as efflorescent crusts and masses as in the limestone caves of Kentucky, and encrusting serpentine and other rocks rich in magnesium.

USES. In medicine and in tanning.

KIESERITE

COMP. Hydrous magnesium sulphate, $MgSO_4.H_2O$.

CHARACTERS AND OCCURRENCE. A white mineral, massive granular or compact, occurring as a saline residue in Stassfurt, Germany, salt deposits (for which see p. 230), and in similar deposits elsewhere.

ZINC MINERALS

Zinc (Zn) is a bluish-white brittle metal, possessing a crystalline structure, and is reported to have been found native in Australia. It melts at 419°C, and has a specific gravity of about 7·15. At a temperature between 100°C and 150°C it may be rolled out into sheets or drawn into wire, but at 200°C it reverts to a brittle condition, and may be readily powdered under the hammer. It becomes superficially tarnished in moist air, and is soluble in dilute acids.

Zinc is obtained by heating in retorts at a high temperature its roasted or calcined ores in admixture with coal or coke; the zinc oxide is reduced to metal, which, being volatile, distils and is condensed; in this process, sulphuric acid resulting from the break-up of the chief zinc ore, the sulphide, is obtained as a by-product. A considerable proportion of the metallic zinc produced is also obtained by an electrolytic process; the roasted zinc sulphide ore concentrates are extracted with sulphuric acid, and the purified solution is electrolyzed between aluminium cathodes and insoluble lead anodes—thus is obtained zinc, 99·9 per cent pure, and cadmium as a by-product.

Metallic zinc, or *spelter*, is used chiefly for coating—galvanizing—iron, and in the manufacture of various alloys, chief of which is brass, others being German Silver and white metal. Zinc is also employed in tubes for containing toothpaste and the like.

Zinc oxide and zinc sulphide are used as pigments, and are sometimes specified in place of white lead; they are less poisonous and retain their colour better than white lead, though their covering power is inferior. Other salts of zinc are industrially important—the chloride is used in soldering, and in preventing decay in wood, and the sulphate is employed in dyeing, glue-making and other processes.

The chief sources of zinc are the sulphide, *blende* or *sphalerite*, ZnS, and the carbonate, *smithsonite*, $ZnCO_3$. Blende often occurs in vast quantities associated with galena, which is an objectionable constituent from the point of view of the zinc smelter. The metallurgical treatment of this 'refractory sulphide ore' has long been a problem, and although great improvements have been effected in mechanically separating the two values by wet dressing, oil-flotation, etc., and various chemical and metallurgical processes have been devised, it does not yet appear to have been solved by a process of universal application. Principally owing to the improved methods of mechanical separation, however, large deposits of zinc-lead sulphide ores of low grade have been opened up.

Smithsonite, the carbonate, is not amenable to wet concentration, and smelters do not readily purchase such ores containing less than 35 per cent zinc. Fluorspar is frequently associated with zinc ores and is an objectionable constituent.

The world production of spelter amounted in 1966 to more than $3\frac{1}{2}$ million tons—the chief producers of the metal being United States, U.S.S.R., Canada, Belgium, Poland, Japan, France, West Germany and Australia. The chief zinc ore-producing country is the United States; the percentage of zinc in the ore mined is frequently very low and often averages only 3 per cent, but this is increased by concentration. The Broken Hill district of New South Wales produces annually over a quarter of a million tons of concentrates averaging over 50 per cent metallic zinc.

The most important primary zinc mineral is *blende*. Deposits carrying blende at depth have often undergone alteration at the surface with the production there of the oxy-salts such as the carbonates, *smithsonite*, $ZnCO_3$, and *hydrozincite*, $Zn_5(OH)_6(CO_3)_2$, the hydrated silicate, *hemimorphite*, $Zn_4Si_2O_7(OH)_2.H_2O$, and sometimes the anhydrous silicate, *willemite*, Zn_2SiO_4, in such oxidized zones the hydrated sulphate, *goslarite*, $ZnSO_4.7H_2O$, often occurs as an efflorescence.

In most occurrences of zinc ore, the blende is accompanied by galena. There are several types of zinc deposits. In one very important type illustrated by the great Tri-State field in the Mississippi Valley, galena and blende occur as metasomatic disseminations or gash, cavity or joint fillings in limestone; the ore is certainly epigenetic, but whether it was derived from below and transported by ascending solutions, or from above and carried down, is a matter of discussion. The important Broken Hill deposits occur in lodes along fault-planes in a series of metamorphosed rocks, and are of hydrothermal origin. Other deposits of hydrothermal origin replace limestone and are exemplified by the Leadville, Colorado, field. Other zinc-lead deposits are found as contact-metamorphic deposits, but these are not very important. The ore of the famous Franklin Furnace deposit of New Jersey is *franklinite*, $(Fe,Zn,Mn)(Fe,Mn)_2O_4$, *willemite* and *zincite*, ZnO, and occurs as bands and lenses in crystalline limestone; this remarkable deposit is interpreted as of pyrometasomatic origin, but may possibly be a hydrothermal zinc deposit which has been subsequently contact-metamorphosed. Finally, the decay of rocks such as limestones in which there are zinc-lead veins and deposits gives rise to residual deposits of these minerals.

TESTS. Zinc minerals heated on charcoal give an encrustation which is yellow when hot, white when cold; this encrustation, moistened with cobalt nitrate and strongly reheated, assumes a fine green colour.

The nomenclature of some of the zinc minerals is rather confused; the sulphide is usually called zinc-blende or blende in Britain, but is known as sphalerite in America; the anhydrous carbonate has been called calamine in Britian, but smithsonite in America; the hydrated silicate is hemimorphite in Britain, but has been called calamine in America. It is recommended that the name calamine should be discarded. The zinc minerals considered here are:

Element	Native zinc (doubtful).
Oxides	Zincite, ZnO.
	Franklinite, $(Fe,Zn,Mn)(Fe,Mn)_2O_4$.

Sulphide	Blende, Sphalerite, ZnS.
Carbonate	Smithsonite, $ZnCO_3$.
Basic Carbonate	Hydrozincite, $Zn_5(OH)_6(CO_3)_2$.
Silicates	Willemite, Zn_2SiO_4.
	Hemimorphite, $Zn_4Si_2O_7(OH)_2.H_2O$.
Sulphate	Goslarite, $ZnSO_4.7H_2O$.

NATIVE ZINC

Native zinc is said to have been found in basalt near Melbourne in Australia, and has also been reported as occurring in the auriferous sands of the Nittamitta River in the same district, associated with topaz and corundum. The existence of native zinc is, however, still doubted. Zinc has been artificially crystallized in hexagonal prisms, with low pyramidal terminations.

ZINCITE, Red Oxide of Zinc, Spartalite

COMP. Zinc oxide, ZnO; usually contains impurities of oxides of manganese, to which it is considered the colour is due, since chemically pure oxide of zinc is white; the manganese impurities have been found to vary from mere traces up to 12 per cent; occasionally traces of iron oxide are present.

CRYST. SYST. Hexagonal, hemimorphic. COM. FORM. Crystals not common; usually found massive, foliaceous, granular, or in disseminated grains. CLEAV. Perfect prismatic, poor basal. COLOUR. Deep red, but when in very thin scales deep yellow by transmitted light; weathers to a white crust of zinc carbonate. STREAK. Orange-yellow. LUSTRE. Subadamantine; translucent to subtranslucent. FRACT. Subconchoidal; brittle. H. 4–4·5. SP. GR. 5·4–5·7.

TESTS. Heated in the closed tube, it blackens, but on cooling reverts to its original colour; dissolves in acids without effervescence; heated alone before the blowpipe, it is infusible; heated in the reducing flame on charcoal, it yields a white encrustation of zinc oxide, which turns green when moistened with cobalt nitrate and reheated in the oxidizing flame; manganese usually present as indicated by reddish-violet colour of the borax bead in the oxidizing flame.

OCCURRENCE. Occurs with franklinite, willemite and calcite in the Franklin Furnace ore deposit in New Jersey, and is of pyro-metasomatic or contact-metamorphic origin (see before on p. 300, and also below.

FRANKLINITE

COMP. Variable; oxide of iron, zinc and manganese, (Fe,Zn, Mn) $(Fe,Mn)_2O_4$; compare the spinel composition on p. 320.

CRYST. SYST. Cubic. COM. FORM. Octahedra, often rounded at the edges; also in rounded grains, and massive. COLOUR. Black. STREAK. Black. LUSTRE. Metallic; opaque. FRACT. Uneven; brittle. H. 5·5–6·5. SP. GR. 5–5·2.

TESTS. Borax bead amethyst coloured in oxidizing flame, due to manganese, and in reducing flame bottle-green, due to iron; sodium carbonate bead is bluish-green; heated with cobalt nitrate on charcoal, greenish mass due to zinc.

OCCURRENCE. Franklinite occurs at Franklin Furnace, New Jersey, associated with willemite, zincite and calcite, an average ore being 50 per cent franklinite, 25 per cent willemite, 5 per cent zincite, and 20 per cent calcite; the zinc minerals occur as rounded grains and lenses in a crystalline limestone and are considered to be the result of pyrometasomatism, but may possibly result from the contact-metamorphism of previously existing hydrothermal zinc deposits.

BLENDE, SPHALERITE, Black Jack

COMP. Zinc sulphide, ZnS; part of the zinc is, however, usually replaced by iron, and a little cadmium is often present, but never reaches 5 per cent.

CRYST. SYST. Cubic, tetrahedral. COM. FORM. Tetrahedra (see Fig. 47) and rhombdodecahedra common; crystals often twinned and modified, the forms being then difficult to determine; also occurs massive and compact, and occasionally botryoidal or fibrous. CLEAV. Perfect parallel to the faces of the rhomb-dodecahedron. COLOUR. Usually black or brown, sometimes yellow or white and, rarely, colourless. STREAK. White to reddish-brown. LUSTRE. Resinous to adamantine; transparent, translucent or opaque. FRACT. Conchoidal; brittle. H. 3·5–4. SP. GR. 3·9–4·2.

TESTS. Heated alone before the blowpipe, infusible, or very difficulty fusible; with sodium carbonate on charcoal in the reducing flame, it colours the flame strongly green; on charcoal, when roasted in the oxidizing flame, and then intensely heated in the reducing flame, it yields an encrustation of zinc oxide, which is yellow when hot and white when cold—this encrustation assumes a green colour when heated with cobalt nitrate solution; some varieties of blende heated on charcoal with sodium carbonate first give a reddish-brown coating of cadmium oxide; blende is soluble in hydrochloric acid, with evolution of sulphuretted hydrogen—the solution gives a white precipitate of zinc sulphide on the addition of ammonium sulphide.

OCCURRENCE. Blende is the common ore of zinc and is found associated with galena in deposits of various types, as mentioned on p. 299; examples of these types are the metasomatic deposits in limestone of the Tri-State field in the United States, hydrothermal lode and vein deposits of Broken Hill, N.S.W., Colorado, Cornwall and Cardiganshire, replacement-deposits in limestone as in Derbyshire, Cumberland, Westphalia and Colorado, and contact-metamorphic deposits as in New Mexico.

USES. Blende is the most important ore of zinc.

SMITHSONITE, (Formerly Calamine in Britain)

COMP. Zinc carbonate, $ZnCO_3$; the zinc is often partly replaced by iron or manganese, and a little lime, magnesia or cadmium oxide is often present.

CRYST. SYST. Hexagonal-trigonal, Calcite type. COM. FORM. Crystals, modified rhombohedra, rare; commonly found massive, reniform, botryoidal, stalactitic, encrusting, granular or earthy. CLEAV. Perfect rhombohedral. COLOUR. White, greyish, greenish, brownish-white. STREAK. White. LUSTRE. Vitreous, inclining to pearly; subtransparent, translucent or opaque. FRACT. Uneven; brittle. H. 4·5. SP. GR. 4–4·5.

TESTS. Heated in the closed tube, the mineral gives off carbon dioxide and turns yellow when hot, white when cold; heated alone before the blowpipe it is infusible; heated on charcoal, moistened with cobalt nitrate and strongly reheated, it assumes a green colour on cooling; heated with sodium carbonate on charcoal it gives zinc vapours and forms the usual encrustation of zinc oxide; soluble in hydrochloric acid with effervescence.

OCCURRENCE. Smithsonite occurs in beds and veins, and is usually associated with blende, hemimorphite, galena, and iron and copper ores; in many cases it is a metasomatic replacement of limestone but it is probable that it is always a secondary mineral resulting from the alteration of primary blende. Localities are, for example, Mendip Hills (Somerset), Matlock (Derbyshire), Alston Moor (Cumberland), Leadhills (Scotland), and in most lead and zinc mining centres.

USES. Smithsonite is an important ore of zinc; commercially, the term calamine includes the zinc silicates as well as the carbonate.

HYDROZINCITE

COMP. Basic zinc carbonate, $Zn_5(OH)_6(CO_3)_2$.

CHARACTERS AND OCCURRENCE. Monoclinic, but usually occurring massive, fibrous or encrusting; white in colour, hardness 2–2·5, specific gravity 3·6–3·8; results from the alteration of blende, and found with smithsonite in the oxidation zones of zinc deposits, as near Santander, Spain, and elsewhere.

WILLEMITE, Wilhelmite

COMP. Zinc silicate, Zn_2SiO_4.

CRYST. SYST. Hexagonal-trigonal. COM. FORM. Prismatic crystals, formed by hexagonal prism with rhombohedral terminations; also massive. CLEAV. Two imperfect cleavages. COLOUR. Green, yellow or brown. LUSTRE. Vitreous to resinous. H. 5–5·5. SP. GR. 4–4·1.

TESTS. Heated on charcoal, moistened with cobalt nitrate and strongly reheated gives a green mass; soluble in hydrochloric acid, the solution gelatinizing when concentrated.

VARIETY. *Troostite* is a manganese-bearing variety occurring in large crystals. OCCURRENCE. Willemite occurs with zincite and franklinite in the deposit at Franklin Furnace, New Jersey, mentioned on pp. 300, 302; also at Moresnet and Vieille Montagne, Belgium, and Raibl, Carinthia.

HEMIMORPHITE (formerly Calamine in America), Electric Calamine, Galmei, Silicate of Zinc

COMP. Hydrous zinc silicate, $Zn_4Si_2O_7(OH)_2.H_2O$.

CRYST. SYST. Orthorhombic, hemimorphic. COM. FORM. Modified orthorhombic prisms, the opposite extremities terminated by dissimilar faces; crystals frequently twinned; also massive, granular, fibrous, mammillated, encrusting, stalactitic, or banded. CLEAV. Perfect prismatic. COLOUR. White, yellowish-brown, sometimes faintly greenish or bluish; sometimes banded in blue and white. STREAK. White. LUSTRE. Vitreous; sub-pearly on basal pinacoid; sometimes adamantine; transparent to translucent or opaque. FRACT. Uneven; brittle. OTHER PROPERTIES. Becomes electrically charged when heated, and phosphorescent when rubbed. H. 4·5–5. SP. GR. 3·45.

TESTS. Heated in the closed tube, it decrepitates, whitens and gives off water; heated alone before the blowpipe, it is almost infusible; heated with sodium carbonate on charcoal, gives an encrustation yellow while hot and white when cold—this encrustation moistened with cobalt nitrate and strongly reheated, assumes a green colour; hemimorphite gelatinizes with acids, and is decomposed even by acetic acid with gelatinization; hemimorphite is soluble in a strong solution of caustic potash.

OCCURRENCE. Accompanies the sulphides of zinc, iron and lead, and is found associated with smithsonite; it is a product in most cases of the oxidation of primary sulphide ore; occurrences are as for blende and smithsonite above.

GOSLARITE, White Vitriol

COMP. Hydrous zinc sulphate, $ZnSO_4.7H_2O$.

CRYST. SYST. Orthorhombic. COM. FORM. Prismatic crystals; usually stalactitic or encrusting. CLEAV. Perfect parallel to the brachypinacoid. COLOUR. White. STREAK. White. LUSTRE. Vitreous; transparent to translucent. TASTE. Astringent, metallic and nauseous. H. 2–2·5. SP. GR. 2·1.

TESTS. Heated in the closed tube, it boils and gives off water, fusing to an opaque white mass; heated on charcoal, fuses with ebullition, and gives an encrustation of zinc oxide, yellow hot, white cold—this encrustation moistened with cobalt nitrate and reheated gives a green mass; goslarite is readily soluble in water, the solution yielding a white precipitate of barium sulphate on the addition of barium chloride solution—presence of sulphate.

OCCURRENCE. Goslarite results from the decomposition of blende, and is found sparingly in some of the Cornish mines and at Holywell, Flintshire; the chief locality is the Rammelsberg Mine, Goslar, Germany.

CADMIUM MINERALS

Cadmium (Cd) is a bluish-white metal having a brilliant lustre, and closely resembling zinc. It is very malleable and ductile, and has a specific gravity of about 8·6 and melts at 320°C. It is found in nature as the sulphide, *greenockite*, but this mineral is of rare occurrence. Cadmium also occurs in small quantities, probably as the sulphide also, in zinc ores, and the metal is obtained as a by-product in the distillation or electrolysis of zinc ores, in which it seldom occurs in greater amount than 0·4 per cent and usually less.

The production of cadmium averages over 12,000 tons annually, the chief producers being those for zinc—U.S.A., Canada, Mexico, South-west Africa, Belgium, U.S.S.R., Japan and the Congo. Cadmium and its compounds are used for a number of purposes. The metal is employed in several important alloys—fusible alloys used in fire-extinguishers, bearing-alloys in motor manufacture, etc.; it is also of importance in some processes of electro-plating and metal spraying, and in the manufacture of electric transmission wires. The cadmium salts are of importance as pigments, giving with certain other materials, such as selenium, brilliant reds and yellows to pigments and glass.

TEST. Cadmium minerals when heated with sodium carbonate

on charcoal, give a reddish-brown encrustation, which is yellow at some distance from the assay.

The only cadmium mineral considered here is:

Sulphide   Greenockite, CdS.

GREENOCKITE

COMP. Cadmium sulphide, CdS.

CRYST. SYST. Hexagonal, hemimorphic—the opposite extremities of crystals being dissimilarly modified. COM. FORM. Short hexagonal crystals not common; often as a coating on zinc ores. COLOUR. Honey, citron or orange-yellow. STREAK. Between orange-yellow and brick-red. LUSTRE. Adamantine, resinous; nearly transparent. H. 3–3·5. SP. GR. 5.

TESTS. Heated in the closed tube, turns carmine-red, and reverts to its original colour on cooling; heated in the open tube, gives off sulphurous fumes; heated with sodium carbonate on charcoal, gives a reddish-brown encrustation—this encrustation is yielded before the zinc encrustation, so that careful observation of the behaviour of samples of blende may indicate the presence of cadmium in them: greenockite dissolves in hydrochloric acid, with the evolution of sulphuretted hydrogen.

OCCURRENCE. Almost invariably occurs associated with zinc ores, on which it forms a coating; localities are—Bishopton (Scotland), Pribram (Czechoslovakia), Freidensville (Pennsylvania).

MERCURY MINERALS

Mercury or quicksilver (Hg) exists native, but as such is an unimportant source of the metal. It is a silver white metal, liquid at ordinary temperatures; it boils at 357°C, and has a specific gravity of 13·59. When pure it is unaffected by dry or moist air. Mercury combines with most metals to form alloys called *amalgams* and these decompose on heating with volatilization of the metallic mercury.

Mercury is usually obtained from its ore, *cinnabar*, HgS, by roasting in an oxidizing atmosphere, whereby the sulphur is oxidized to sulphurous acid, and the freed metal volatilized and condensed. It may also be obtained by distillation in retorts in the presence of lime or iron which, combining with the sulphur, liberates the mercury. The presence of antimony in the ore is not uncommon, and renders the collection of the mercury difficult.

The only source of the metal is the sulphide, cinnabar, HgS. Many of the ores treated are extremely low in this mineral; in California the average yield amounts to between 0·6 and 0·7 per cent of metal. Cinnabar is not easily altered and the oxidized zone as such does not exist; sometimes there is a small development of the native metal, the chloride *calomel*, and various oxy-chlorides, from the sulphide. Some deposits of tetrahedrite and related copper ores (see p. 248) carry a small percentage of mercury, and such deposits have been worked for mercury on a small scale.

Cinnabar is being deposited from the waters of hot springs at the present day, and it is probable that most cinnabar deposits have been formed in connection with Tertiary or recent volcanic activity.

Mercury is sold in flasks containing 76 pounds. The world production in 1959 was about 8,000 tons, the chief producers being Spain, Italy, U.S.A., U.S.S.R., China, Mexico, Japan and Yugoslavia, with a few less important sources. The chief uses of mercury are in the manufacture of drugs and chemicals, such as the chlorides—corrosive sublimate, and calomel—of chlorine and caustic soda, of vermilion pigments, and anti-fouling paints. It is used in the electrical industry for rectifiers, automatic switches, mercury vapour lamps, etc., and in the instrument industry in the construction of thermometers, barometers, etc. Mercury compounds are used in some quantity as insecticides and fungicides in agriculture and industry, and dental preparations. At one time a considerable amount of mercury was used in the extraction of gold and silver from their ores.

TESTS. Mercury compounds, heated in the closed tube with sodium carbonate and charcoal, give a metallic mercury mirror. Heated with potassium iodide and sulphur, mercury compounds give a greenish-yellow encrustation and greenish-yellow fumes.

The mercury minerals here considered are:

Element	Native Mercury, Hg.
Amalgam	Native Amalgam, Hg_xAg_y.
Sulphide	Cinnabar, HgS.
Chloride	Calomel, Hg_2Cl_2.

NATIVE MERCURY, Quicksilver

COMP. Pure mercury, Hg; a little silver is sometimes present.

CRYST. SYST. Crystallizes when frozen, the crystals showing a rhombohedral structure by X-rays; freezes at $-39\,^{\circ}$C. COM. FORM. Occurs as small fluid globules disseminated through the matrix in which it occurs. COLOUR. Tin-white. LUSTRE. Metallic; opaque. SP. GR. 13·59.

TESTS. Dissolves readily in nitric acid; heated before the blowpipe, mercury volatilizes with little or no residue—should any residue be left, the presence of silver may be shown by fusion with sodium carbonate on charcoal and the production of a silver bead.

OCCURRENCE. Native mercury occurs as fluid globules scattered through cinnabar, HgS, as at Almaden (Spain), Idria (Italy), etc., and is sometimes found in some quantity filling cavities. It is a rare mineral, and is of secondary origin. It is deposited with cinnabar from the waters of certain hot springs.

NATIVE AMALGAM, Silver Amalgam

COMP. Mercury and silver in varying proportions.

CRYST. SYST. Cubic. COM. FORM. Rhombdodecahedron; also occurs massive. COLOUR. Silver-white. STREAK. Silver-white. LUSTRE. Metallic opaque. FRACT. Conchoidal, uneven; brittle and grates under the knife when cut. H. 3–3·5. SP. GR. 10·5 to 14, depending on the composition.

TESTS. Heated in the closed tube, the mercury sublimes and condenses on the cold portion of the tube, leaving a residue of silver; heated before the blowpipe, the mercury volatilizes, and a globule of silver is left; amalgam is soluble in nitric acid; when rubbed on copper amalgam imparts a silvery lustre.

VARIETY. *Arquerite* is a silver-rich variety, soft and ductile, found at the mines of Arqueros, Coquimbo, Chile.

OCCURRENCE. Occurs as scattered grains with cinnabar, as at Almaden, Spain, or in the oxidation zone of silver deposits where it is associated with cerargyrite.

CINNABAR

COMP. Mercury sulphide, HgS; usually contains impurities of clay, bitumen, etc.

CRYST. SYST. Hexagonal, trigonal-trapezohedral, Quartz Type.

COM. FORM. Occurs in rhombohedra or prisms, the crystals often being tabular; usually massive, granular, and sometimes forming crusts. CLEAV. Perfect prismatic. COLOUR. Cochineal-red, sometimes brownish or dark-coloured. STREAK. Scarlet. LUSTRE. When massive, often dull; of crystals, adamantine; subtransparent to opaque. FRACT. Subconchoidal, or uneven; sectile. H. 2–2·5. SP. GR. 8·09.

TESTS. Heated in the open tube, yields a sublimate of metallic mercury, also a black one of mercury sulphide and fumes of sulphur dioxide; heated in the closed tube, cinnabar gives a black sublimate, which becomes red if detached and rubbed on a streak plate; heated with sodium carbonate and charcoal in the closed tube, gives metallic mercury; heated on charcoal with potassium iodide and sulphur, gives greenish fumes and a slight greenish encrustation.

VARIETY. *Hepatic Cinnabar* is a compact variety of a liver-brown colour, and sometimes giving a brownish streak.

OCCURRENCE. Cinnabar occurs as disseminations, impregnations and stockworks in a variety of rocks, but in many cases under circumstances that indicate that it is the result of volcanic activity. It is deposited by certain hot springs in volcanic areas; it occurs in small quantity in some gold-quartz veins. The common associates of the first type of deposit—the most important economically—are chalcopyrite, pyrite, realgar, stibnite, quartz and opal, calcite, and often bitumen. The most important locality is Almaden in Spain, where the mineral occurs as impregnations or small veins in quartzite; other important localities are Idria in Italy, and the western states of the United States, especially California. Small productions come from Russia, Mexico, China, Algeria, Czechoslovakia, and the mineral has been worked in British Columbia, Queensland, New South Wales and New Zealand.

USES. Cinnabar supplies practically all the mercury of commerce. The paint, vermilion, which has the same composition, is prepared from this ore.

CALOMEL, Horn Quicksilver

COMP. Mercury chloride, Hg_2Cl_2.

CHARACTERS AND OCCURRENCE. Forms small crystals, whitish, greyish, or brownish in colour, with a hardness 1–2, and specific gravity 6·48, and found associated with cinnabar at Idria, Italy, and Almaden, Spain, and elsewhere.

GROUP 3B
BORON, ALUMINIUM

BORON MINERALS

Boron does not occur native, but may be procured as a grey amorphous powder by reduction of boric acid.

Boron occurs in combination in several silicates, the chief of these being *tourmaline*, *axinite* and *datolite*—the first two of these minerals are described with the other rock-forming silicates on pp. 376, 379.

The principal forms in which boron occurs in nature are, however, as *borates*, and these are of great industrial importance. The chief borates are as follow, and these are dealt with in this book:

Boric acid	Sassoline, H_3BO_3.
Hydrous Sodium Borates	$\begin{cases} \text{Borax,} & Na_2B_4O_7.10H_2O. \\ \text{Kernite,} & Na_2B_4O_7.4H_2O. \end{cases}$
Hydrous Calcium Borate	Colemanite, $Ca_2B_6O_{11}.5H_2O$.
Hydrous Sodium Calcium Borate	Ulexite, $NaCaB_5O_9.8H_2O$
Magnesium Borate and Chloride	Boracite, $Mg_3B_7O_{13}Cl$.

The borates occur in two chief ways: (1) as deposits from volcanic emanations and from the waters of hot springs of volcanic areas—such deposits being dominantly sassoline or borax, (2) as a result of the drying-up of enclosed bodies of water. To this type belongs the boracite in the Stassfurt saline residues described on p. 208, and the *playa* (dried-up shallow basin) deposits and lake-deposits of Tertiary age in the Western United States—these deposits containing borax, kernite, ulexite, colemanite. From the Tertiary lake-deposits have been leached beds of borates constituting the borax marsh deposits which were formerly of great economic importance. Not all the borates listed above are of primary sedimentary origin, as it is probable that colemanite, for example, has been formed by the leaching of ulexite with sodium chloride solutions.

Borates are used for a great variety of industrial purposes. Boric acid, borax, and other borates are used as fluxes for the manufacture of artificial gems, glasses, and enamels, especially in the pottery and enamelled iron trades. Borax is also used in the soap and glue industries, and in cloth manufacture and tanning. Large quantities are employed as preservatives, antiseptics, and in paint-driers. The compound boron carbide, produced in the electric furnace, ranks close to diamond in hardness, and appears to have an extensive application as an abrasive and resistant material. The chief producer of borates is the United States, chiefly from the kernite and colemanite deposits of California, but also as a by-product of the Californian potash industry; Chile, Argentina and Turkey produce increasing amounts from ulexite and colemanite deposits, and Italy exploits sassoline deposits of volcanic origin.

TESTS. Boron minerals give a rather nondescript yellow-green flame when heated before the blowpipe. Fusible borates, heated in the oxidizing flame on charcoal, moistened with cobalt nitrate and strongly reheated, give a blue glassy residue—a similar residue is given, however, by fusible silicates and phosphates. Boron minerals dissolved in dilute hydrochloric acid, if need be after fusion with sodium carbonate, give a solution which has a characteristic effect on turmeric paper. This, moistened with the solution and dried at 100°C by placing it on a flask containing boiling water, assumes a reddish-brown colour which changes to inky-black by moistening with ammonia.

SASSOLINE, Native Boric Acid

COMP. Boric acid, H_3BO_3.

CRYST. SYST. Triclinic. COM. FORM. Sometimes in prismatic crystals; usually as small glistening scales associated with sulphur. CLEAV. Perfect basal. COLOUR. White, greyish; sometimes yellow, from the presence of sulphur. LUSTRE. Pearly; translucent to transparent. FEEL. Smooth and unctuous. TASTE. Acidulous, slightly saline and bitter. TENACITY. Sectile and flexible. H. 1. SP. GR. 1·48.

TESTS. Fuses easily in the blowpipe flame, tingeing the flame green; soluble in water and in alcohol; dissolved in alcohol, colours flame green by presence of boron; test with turmeric paper as given above.

OCCURRENCE. Sassoline occurs with sulphur in the crater of Vulcano, Lipari; also found around fumaroles (small vents or outlets of sulphurous emanations), and in the steam or vapours which rise from the bottom of the small hot lakes or lagoons of Tuscany; it is condensed in the water, and afterwards separates out in large flakes, which contain about 50 per cent of boric acid; also occurs in the natural waters of Clear Lake, California.

BORAX, Tincal

COMP. Hydrous sodium borate, $Na_2B_4O_7.10H_2O$.

CRYST. SYST. Monoclinic. COM. FORM. Occurs in prismatic crystals resembling in shape the crystals of augite; also in lumps and masses. CLEAV. Parallel to the ortho-pinacoid (100) and to the prism (110). COLOUR. White, sometimes with tinges of blue, green, or grey. STREAK. White. LUSTRE. Vitreous, sometimes earthy. TASTE. Sweetish, alkaline. FRACT. Conchoidal; soft and brittle. H. 2–2·5. SP. GR. 1·7.

TESTS. Heated before the blowpipe, bubbles up and fuses to a clear glassy bead; borax is soluble in water, producing an alkaline solution; colours the flame yellow, due to sodium: with sulphuric acid gives a green flame due to boron: test with turmeric paper as detailed in the introduction to boron minerals.

OCCURRENCE. Borax, together with other borates, ulexite and colemanite, occurs in playa-deposits, 'alkaline flats' and borax marshes formed by the drying-up of saline lakes; deposits of this type are well developed in California, and here the borates have been leached out from bedded colemanite deposits of Tertiary age.

Borax accompanies other borates in the lake-deposits of Tertiary age in the same area. Borax also occurs in Tibet, on the shores and in the waters of lakes, and is there called 'tincal'.

KERNITE, Rasorite

COMP. Hydrous sodium borate, $Na_2B_4O_7.4H_2O$.

CRYST. SYST. Monoclinic. COM. FORM. Massive. CLEAV. Perfect parallel to the orthopinacoid and to the basal pinacoid. COLOUR. White. LUSTRE. Pearly to vitreous; transparent to translucent. H. 3. SP. GR. 1·95.

TESTS. As for borax.

OCCURRENCE. Kernite is now the most important source of industrial borates; the important deposits are beds resulting from the drying-up of a Miocene salt-lake, and are worked in California.

COLEMANITE

COMP. Hydrous calcium borate, $Ca_2B_6O_{11}.5H_2O$.

CRYST. SYST. Monoclinic. COM. FORM. Short prismatic crystals usually projecting into cavities; also massive crystalline, and granular. CLEAV. Perfect parallel to the clinopinacoid, less good parallel to the basal pinacoid. COLOUR. Colourless, white or greyish. STREAK. White. LUSTRE. Vitreous to adamantine; transparent to translucent. FRACT. Hackly. H. 4–4·5. SP. GR. 2·42.

TESTS. Colemanite resembles feldspar, calcite and some other spars in appearance, but may be readily distinguished from these by blowpipe tests; yields water on heating; heated before the blowpipe decrepitates and colours the flame yellowish-green; heated on charcoal it becomes white, and, moistened with cobalt nitrate and reheated, it turns blue; colemanite is soluble in hydrochloric acid, with separation of boric acid on cooling; test with turmeric paper as described in the introduction to the boron minerals.

OCCURRENCE. Colemanite occurs in deposits of Tertiary age in San Bernadino, Los Angeles, Kern and Inyo counties, California;

the colemanite is present as nodules in clays, and also in beds 10 to 50 feet thick, resting on rhyolitic tuffs; from these deposits have been derived the lake deposits of borax and ulexite, to which reference is made under the descriptions of these minerals. It is considered that the colemanite deposits result from the leaching of ulexite beds by sodium chloride solutions.

OTHER HYDRATED CALCIUM BORATES. *Priceite* and *Pander-mite* are hydrated calcium borates related to colemanite; priceite is a soft white earthy mineral and pandermite is a somewhat harder mineral.

ULEXITE, Boronatrocalcite, Natroborocalcite

COMP. Hydrous sodium calcium borate, $NaCaB_5O_9.8H_2O$.

CRYST. SYST. Triclinic. COM. FORM. Globular or reniform masses, which, when broken open, exhibit a fibrous structure. COLOUR. White. LUSTRE. Silky internally; opaque. TASTE. None. H. 1. SP. GR. 1·9.

TESTS. Gives water on heating; fuses to a clear glass, colouring the flame yellow; moistened with sulphuric acid, colours the flame green for an instant; test with turmeric paper as given in the introduction to the boron minerals.

OCCURRENCE. Ulexite occurs associated with borax in the lake deposits of California mentioned above; also in Chile and Argentina, as white reniform masses, in lagoon deposits associated with gypsum and rock-salt.

BORACITE, Stassfurtite

COMP. Borate and chloride of magnesium, $Mg_3B_7O_{13}Cl$.

CRYST. SYST. Cubic, tetrahedral. COM. FORM. Crystals cubic or octahedral in shape (see Fig. 48B); also massive and columnar, or granular. CLEAV. Very imperfect parallel to the two tetrahedra. COLOUR. Colourless, white, yellow, greenish or greyish. STREAK. White. LUSTRE. Vitreous; subtransparent to subtranslucent. FRACT. Uneven, conchoidal; brittle. H. 7. SP. GR. 2·95.

TESTS. Heated on charcoal, fuses and forms a bead which solidifies on cooling to a crystalline mass; gives the green flame of boron; heated on charcoal, moistened with cobalt nitrate and reheated, yields a pink mass, due to magnesium; chloride given by the copper oxide-microcosmic salt bead test; boracite is insoluble in water, but soluble in hot hydrochloric acid.

OCCURRENCE. Boracite occurs in the Stassfurt saline deposit in Germany, associated with rock-salt, gypsum and anhydrite; these deposits are the result of the evaporation of sea-water (see p. 208), and in them the boracite occurs as small crystals or concretions; boracite is also found at Panderma in Turkey, at Kalkberg and Schildstein in Hanover, and Luneville, La Meurthe, France, where it is associated with the same set of minerals as at Stassfurt.

ALUMINIUM MINERALS

Aluminium (Al) is not found in a free state, but in combination constitutes 8 per cent of the earth's crust and is the most abundant of metals. It is an essential constituent of the *clay minerals*, and of a large number of important silicates such as the *feldspars*, *micas*, *sillimanite*, etc. The chief industrial sources of aluminium and its compounds are *bauxite* (hydrated oxides) and to a less extent, *cryolite*, Na_3AlF_6, *alunite*, $KAl_3(SO_4)_2(OH)_6$, *leucite*, $KAlSi_2O_6$, and *alum shales*. Such industrial minerals as potter's clay, china clay or kaolin, fuller's earth, feldspar, garnet, mica, etc., are aluminium silicates, while aluminium oxides occur as bauxite, corundum and emery.

The metal aluminium is produced in the electric furnace by the reduction of alumina obtained from bauxite. It is a silver-white durable metal, capable of taking a high polish. Owing to its low specific gravity, 2·58, it is of great value in the manufacture of many articles where lightness is of importance. It melts at 658°C, and alloys with most metals and some non-metals—light alloys of importance being those with zinc, copper or magnesium. Aluminium is also employed in the manufacture of household articles, of wrapping foil, cans, building-sheet, etc., in electrical equipment, in car and ship construction and in metallurgical and chemical processes. It is produced especially in localities where hydroelectric installations are practicable. The world production of aluminium metal was over 7 million tons in 1966. The main producers, all with over 100,000 tons annually, were in order of importance U.S.A., with over $2\frac{1}{2}$ million tons, U.S.S.R. with 1 million tons, Canada, France, Japan, West Germany and Italy. The rare element *gallium* (Ga), important in the electronics industry, is obtained from commercial aluminium.

The energetic action of finely divided aluminium on a metallic oxide when heated together is utilized in the 'Thermit' process for the production of metallic chromium, manganese, molybdenum, tungsten, uranium, etc., and in the welding of rails, etc.; the aluminium combines directly with the oxygen of the oxide, and the heat evolved by this reaction is sufficient to promote the fusion of the reduced metal.

There are a number of industrially important silicates that contain aluminium—such as feldspar, muscovite mica, a group of hydrated silicates illustrated by kaolin, fuller's earth, etc., the aluminium silicates proper—sillimanite, andalusite and kyanite—and garnet; most of these silicates are important rock-forming minerals and, as they are not worked for aluminium or its salts, their industrial value depending mainly upon their physical properties, it is best to deal with them under the silicates.

TEST. Most aluminium minerals, when finely powdered and heated before the blowpipe, moistened with cobalt nitrate solution and strongly reheated, give a mass showing a fine blue colour. The non-silicate aluminium minerals described here are:

Oxides	Corundum, Al_2O_3.
	Spinel, $MgAl_2O_4$.
	Chrysoberyl, $BeAl_2O_4$ (described) on p. 293 with the *Beryllium* minerals).
Hydroxides	Diaspore, $HAlO_2$.
	Boehmite, $AlO(OH)$.
	Gibbsite, $Al(OH)_3$.
	Bauxite, mixture of Al hydroxides.
Sulphates	Websterite, Aluminite, $Al_2O_3.SO_3.9H_2O$.
	Alunogene, $Al_2(SO_4)_3.16H_2O$.
	Alum, $KAl(SO_4)_2.12H_2O$.
	Alunite, $KAl_3(SO_4)_2(OH)_6$ (described on p. 239 with the *Potassium* minerals).

Fluoride	Cryolite, Na_3AlF_6.
Phosphates	Turquoise, $CuAl_6(PO_4)_4.(OH)_8.4H_2O$.
	Wavellite, $Al_6(PO_4)_4(OH)_6.9H_2O$.

CORUNDUM

COMP. Aluminium oxide, Al_2O_3.

CRYST. SYST. Hexagonal-trigonal, Calcite type. COM. FORM. Occurs mostly in barrel-shaped or pyramidal crystals, shown in Fig. 123, due to the presence of various bipyramids and the basal pinacoid; also in steep hexagonal bipyramids, illustrated also in Fig. 123; the crystals from alluvial deposits are usually much water-worn and rounded; corundum also occurs massive and granular. CLEAV. None, but separation-planes parallel to the basal pinacoid are common. COLOUR. Of common varieties, grey, greenish or reddish, and dull; sometimes colourless; the well-known red colour of the *ruby*, and the blue colour of the *sapphire* serve to distinguish and fact to constitute these varieties; *oriental amethyst*, *oriental emerald*, and *oriental topaz* are purplish, green and yellow respectively. LUSTRE. Vitreous; crystal faces frequently dull. FRACT. Conchoidal or uneven. H. 9, next to diamond. SP. GR. 3·9–4·1.

OPT. PROPS. Occurs in crystals somewhat elongated and irregular, showing the basal parting in suitable sections; usually clear, and, if coloured, the colour is often patchy; the refractive index is very high ($\omega = 1·768$, $\varepsilon = 1·759$), and the birefringence is about that of quartz, so that low-order greys are given between crossed nicols; uniaxial negative; the crystals show negative elongation; the high refractive index distinguishes corundum from quartz and apatite.

TESTS. Not acted on by acids; the hardness and physical properties are usually distinctive; finely powdered corundum heated with cobalt nitrate on charcoal assumes a fine blue colour.

VARIETIES. (1) *Corundum Gemstones*—the *Ruby, Sapphire, Oriental Amethyst, Oriental Emerald*, and *Oriental Topaz* are varieties of corundum coloured red, blue, purple, green, and yellow respectively, and are used as gemstones. (2) Under the name *Corundum* the ordinary types not of gem-quality are included.

(3) *Emery* is a greyish-black variety of corundum containing much admixed magnetite and hematite; it is crushed, powdered and sifted, and the powder used for polishing hard surfaces.

OCCURRENCE. Corundum occurs in several ways. It is produced by the contact-metamorphism of shales, as in the silica-poor

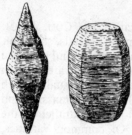

hornfelses. It occurs as veins and segregations associated with peridotites in the Appalachian belt in the eastern United States. It is an original constituent of various igneous rocks, such as syenite, as in Canada, and anorthosite, as in India. It occurs as a result of the contact-metamorphism of limestone, as in Burma. Corundum was formerly mined at several localities, but now its production is from Rhodesia,

FIG. 123. Corundum

India, South Africa and U.S.S.R. The corundum gemstones occur either as isolated crystals in crystalline limestone, or as rounded pebbles in alluvial deposits derived from such rocks; important producers are Burma, Ceylon and Siam. Emery occurs as segregations in igneous rocks or as masses in granular limestone and gneiss; the important producing localities are Naxos in Greece and Kayabachi in Turkey; the Peekshill emery of New York is largely spinel.

USES. Corundum is, with the exception of diamond, the hardest mineral known, and is used as an abrasive. Grinding 'wheels' are made by the incorporation of a binding material such as shellac, with crushed corundum. Artificial corundum, marketed under the names of 'Alundum' and 'Aloxite', is made by fusing bauxite in an electric furnace. Emery is similarly used as an abrasive, and as a refractory material. The coloured varieties of corundum are used as gemstones.

SPINEL

COMP. Magnesium aluminium oxide, $MgAl_2O_4$; spinel is a member of the Spinel Group of minerals which have the general formula, AB_2O_4, where A^{2+} is Mg,Fe,Zn,Mn, and B^{3+} is Al,Fe,Cr,Mn—other examples of this group being magnetite,

Fe_3O_4, and chromite, $FeCr_2O_4$; members of the Spinel Group are isomorphous and isostructural, so that spinel proper usually has iron oxide, manganese oxide or chromium oxide replacing in small amount the magnesia or alumina.

CRYST. SYST. Cubic. COM. FORM. Crystals commonly octahedra, more rarely rhombdodecahedra; often twinned on the face of the octahedron. COLOUR. Red, brown or black, sometimes green or blue. LUSTRE. Vitreous; dark-coloured crystals usually opaque. FRACT. Conchoidal. H. 8. SP. GR. 3·5–4·1, depending on the composition.

OPT. PROPS. Gives four or six-sided sections, or irregular or rounded grains; colourless, or green (*pleonaste*), red or brown (*picotite*); the refractive index is high, for pure spinel being 1·718, for pleonaste 1·77 and for picotite 2·05; between crossed nicols, isotropic.

TESTS. Infusible; when black resembles magnetite, but is not magnetic; when red, resembles garnet, but is not fusible; when brown, resembles zircon, but is harder; form, hardness and infusibility are characteristic.

VARIETIES. *Ruby-Spinel* or *Magnesia Spinel* is the clear red variety; *Spinel-Ruby*, *Balas-Ruby*, *Rubicelle*, are deep-red, rose-red and yellow varieties, used as gemstones. *Pleonaste* is a dark-green spinel, containing iron; *Picotite* is a brown spinel containing iron and chromium; *Hercynite* is an iron-spinel, approaching $FeAl_2O_4$, and black in colour; *Gahnite* is a zinc spinel, $ZnAl_2O_4$, dark green in colour.

OCCURRENCE. The variety picotite occurs as an original constituent of basic and ultrabasic igneous rocks, and in their metamorphic derivatives; pleonaste occurs in igneous rocks, but is especially characteristic of contact-metamorphosed shales, as in the silica-poor hornfelses, and in crystalline limestones of regional and contact metamorphic origin; spinel proper, magnesia spinel, occurs in crystalline limestones and schists; spinel occurs also in alluvial deposits resulting from the degradation of the parent rocks. The gem varieties come from Ceylon, Burma, Thailand and Afghanistan.

DIASPORE

COMP. Hydrogen aluminium oxide, $HAlO_2$.

CHARACTERS AND OCCURRENCE. Occurs in orthorhombic prismatic crystals, foliaceous and scaly forms, of a white colour, with hardness nearly 7 and specific gravity 3·5; it is found with corundum and emery, and probably results from the alteration of these; it occurs also in bauxite deposits.

BOEHMITE

COMP. Aluminium hydroxide, $AlO(OH)$, like diaspore but has a different structure.

CHARACTERS AND OCCURRENCE. Occurs in tiny flakes or aggregates, white in colour, with specific gravity about 3: yields water in the closed tube test and when heated on charcoal, moistened with cobalt nitrate and strongly reheated, gives a blue unfused residue. Boehmite is an important constituent of most bauxites (see below).

GIBBSITE, Hydrargillite

COMP. Aluminium hydroxide, $Al(OH)_3$.

CRYST. SYST. Monoclinic. COM. FORM. Crystals rare, usually as concretions. COLOUR. White. H. 3. SP. GR. 2·35.

TESTS. Gives water when heated in the closed tube; heated with cobalt nitrate gives a blue residue.

OCCURRENCE. In deposits of bauxite, and as an alteration product of aluminium silicates.

BAUXITE

COMP. A mixture of the aluminium hydroxides, diaspore boehmite and gibbsite, in different amounts, together with impurities of iron oxide, phosphorus compounds and titania, the latter sometimes amounting to 4 per cent.

COM. FORM. Amorphous in earthy granular or pisolitic masses. COLOUR. Dirty white, greyish, brown, yellow, or reddish-brown.

TESTS. Forms and general characters distinctive; gives the aluminium reaction when heated with cobalt nitrate; does not give a silica skeleton in the microcosmic salt bead.

OCCURRENCE. Bauxite results from the decay and weathering of aluminium-bearing rocks, often igneous but not necessarily so, under tropical conditions; it may form residual deposits replacing the original rock, or it may be transported from its place of origin and form deposits elsewhere. It occurs in France, in pockets in Cretaceous limestone and is there the result of pre-Tertiary tropical weathering.

USES. While the principal use of bauxite is for the manufacture of aluminium, considerable quantities are used as abrasives and in the manufacture of aluminium compounds. Lower grades of bauxite are used as refractories, as refractory bricks and for furnace and converter linings. There are, commercially speaking, two kinds of bauxite, red and white. For chemical purposes the white bauxite containing only a trace of iron is used, but for the manufacture of metallic aluminium, iron is not harmful, although the presence of more than 3 per cent of silica or of titania is objectionable. The world production of bauxite is about 50 million tons annually, the chief producers in order of importance in 1965 being Jamaica, Surinam, U.S.S.R., Guyana, France, U.S.A., Yugoslavia, Guinea, Hungary, Greece and Australia, all with over 1 million tons.

WEBSTERITE, Aluminite

COMP. Hydrous aluminium sulphate, $Al_2O_3.SO_3.9H_2O$.

COM. FORM. An earthy material occurring in veins and in reniform or tuberose masses. COLOUR. White and yellowish. LUSTRE. Dull and opaque. FRACT. Earthy; adheres to the tongue, and yields to the fingernail. H. 1–2. SP. GR. 1·66.

TESTS. Heated in closed tube, gives water; heated on charcoal with cobalt nitrate gives blue mass; heated on charcoal with sodium carbonate, mass transferred to a silver coin and moistened, gives a black stain; the physical properties are distinctive.

OCCURRENCE. Usually found in clays of Tertiary age; sometimes in clay-filled pipes or pot-holes in the surface of the Chalk, as at Newhaven in Sussex, and elsewhere.

ALUNOGENE

COMP. Hydrous aluminium sulphate, $Al_2(SO_4)_3.16H_2O$.

CRYST. SYST. Monoclinic(?). COM. FORM. Crystals uncommon; usually occurs in masses composed of delicate and closely packed fibres; also massive and in crusts. COLOUR. White, sometimes yellowish or reddish. LUSTRE. Vitreous and silky; subtransparent to subtranslucent. TASTE. Of common alum; soluble in water. H. 1·5–2· SP. GR. 1·6–1·8.

TESTS. Heated in closed tube, gives water; gives a blue mass when heated with cobalt nitrate on charcoal; soluble in water.

OCCURRENCE. Occurs in the neighbourhood of volcanoes; and in shales, especially alum shales, where pyrites is decomposing.

ALUM, Potash Alum, Kalinite

COMP. Hydrous aluminium potassium sulphate, $KAl(SO_4)_2.12H_2O$.

CHARACTERS AND OCCURRENCE. Crystallizes in the cubic system, generally in small octahedra, and occurs often in shales, *alum-shales*, which contain pyrites and are undergoing decomposition; such shales occur at Whitby in Yorkshire and elsewhere; alum also occurs in the neighbourhood of volcanoes; but in neither mode of occurrence is the mineral sufficiently plentiful to be of much economic value nowadays; alum is readily soluble in water, and has a characteristic taste; its hardness is 2–2·5, and its specific gravity 1·75.

CRYOLITE

COMP. Fluoride of aluminium and sodium, Na_3AlF_6.

CRYST. SYST. Monoclinic. COM. FORM. Crystals not common; usually found massive and cleavable, with a lamellar structure. CLEAV. Parting parallel to the basal pinacoid, less good parallel to the prism and orthodome. COLOUR. Colourless, snow-white, reddish, brownish, brick-red, and even black. LUSTRE. Vitreous; subtransparent to subtranslucent. FRACT. Uneven; brittle. H. 2·5. SP. GR. 2·97.

Tests. Cryolite becomes practically invisible when immersed in water, since its refractive index equals that of water; heated alone before the blowpipe, cryolite fuses easily, colouring the flame intense yellow from sodium; the residue from heating on charcoal, when moistened with cobalt nitrate and strongly reheated, gives a blue mass from aluminium; heated with sulphuric acid, greasy bubbles of hydrofluoric acid are evolved, indicating a fluoride.

Occurrence. Cryolite occurs in a pegmatite vein in granite at Ivigtut in West Greenland, associated with galena, blende, siderite, fluorspar, etc. This deposit was worked, and the cryolite has been proved to a depth of 150 feet.

Uses. Cryolite is used in the manufacture of aluminium, for making sodium and aluminium salts, and in the manufacture of a white porcellanous glass. Its use has largely been replaced by that of synthetic cryolite.

TURQUOISE

Comp. Basic hydrous phosphate of aluminium and copper possibly $CuAl_6(PO_4)_4.(OH)_8.4H_2O$.

Cryst. Syst. Triclinic. Com. Form. Massive, reniform, stalactitic, or encrusting. Fract. Conchoidal; brittle. Colour. Turquoise-blue or bluish-green. Lustre. Rather waxy, internally dull; feebly translucent to opaque. H. 6. Sp. Gr. 2·6–2·8.

Tests. Gives water when heated in the closed tube; gives reactions for copper; soluble in hydrochloric acid after ignition.

Occurrence. Turquoise occurs in thin veins, patchy deposits and seams in rocks, such as trachytes and other igneous rocks, that have been profoundly altered; it is considered that the phosphate is derived from apatite; the gem production comes from Persia, United States, Egypt and U.S.S.R.

Uses. Turquoise is used in jewellery; fossil bones and teeth, coloured by phosphate of iron, *vivianite*, and termed *odontolite* or *bone turquoise*, are frequently cut and polished for the same purpose.

WAVELLITE

COMP. Hydrous phosphate of aluminium, $Al_6(PO_4)_4(OH)_6.$ $9H_2O$; some analyses show the presence of a little fluorine and of iron oxide.

FIG. 124. Wavellite

CRYST. SYST. Orthorhombic. COM. FORM. Crystals rare; usually occurs in small spheres as shown in Fig. 124, having a radiating structure which is well displayed when they are broken across—the spheres usually range from half an inch in diameter to less. COLOUR. White, yellowish or brownish. LUSTRE. Vitreous, inclining to pearly or resinous; translucent. H. 3·5–4. SP. GR. 2·33.

TESTS. Structure quite characteristic; yields water when heated in the closed tube; gives a blue mass when heated with cobalt nitrate on charcoal; also usually gives reactions for fluorine.

OCCURRENCE. Occurs in residual deposits formed from igneous rocks, as at St. Austell in Cornwall; also occurs as nodular masses associated with manganese ores and limonite, as at Holly Springs, Pennsylvania.

GROUP 4A
TITANIUM, ZIRCONIUM, CERLUM, THORIUM

TITANIUM MINERALS

Titanium has not been found in a free state in nature. It is a greyish metal and resembles tin in its chemical properties and, like that metal, is capable of forming two oxides, TiO and TiO_2. Only the latter occurs in nature; it supplies an example of trimorphism, constituting the three distinct minerals, *rutile*, *anatase* and *brookite*. Titanium oxide also enters into the composition of *ilmenite*, the oxide of titanium and iron, $FeTiO_3$, and many samples of magnetite contain varying amounts of titanium, giving the *titaniferous magnetites*. At the present time, ilmenite is the chief source of the titanium required in industry, whilst the employment of this mineral as an iron ore proper is not yet possible—accordingly ilmenite is described here with the other titanium minerals. Titanium occurs in a number of rock-forming silicates, the chief of which is *sphene* or *titanite*, $CaTiSiO_5$, described on p. 363.

The titanium minerals of economic importance are ilmenite and rutile; primary ilmenite deposits are magmatic segregations or veins derived therefrom; primary rutile occurs in a number of ways, segregations in igneous rocks such as syenite, anorthosite and gabbro, and in pegmatitic rocks of various kinds. Most of the exploited deposits of both minerals, however, are of detrital character, such as beach-sands. The world production of titanium concentrates approaches two million tons annually, the chief producers of ilmenite being U.S.A., U.S.S.R., Canada, India, Norway and Finland, and of rutile, predominantly Australia, followed by Sierra Leone.

Titanium is used in the production of certain special alloys, such as with iron or iron and carbon, and certain non-ferrous alloys are becoming important. Such alloys are employed in the aircraft industry. Some titaniferous iron-ores are smelted in the electro-furnace direct for the production of ferro-titanium alloys. Titanium-carbide is an extremely obdurate material, and can be employed for cutting-tools. The oxide is used for imparting an ivory tint to artificial teeth, and as a yellow glaze in pottery manufacture. Titanium white pigments are of considerable importance, and their production absorbs the greater part of the output of titanium ores.

TESTS. Titanium minerals colour the microcosmic bead yellow when hot, colourless when cold, in the oxidizing flame; and yellow hot, violet cold, in the reducing flame. Titanium minerals, when fused with sodium carbonate on charcoal, dissolved in hydrochloric acid, and a few grains of tin added, give a violet solution. The best test for titanium, however, is performed as follows: the mineral is heated with sodium carbonate on charcoal, and the resulting mass dissolved in sulphuric acid to which has been added an equal quantity of water; the solution is cooled, and diluted with water; a drop of hydrogen peroxide is added to the solution, and the formation of an amber colour indicates the presence of titanium.

The titanium minerals dealt with here are:

Oxides
$\begin{cases} \text{Rutile, } TiO_2, \text{ tetragonal.} \\ \text{Anatase, } TiO_2, \text{ tetragonal.} \\ \text{Brookite, } TiO_2, \text{ orthorhombic.} \end{cases}$

Oxide of Titanium and Iron   Ilmenite, $FeTiO_3$.

RUTILE

COMP. Titanium dioxide, TiO_2.

CRYST. SYST. Tetragonal, axial ratio, 0·644. COM. FORM. Crystals often tetragonal prisms terminated by pyramids, as shown in Fig. 56B; ditetragonal forms are sometimes developed; crystals are frequently acicular, and radiately grouped in the interior of other minerals, especially quartz; at times, crystals are twinned on the second order pyramid (101), giving forms bent at a sharp angle, called geniculate or knee-shaped twins, as illustrated in Fig. 87B and described on p. 153—the twinning may be repeated until wheel-shaped multiple twins result; the crystals often show a longitudinal striation. CLEAV. Poor parallel to the prisms (110) and (100). COLOUR. Reddish-brown, red, yellowish or black; in thin section, foxy-red. STREAK. Pale brown. LUSTRE. Metallic, adamantine; opaque or subtransparent. FRACT. Subconchoidal or uneven; brittle. H. 6–6·5. SP. GR. 4·2.

TESTS. Heated alone, infusible; the microcosmic salt bead is, in the oxidizing flame, yellow hot, colourless cold, and, in the reducing flame, yellow hot, violet cold; reacts to the special tests for titanium described above in the introduction to the titanium minerals.

OCCURRENCE. Rutile occurs as an accessory constituent of igneous rocks of many kinds, granites, diorites, etc., and their metamorphic derivatives, such as gneisses and amphibolites; it forms fine needles, *clayslate needles*, in some slates and phyllites, where it results from the decay of titanium-bearing micas; decayed biotite often shows a lattice-like collection of rutile needles, *sagenite*—the lattice-arrangement arising by twinning; it forms acicular needles in quartz and feldspar; the economically important rutile deposits are first, segregations in igneous rocks— syenite, gabbro, anorthosite, etc., as in Virginia, Canada, Norway, second, a peculiar type of dyke, probably of pegmatitic character, composed of rutile, apatite and ilmenite, and known as *nelsonite* and occurring associated with the rutile segregations of Virginia, and third, as an important constituent of beach sands resulting from the denudation of rutile-bearing rocks, as in Australia, Florida and India.

USES. As a source of titanium.

ANATASE, Octahedrite

COMP. Titanium dioxide, TiO_2.

CRYST. SYST. Tetragonal, axial ratio, 2·514. COM. FORM. In crystals of two habits, either in slender acute tetragonal bipyramids, whence the name octahedrite, or else in tabular crystals with basal plane prominent. CLEAV. Perfect basal, and pyramidal.

COLOUR. Brown, indigo-blue or black. STREAK. Colourless. LUSTRE. Adamantine; transparent to opaque. H. 5·5–6. SP. GR. 3·82–3·95.

TESTS. As for rutile; form, and blue colour, when present, are distinctive.

OCCURRENCE. Usually results from the alteration of other titanium-bearing minerals, but also formed in veins of hydro-thermal origin.

BROOKITE

COMP. Titanium dioxide, TiO_2.

CRYST. SYST. Orthorhombic. COM. FORM. Thin crystals, tabular parallel to the front pinacoid. COLOUR. Hair-brown, reddish, iron-black. STREAK. Colourless. FRACT. Metallic; brittle. H. 5·5–6. SP. GR. 4.

TESTS. As for rutile.

OCCURRENCE. Brookite occurs as an alteration product of other titanium-bearing minerals, as in the rotten dolerites of Tremadoc, North Wales, and in other decomposed rocks; it may be of contact-metamorphic origin, as in Arkansas, and it occurs also in mineral druses in the Alps.

ILMENITE, Menaccanite

COMP. Iron titanium oxide, $FeTiO_3$; specimens vary in compo-sition, especially in the ratio of titanium to iron, a variation due to the intergrowth of magnetite or hematite with the ilmenite; a little magnesia is often present.

CRYST. SYST. Hexagonal-trigonal. COM. FORM. Often occurs in thin plates or scales, and massive, and also as sand. COLOUR. Iron-black. STREAK. Black to brownish-black. LUSTRE. Sub-metallic; opaque. FRACT. Conchoidal. H. 5–6. SP. GR. 4·5–5. OPT. PROPS. In thin sections, ilmenite is black and opaque; in many cases when viewed by reflected light, it is seen to be altered along three directions into a white substance, called *leucoxene*, which may be a variety of sphene.

TESTS. Heated before the blowpipe, ilmenite is infusible, or nearly so; gives iron reactions with the fluxes; heated on charcoal with sodium carbonate, dissolved in hydrochloric acid, gives a violet colour on the addition of a small particle of tin; heated on charcoal with sodium carbonate, the resulting mass dissolved in sulphuric acid, and cooled—the solution diluted with an equal bulk of water gives an amber-coloured solution on the addition of a drop of hydrogen-peroxide.

VARIETIES. *Menaccanite* is a variety of ilmenite occurring as a sand at Menaccan, Cornwall; *Iserine*, a variety occurring mostly in the form of loose granules or sand, and frequently in octahedral crystals, which are probably pseudomorphs, is most likely an iron-rich variety of rutile; *Kibdelophane* is a variety rich in titanium.

OCCURRENCE. Ilmenite occurs as an accessory constituent in the more basic igneous rocks, especially in gabbros and norites; associated with such rocks it forms magmatic segregations of great size, as at Taberg and Ekersund in Norway, in Quebec and Ontario, and in the Adirondacks, United States: it also occurs in dyke-like bodies derived from such magmatic segregations, and as a minor constituent of certain copper veins. Many important deposits of ilmenite are of detrital character, chiefly beach-sands; deposits of this type are worked at Travancore in India, in Australia, Senegal, Florida, Tasmania and elsewhere.

ZIRCONIUM MINERALS

Zirconium (Zr) does not occur free in nature, but it enters into the composition of a number of complex silicates most of which are rare. The simple silicate, $ZrSiO_4$, is, however, of widespread occurrence as the mineral *zircon*, and this mineral is the source of the metal and its compounds used in industry. The metal zirconium is allied to titanium in its properties, and may be prepared by reduction of zirconium oxide by magnesium; the oxide is produced by fusing zircon with acid potassium fluoride, treating with hydrochloric acid and precipitating the oxide with ammonia. The metal has a specific gravity of 4·08, melting point of 1300°C, and is produced either as crystals or in powder, the latter burning readily in air.

Zirconium and its compounds are becoming of greater industrial importance. The metal is used in alloys with iron, silicon, tungsten, etc., in nuclear reactors and for removing oxides and nitrides from steel; a small quantity is employed in flashlamps of various kinds. Zirconia, the oxide, is used as a refractory, in abrasives, and in enamels, etc. The source of the metal and its compounds is the silicate, zircon, which itself is employed as a refractory and as a gemstone. Zircon occurs as a constituent of acid igneous rocks, and pegmatitic deposits of the mineral, or concentrations derived from the decay of such deposits, are worked. The chief producer is New South Wales, followed by the United States, Senegal, South Africa and Brazil. The gem varieties come chiefly from Ceylon and Burma.

The rare element *hafnium* (Hf) replaces the zirconium in zircon up to about 1 per cent; hafnium is being increasingly employed in the nuclear energy industry and in space-rockets.

TESTS. The mineral zircon may be recognized by its physical properties. Zirconium minerals, fused with sodium carbonate, and dissolved in hydrochloric acid, give a solution which turns turmeric paper an orange colour—the paper, dried by gently heating, assumes a yellow-red colour.

The only zirconium mineral dealt with here is:
    Silicate   Zircon, $ZrSiO_4$.

ZIRCON

COMP. Zirconium silicate, $ZrSiO_4$.

CRYST. SYST. Tetragonal, Zircon Type (see p. 114). COM. FORM. Crystals usually prismatic, consisting of the tetragonal prism and tetragonal bipyramid, as shown in Fig. 56A, p. 119; also in rounded detrital grains. CLEAV. Parallel to the faces of the prism (110), indistinct, and parallel to the bipyramid (111), still less distinct. COLOUR. Colourless, grey, pale yellow, greenish, reddish-brown. STREAK. Colourless. LUSTRE. Adamantine; transparent to opaque. FRACT. Conchoidal. H. 7·5. SP. GR. 4·7.

OPT. PROPS. In thin sections of rocks, zircon occurs in small colourless prismatic crystals, showing square cross-sections, or elongated longitudinal sections, the former being isotropic, and the latter polarizing in very high colours. The refractive index is very high, $\omega = 1.926$ to 1.936, $\varepsilon = 1.985$ to 1.991, so that the mineral has an intense black border. The mineral is optically positive, and the vibrations parallel to the length of the crystals are slow. Zircon is distinguished from apatite by its higher refractive index and birefringence.

VARIETIES. *Hyacinth* is a gem variety, red in colour and transparent; *Jargoon* is another gem variety, colourless and smoky tinted; *Zirconite* is a grey or brown variety.

TESTS. Heated before the blowpipe alone, zircon is infusible; the coloured varieties, however, become colourless and transparent when heated; when zircon is powdered and fused with sodium carbonate, and this product dissolved in dilute hydrochloric acid, a solution is obtained which turns turmeric paper orange.

OCCURRENCE. Zircon occurs as a primary constituent of igneous rocks, especially the more acid—granite, nepheline-syenite, etc., and corresponding pegmatitic forms; certain decomposed pegmatites which contain large zircon crystals have been worked for the mineral in Madagascar, Brazil, and elsewhere. Zircon is also found in crystalline limestones, gneisses and other metamorphic rocks, and the detrital deposits containing zircons used as gems are derived from such primary deposits, as in Ceylon, Burma, etc. Zircon, being an obdurate mineral, is a common constituent of the heavy residues of various sedimentary rocks, such as sandstones; concentrations of zircon, associated with ilmenite, rutile, and monazite occur as beach-sands in Australia, at Travancore in India, Florida, Brazil, etc., and such deposits have been worked.

CERIUM AND THORIUM MINERALS

Cerium (Ce) is an important member of the group of metals called the Rare Earth Metals, which includes such metals as lanthanum, erbium, yttrium, europium and others. Thorium is closely associated in nature with this group, and the source of thorium salts is the cerium mineral, *monazite*, so that minerals of cerium and thorium are conveniently dealt with together.

Cerium metal is produced by the electrolysis of the fused chloride; it is iron-grey in colour and has a metallic lustre. It has few industrial applications; a small quantity is used in the manufacture of an iron-cerium alloy employed for producing the spark in lighters. Cerium salts, and salts of others of the rare earths, are used in the manufacture of gas-mantles, and for certain processes in industrial chemistry. The demand for rare earths is likely to be increased by the development of colour-television.

There are many minerals containing rare earths, but two only are considered in this book. Cerium enters into the composition of the silicate, *orthite* or *allanite*, a cerium-bearing member of the epidote group of rock-forming silicates described on pp. 372–375. *Monazite* is a phosphate of the cerium metals, but is industrially important as a source of thorium compounds, as it contains a small percentage of thorium oxide or thorium silicate.

Thorium (Th) is a metal related to titanium; its specific gravity is 11. The metal is used in magnesium-thorium and other thorium alloys and in electrical apparatus. The oxide is of considerable commercial importance. It is used chiefly in the manufacture of gas-mantles; it also enters into the construction of Nernst lamps, and it is used in medicine. Thorium is a radioactive metal and a possible source of atomic energy.

As already stated, an important source of thorium is the cerium phosphate, monazite, which invariably contains a certain proportion of thorium oxide. Thorium is also produced from conglomerates containing uranium, thorium and rare earths in Ontario. The production of monazite comes from Australia, Malaysia, India, Ceylon, Nigeria and the United States. The monazite mostly occurs in beach-sands derived from the denudation of monazite-bearing acid igneous rocks. When found in quantity, *thorianite*, an oxide of thorium and uranium, is a valuable source of thoria; it has been produced on a commercial scale from beach-deposits in Madagascar, Ceylon and elsewhere.

TESTS. Satisfactory tests for cerium and thorium are complicated chemical ones, and are beyond the scope of this book. Rare Earth metals can be detected by spectroscopic methods. Under the microscope, the identity of grains of monazite can be established by use of the spectroscopic eye-piece, whereby characteristic absorption bands are observed.

The cerium and thorium minerals considered here are:

Phosphate of the Rare Earths	Monazite (Ce,La,Yt)PO_4, with Thorium either as ThO_2 or $ThSiO_4$.
Silicate	Thorite, $ThSiO_4$.
Oxide	Thorianite, $ThO_2.U_3O_8$.

MONAZITE

COMP. Phosphate of the rare-earth metals, (Ce,La,Yt)PO_4, with thoria, ThO_2, and silica, SiO_2, present, often in the right proportion to form thorium silicate.

CRYST. SYST. Monoclinic. COM. FORM. Complex crystals commonly flattened parallel to the orthopinacoid (100), or elongated along the ortho-axis; crystals sometimes twinned on the orthopinacoid, both contact and penetration twins being known; monazite also occurs massive or as rolled grains. CLEAV. Basal, usually imperfect. COLOUR. Pale-yellow to dark reddish-brown. STREAK. White. LUSTRE. Resinous. FRACT. Conchoidal or uneven. H. 5·5. SP. GR. 5·27

OPT. PROPS. In thin sections of granites, etc., monazite is often seen as small grains, sometimes showing good crystal faces, of a pale honey-yellow colour; the refractive index is high, about 1·8, and the double refraction very high; thick sections of the mineral are pleochroic; optically positive.

TESTS. Infusible before the blowpipe; fused with sodium carbonate, mass dissolved in nitric acid, a few drops of ammonium molybdate added to the solution gives a white precipitate indicating phosphate.

OCCURRENCE. Primary monazite occurs as an accessory component of acid igneous rocks, such as granites, and large crystals and masses have been found in pegmatites. It is found as a heavy residue in sediments, and it is obtained on a commercial scale from sands where natural concentration has gone on, the source of the monazite being a neighbouring monazite-bearing granitic rock. The mineral occurs as a constituent of the sea-shore sands, or monazite sands, near Prado, in the south of the State of Bahia, Brazil, and at various parts of the coasts of the states of Espirito

Santo and Rio de Janeiro; the concentrate from such sands contains 5 to 7 per cent of thoria; less important inland deposits yield monazite with 4 to 5·7 per cent thoria. In recent years, Travancore, Madras, India, has become an important producer of monazite, the mineral occurring with ilmenite, rutile, zircon, etc., as beach-sand; the mineral from this locality is richer in thoria than that of Brazil, the oxide rising to 14 per cent in selected specimens. The mineral is also worked in Ceylon, and has been recorded from Nigeria, Nyasaland, and Malaysia, and these localities may become commercially important. In Nigeria and Malaysia, it forms a by-product in the final dressing of alluvial tin-ore. Monazite sands usually consist of monazite naturally concentrated with other heavy obdurate minerals such as garnet, magnetite, rutile, ilmenite, zircon, etc.; separation is effected by electro-magnetic separators, the magnets of which are adjusted to varying intensities—magnetite and ilmenite are first removed, and monazite, usually the most feebly magnetic, last; rutile, zircon and siliceous matter pass into the reject, and are further treated for the recovery of rutile and zircon. The manufacture of the thoria from the separated monazite is a complicated and purely chemical operation.

THORITE

COMP. Thorium silicate, $ThSiO_4$.

CRYST. SYST. Tetragonal. COM. FORM. Crystals like those of zircon in form. CLEAV. Prismatic. COLOUR. Black; orange-yellow in the variety *orangite*. STREAK. Dark-brown. LUSTRE. Vitreous on fresh surfaces. FRACT. Conchoidal. H. 4·5. SP. GR. 5·3.

TESTS. Usually hydrated, and yields water on heating; soluble in hydrochloric acid with gelatinization on concentration of the solution; specific gravity noticeable.

OCCURRENCE. As large crystals in the syenite-pegmatites of the Langesundfiord district of Norway, and in similar rocks elsewhere.

THORIANITE

COMP. Oxide of thorium and uranium, $ThO_2.U_3O_8$.

CRYST. SYST. Cubic. COM. FORM. Cube. COLOUR. Black. STREAK. Black. H. 6·5. SP. GR. 9·3.

OCCURRENCE. In the alluvial gem-deposits in Ceylon, associated with zircon, orthite, etc.

GROUP 4B
CARBON, SILICON, TIN, LEAD

CARBON MINERALS

Carbon (C) is known in three different conditions—transparent and crystallized as diamond, scaly and crystalline as graphite, and amorphous as lamp-black, charcoal, soot, etc. These different forms, though chemically identical, vary in hardness, specific gravity and other physical properties. Analysis of the crystal structure of diamond and graphite by means of X-rays shows the different atomic arrangement in the two minerals (p. 19 and Fig. 6). In graphite the atoms are arranged in layers that are further apart than in diamond, and this is held to account for the scaly nature of graphite—in each layer, however, the atoms are closely linked so that each flake of graphite breaks up into smaller and smaller flakes, a property on which the value of the mineral in lubrication is based. The diamond structure, on the other hand, has great strength and compactness which are reflected in the physical properties of the mineral.

Native carbon occurs as two important minerals, *diamond* and *graphite*, both being of great industrial importance. Amorphous carbon enters largely into the composition of *coals*, which, though not minerals in the strict sense, are considered here in this book. Again, carbon forms with oxygen and hydrogen many series of compounds known as the hydrocarbons; naturally occurring hydrocarbons constitute the very important *bitumens* and *petroleums*, and these substances, though again not strictly minerals, are also dealt with here; with the hydrocarbons are also considered *amber* and the natural *resins*.

Carbonic acid is a combination of carbon, oxygen and hydrogen, and has the chemical formula H_2CO_3. The salts (see p. 29) of this acid are called *carbonates*, and are very common minerals, existing in the earth's crust in enormous quantities. For example, the common limestones are composed mainly of calcium carbonate, $CaCO_3$; the rock dolomite is a carbonate of calcium and magnesium, $CaMg(CO_3)_2$; the important iron ore, siderite, is iron carbonate, $FeCO_3$, and a host of other economically important carbonates is known. The carbonates are described under the headings of the metallic elements occurring in them.

DIAMOND

COMP. Pure carbon, C.

CRYST. SYST. Cubic. CRYST. STRUCT. See p. 19 and Fig. 6.
COM. FORM. Octahedral crystals, shown by X-ray examination to possess normal cubic symmetry (see p. 104 and Fig. 6A); crystals often have curved faces and are commonly twinned; also in water-worn grains in alluvial deposits. CLEAV. Perfect parallel to the octahedron. COLOUR. White or colourless, sometimes yellow, red, green, or very rarely blue or black; those most free from colour are termed diamonds of the first water, and are the most valuable. LUSTRE. Brilliantly adamantine; transparent; when dark coloured, translucent. FRACT. Conchoidal. H. 10. SP. GR. 3·52.

OPT. PROPS. The refractive index is very high, 2·417 for sodium light, and the dispersion is also very high, and these two properties cause the diamond to sparkle and have 'fire'.

VARIETIES. *Bort* or *Bortz* and *Carbonado* (a black diamond) are compact varieties of diamond occurring in granular or rounded aggregates, and though of no value as gems, are used extensively for abrasive and cutting purposes in gem-cutting, for the cutting-edges of diamond drills, and for dressing emery wheels. Much *Diamond* not of the first water is also employed for these purposes, shaped diamonds being used as turning-tools, etc.

OCCURRENCE. Diamond occurs in two types of deposits, in igneous rocks of ultrabasic composition, or else in alluvial deposits derived from these primary sources. In the Kimberley diamond fields the diamond occurs in an ultrabasic igneous breccia (*blue ground*) which forms pipes in black shale; it is considered that the diamond is from a deep-seated crustal layer though it has also been suggested that it is formed by the action of the magma on enclosed pieces of carbonaceous shale. Diamonds have been found in igneous rocks of related types elsewhere in South Africa, U.S.S.R., Brazil and Arkansas, United States; they have been reported in dolerite in New South Wales. Diamond occurs in alluvial deposits associated with other minerals of high specific gravity and extreme obduracy, as in South and South-West Africa, in Brazil, India, etc.

USES AND PRODUCTION. The uses of diamond and its varieties for gems and abrasives have been noted above. The main producers of gem diamonds are South and South-West Africa, U.S.S.R., Ghana, Congo, Sierra Leone, Angola, Liberia, Tanzania and Brazil. Industrial diamond is supplied by the Congo, Ghana, Sierra Leone, U.S.S.R., South Africa, Angola and Tanzania.

GRAPHITE, Plumbago, Black Lead

COMP. Pure carbon, C; sometimes contaminated with a small amount of silica, iron-oxides, clay, etc.

CRYST. SYST. Hexagonal. COM. FORM. Crystals uncommon; usually occurs in scales, laminae, or columnar masses; sometimes granular and rarely earthy. CLEAV. A perfect basal cleavage parallel to the large surface of the scales. TENACITY. Thin laminae are flexible, and the mineral is sectile. COLOUR. Iron-grey to dark steel-grey. STREAK. Black and shining. LUSTRE. Metallic. FEEL. Feels cold like metal when handled, owing to its being a good conductor of heat. H. 1–2. SP. GR. 2–2·3, depending upon the purity of the material.

SPEC. PROP. Graphite resembles molybdenite in most of its physical properties, but is distinguished by its jet-black streak, whereas the streak of molybdenite is greenish-black.

OCCURRENCE. Graphite has been considered to occur as a primary constituent of igneous rocks, but it is probable that the mineral in these cases has been derived from the adjacent country-rock. The main occurrences are of three types: (1) Veins of a true fissure character as in Ceylon, Irkutsk, and at Borrowdale in the Lake District, (2) bedded masses of a lenticular and patchy nature in gneiss, crystalline limestone, etc., as in Eastern Canada, and (3) disseminations through the country-rock, often near veins or associated with the contact of an igneous rock, as in Eastern America, Germany, etc.; graphite deposits occur in metamorphosed rocks of either regional or contact metamorphic origin, and it is probable that many of the deposits result from the metamorphism of carbonaceous material of sedimentary origin. PRODUCTION. Over 600,000 tons of graphite are produced annually, the chief sources being Korea, Austria, U.S.S.R., China, Mexico, West Germany and Ceylon.

USES. The chief uses of graphite are for facings in foundry-moulds, for paint, and for crucibles; additional important uses are as a lubricant, in commutators, as a stove-polish, for lead-pencils, in electro-plating to provide a conducting surface on non-metallic substances which are to be plated, and as electrodes for the electric furnace. Different grades of the material are suitable for different purposes; low-grade graphite, with perhaps no more than 40 per cent of the element, is useful for paint manufacture; for crucibles the flaked graphite is almost essential, and the Ceylon graphite of this quality is eminently suitable; for electrodes purity is of great importance, and artificial graphite prepared in the electric furnace is preferred for electrodes and many other purposes; often the physical properties and the freedom from grit are more important than the amount of carbon actually present.

Hydrocarbons: Coal and Bitumen

In this group is considered a variety of substances which differ very widely among themselves, in mode of occurrence, physical properties and chemical composition, but which all agree in consisting, in the main, of carbon with hydrogen and oxygen. Such substances are here considered in two groups, the *Coals* and the *Bitumens*.

COALS

The name coal is applied to a number of different substances, largely made up of carbon, oxygen and hydrogen, which have this point in common, that they are more or less altered remains of old land vegetation, forest growth, peaty or swampy material, etc., transformed by slow chemical changes (principally the elimination of hydrogen and oxygen from the original woody tissue) into a material richer in carbon. Two theories have been put forward to explain the origin of coal. The first of these, the *growth-in-place theory*, supposes coal to be the result of the geological entombment of vegetable matter decaying in the place in which it grew—this explanation is held to fit the origin of pure well-bedded extensive coals, such as common coal. But a second theory, the *drift theory*, has been advanced to explain the formation of impure current-bedded local coal, such as some forms of cannel, which is thought to be the result of the burying of drifted vegetable matter in a delta. According to the vicissitudes which it has undergone, coal has varying amounts of carbon, hydrogen, oxygen and nitrogen—the less-altered varieties, such as lignite, containing large amounts of the gases, and the highly altered varieties, such as anthracite, containing as much as 95 per cent of carbon. Hence, several varieties of coal can be distinguished. Typical analyses of the main varieties are given in the Table and the varieties are described below. Whatever may be the origin of any particular coal, the processes of organic decay have largely influenced its final character.

COMPOSITION OF COALS, ETC.[1]

	Proportions as percentages				Proportions recalculated with carbon as 100			
	C	H	N	O	C	H	N	O
Wood	49·65	6·23	0·92	43·20	100	12·5	1·8	87·0
Peat	55·44	6·28	1·72	35·56	100	11·3	3·1	64·1
Lignite	72·95	5·24	1·31	20·50	100	7·2	1·8	28·1
Bituminous Coal	84·24	5·55	1·52	8·69	100	6·6	1·8	10·3
Anthracite	93·50	2·81	0·97	2·72	100	3·0	1·3	2·9

[1]F. W. Clarke, 'Data of Geochemistry,' Bull. 695, *U.S. Geol. Surv.* 1920. p. 763.

Coal occurs in true beds of various geological ages. The Carboniferous System is by far the most important of the coal-bearing groups, and the coal of Great Britian, Pennsylvania, Germany, India, etc., is of this age. In Britain, the Carboniferous rocks have been gently folded, and, though the coal-bearing division has in most cases been removed from the crests of these folds by denudation, it has been preserved in the troughs or coal-basins. Examples of such basins are afforded by the Lancashire coalfield and the Yorkshire coalfield, which occupy the downfolded troughs on either side of the crest of the Pennines.

There are, however, extensive deposits of coal at geological horizons other than the Carboniferous, though these deposits are usually less valuable both in quality and thickness. Cretaceous coals occur extensively in the United States and Europe, and more recent deposits are also worked.

Coals are classified in various ways. That adopted in the *Coal Resources of the World*, 1913, depended on the following factors:
1. *Fuel Ratio*. The ratio of fixed carbon to volatile matter.
2. *Calorific value*. The amount of heat produced by the complete burning of a standard weight of coal. It is measured by the British Thermal Unit, which is the amount of heat required to raise a similar standard weight of water 1°F in temperature.
3. *Carbon content, content of volatiles, amount of moisture*.
4. *Nature of the coke*, produced by the coal.

On this basis, the following types of coal can be established:[1]

Lignite. Moisture over 20 per cent; calorific value, 7,000–11,000 B.T.U.; Carbon, 45–65 per cent.

Lignitic or Semi-bituminous. Moisture over 6 per cent, and up to 20 per cent; calorific value, 10,000–13,000 B.T.U.; carbon, 60–75 per cent.

Cannel. Calorific value, 12,000–16,000 B.T.U.; coke very porous; yields 30–40 per cent volatile matter on distillation.

Low-carbon Bituminous. Moisture occasionally reaches 6 per cent; volatile matter, up to 35 per cent; calorific value, 12,000–14,000 B.T.U.; carbon, 70–80 per cent; coke porous and friable.

Bituminous. Fuel ratio, 1·2 to 7; calorific value, 14,000–16,000 B.T.U.; carbon, 75–90 per cent; volatiles, 12–26 per cent; generally cokes.

[1] See also *Coal*, 'Monographs on Mineral Resources with Special Reference to the British Empire.' *Imperial Institute,* 1920, pp. 3–4.

Anthracitic and High-carbon Bituminous. Fuel ratio, 4–7; calorific value, 15,200–16,000 B.T.U.; carbon, 80–90 per cent; volatiles, 12–15 per cent; does not readily coke.

Semi-anthracite. Fuel ratio, 7–12; calorific value, 15,000–15,500 B.T.U.; carbon, 90–93 per cent; volatiles, 7–12 per cent.

Anthracite. Fuel ratio more than 12; calorific value, 14,500 to 15,000 B.T.U.; carbon, 93–95 per cent; volatile combustible matter, 3–5 per cent.

VARIETIES OF COAL

Peat. Peat results from the accumulation of vegetable matter, chiefly mosses and other bog plants, and forms extensive beds in Ireland, Russia, Canada, and elsewhere. The organic nature of the deposit is evident throughout the entire mass, although the bottom layers may become compressed into a compact homogeneous substance, a change leading to an increase in carbon content.

Lignite or Brown Coal. A further stage in the alteration of a vegetable deposit is marked by lignite, which, though compact and having a brilliant lustre, shows nevertheless distinct traces of its origin by containing impressions and remains of vegetable fragments, leaves, etc. Lignites contain a good deal of moisture, and often go into powder on drying. Lignite occurs in beds at many horizons in the more recent geological formations, as in Germany, Hungary, Western Canada, and the Mississippi Valley. The name *brown coal* is often restricted to a coal of which the vegetable origin is not so evident as that of lignite. *Jet* is a resinous, hard, coal-black variety of lignite, capable of taking a high polish, and hence suitable for ornaments. It is found at Whitby in Yorkshire, and elsewhere.

Cannel Coal. Cannel is a variety of coal which ignites in the candle flame and burns with a smoky flame. It is one of the bituminous coals, but differs from the usual types of such coals in its texture, lustre, fracture, colour and composition. It is dense, has no lustre, its fracture is conchoidal, its colour dull-grey or black, and it contains a large amount of gas. Microscopic examination shows that cannel is typically composed of spore and pollen remains with an abundance of those of oil-bearing algae. On distillation, cannel produces a large amount of volatile material, and is of value for the production of oils of various types. *Torbanite* or *Boghead Coal* is a variety of cannel considered to arise by the deposition of vegetable matter in lakes. It is found at Torban, Boghead, and at other localities in Linlithgowshire, Scotland, and forms also lenticular deposits in New South Wales. Its exploitation in Scotland gave rise to a celebrated lawsuit which involved an accurate definition of the term 'coal'.

Bituminous Coals. Bituminous coals vary considerably in character, but they all burn with a smoky flame, and during combustion soften and swell up in a manner resembling the fusion of pitch or bitumen. This, however, is only a first step in their destructive distillation, and there is no actual bitumen present. These coals have a bright pitchy lustre, and their specific gravity varies from 1·14 to 1·40. Varieties of bituminous coals are distinguished by their manner of burning, for example, caking coal and non-caking coal. Bituminous coal, exemplified by most ordinary house-coal, is usually banded parallel with its bedding, the bedding-planes being marked by soft powdery charcoal-like material which is called *fusain*. Other bands parallel to the fusain layers show different types of coal, notably *durain*, which is hard and dull, *clarain* which is brighter in lustre and *vitrain*, bright glassy-looking streaks. A piece of coal should be examined and these different layers identified.

Anthracite. Anthracite is a black or brownish-black, sometimes iridescent variety of coal. Its streak is black, and it does not soil the fingers. Its lustre is usually brilliant, and it breaks with a conchoidal or uneven fracture. The hardness varies from 0·5 to 2·5, and its specific gravity from 1 to 1·8. Anthracite contains up to 95 per cent of carbon, the constituents hydrogen, oxygen and nitrogen, being present in very small quantity. It is less easily kindled than other coals, and burns with but little flame, and during combustion gives out much heat. Passages from anthracite into ordinary coal have been observed, and this variety usually occurs where coal-bearing strata have been subjected to considerable pressures or raised temperatures; but there are exceptions to this rule, and the anthracite may be due in some cases to alteration of the original vegetable material before entombment. Anthracite occurs locally in many coalfields, as in South Wales, Scotland, and Pennsylvania. It is used where a smokeless fire is required.

BITUMENS

Bitumens are essentially hydrocarbons of the paraffin series, C_nH_{2n+2}, and of the naphthene series, C_nH_{2n}. Different bitumens have different proportions of these two series, and also have subordinate amounts of allied series. The bitumens, in the widest sense of the term, include members ranging from a very liquid, light-yellow, oil, of specific gravity 0·771, through gradations to solid bitumens, such as asphalt, and to waxy substances such as ozokerite.

Crude Petroleum, Naphtha, Mineral Oil. Under the name of petroleum are included liquids of a brown or blackish colour, often with a greenish tinge, generally somewhat lighter than water, and usually possessing a powerful and disagreeable odour. By fractional distillation various oils, known as petroleum ether, petroleum spirit, benzine, etc., are separated, and are extensively used in internal combustion engines. The intermediate fractions of distillation are used for illumination, and the heavy products provide lubricating and fuel oils. Petroleum is the basis of the petrochemical industry.

Petroleum is usually found in sandstones and dolomites which are bent into gentle folds or domes, and the most important oil-bearing strata are of a geologically recent date. The chief oil-fields, arranged according to the age of the rocks containing the oil, are: Tertiary—The Persian Gulf area, California, Russia, Gulf States of United States, Rumania, Venezuela, Burma; Cretaceous—Texas, Wyoming, Galicia; Carboniferous and Devonian—Texas, Oklahoma, Kansas, Pennsylvania, Illinois, Canada; Ordovician—Indiana. The presence of petroleum may be shown at the surface by seepage of oil, or by pitch or bitumen deposits, the latter due to the evaporation and oxidation of the volatile hydrocarbons.

Two types of views are held concerning the origin of petroleum. By some it is considered to be of inorganic origin, resulting either from volcanic action, or from the decomposition of carbides in the interior of the earth on the accession of waters coming from the surface. The more generally accepted view regards the petroleum as of organic origin, derived from the decay of accumulations of organic material, possibly dominantly of vegetable origin.

Asphalt, Asphaltum, Mineral Pitch. Asphalt is a mixture of different hydrocarbons, usually occurring in the form of a black or brown pitchy substance, soft, but occasionally solid and then maybe with a conchoidal fracture. By the action of suitable solvents, such as carbon bisulphide, asphalt may be dissolved on the one hand into various hydrocarbons, and on the other into a non-bituminous organic matter approximating to coal, with this latter being mixed any inorganic substances present in the crude material. Asphalt occurs in quantity in the celebrated pitch-lake of Trinidad, and in Venezuela, Cuba and Alberta; in these localities the asphalt deposits result from the oxidation of petroleum. In England asphalt has been found at Castleton in Derbyshire, Pitchford near Shrewsbury, and at Stanton Harold, Leicestershire, where it encrusts crystals of galena and copper pyrites; these occurrences are of no commercial value whatever.

Porous sedimentary rocks, such as sandstones or dolomites, may become impregnated with up to 15 per cent asphalt, and are then worked for natural paving material, or for the extraction of asphalt. Such asphalt rocks occur at Neufchatel, France, and in Kentucky and Oklahoma, and near all asphalt deposits.

The various grades of asphalt are mixed with different quantities of rock-chips to form paving and road materials, and the purer varieties are suitable for waterproofing masonry, as insulating material, etc.

Elaterite, Elastic Bitumen, Mineral Caoutchouc. Elaterite is a soft and elastic brownish solid bitumen, much like india-rubber in its physical properties. It has been reported from Castleton in Derbyshire, Neufchatel, and elsewhere.

Albertite, Gilsonite, Grahamite, Uintaite, Wurtzillite. All these substances are varieties of solid bitumen, differing slightly in their chemical and physical properties. Such materials usually occur filling fissures, as in New Brunswick, and have been derived from petroleum-bearing rocks.

Ozokerite. Ozokerite resembles beeswax in appearance, and is a dark yellow or brownish substance, often with a greenish opalescence. It is found associated with petroleum in Utah, Moldavia, and in Galicia, where it is mined. In Galicia, ozokerite occurs squeezed up into fractures, so as to form vein-like bodies; when these 'veins' have been worked out, they are slowly filled by the rising of fresh material from below. Ozokerite is purified to form ceresine, which is used for making candles.

Hatchettine. Hatchettine is a colourless or yellowish, soft waxy substance, resembling ozokerite. It has been found in the cracks of ironstone nodules at Merthyr Tydvil, South Wales.

Amber. Amber is a fossil resin much used for the mouthpieces of pipes, and for beads and ornaments—this latter use being of great antiquity. Amber varies in colour from deep orange-yellow to very pale yellowish tints, and is sometimes even white. It is frequently clouded, and often contains fossil insects, etc., which were enclosed in the gum when it was exuded from the coniferous trees of the period during which the deposits containing it were laid down. It has a hardness from 2 to 2·5, and breaks with a conchoidal fracture. Its specific gravity is about 1·1, and its refractive index 1·54. When heated amber leaves a black residue used in the manufacture of the finest black varnishes. Amber occurs as irregular nodular fragments in strata of recent geological age, deposited under estuarine conditions, and is extensively worked on the southern coast of the Baltic.

Copalite, Highgate Resin. Copalite is a pale yellow or brownish waxy substance, found in small films or fragments in the London Clay, at Highgate Hill, London. It burns easily with a smoky flame, and leaves little ash.

Gum Copal. Gum copal is resin found buried in modern sands, as in New Zealand, and is an inferior kind of amber.

SILICON MINERALS

Silicon (Si) does not occur in a free state in nature, but its compounds are extraordinarily abundant. It constitutes about 28 per cent of the earth's crust. The oxide, quartz, and the great group of the silicates are the most important rock-forming minerals.

Silicon is used as a semi-conductor, a deoxidizer in steel-making, as ferrosilicon alloy and in aluminium-casting and other metal-lurgical processes. Its organic compounds, especially the silicones, are exceedingly important.

Silica, SiO_2, is the only oxide of silicon. It occurs in the form of quartz, chalcedony, agate, flint, etc. Sand is usually made up mostly of small grains of quartz, more rarely of flint, and con-solidated sands provide the important sedimentary rock, sand-stone. Opal is a hydrated form of silica.

The large number of minerals known as the *silicates* were formerly considered as salts of various theoretical silicic acids, but are now classified on the basis of the structural arrangements of their constituent atoms. These structures have been elucidated by the methods of X-ray analysis, discussed on p. 74 and are now described.

Silicate Structures. The fundamental unit in the building of silicate minerals is the SiO_4-tetrahedron in which the silicon atom (or, more strictly, cation) is situated at the centre of a tetrahedron whose corners are occupied by four oxygen atoms (Fig. 125a). The average distance between the centres of two adjacent oxygens is $2 \cdot 7$Å, i.e., almost twice the radius of the oxygen. Classification of the silicates is based on the different ways in which the SiO_4-tetrahedra occur, either separately or linked together. They are as follow:

(a) *Separate SiO_4-groups, nesosilicates.* The separate tetra-hedra are stacked together in a regular manner throughout a crystal structure, and are linked together only through the medium of other cations which lie between them. An example of a mineral built up in this way is olivine, Mg_2SiO_4. Considering the valencies of the elements composing the SiO_4-tetrahedron, we see that silicon has four positive, and each oxygen two negative valencies, *i.e.* there are eight negative valencies in all; the group as a whole therefore has four negative valencies in excess. In the olivine structure, cations (mainly Mg) lie between the tetrahedral groups and contribute the necessary positive charges to make the structure electrically neutral. There are, in effect, two Mg ions to every SiO_4-group, as in the formula Mg_2SiO_4. Some of the Mg is usually replaced by Fe in olivine, thus giving the mineral its

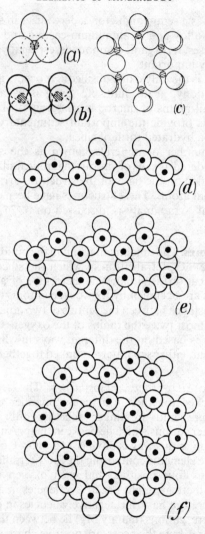

FIG. 125. Silicate Structures. Open circles represent oxygen atoms; silicon is shown by dots or shaded circles.

typical green colour; the proportion of Fe to Mg is variable, and hence the formula of olivine is usually written $(Mg,Fe)_2SiO_4$. The name *nesosilicate* is derived from the Greek for island since the SiO_4-groups remain separate units in the structure.

(b) *Structures with Si_2O_7-groups, sorosilicates*. In some minerals the SiO_4-tetrahedra occur in pairs, in which one oxygen is shared between the two silicons and is inert (Fig. 125b). The composition of each pair of tetrahedra is then Si_2O_7, and these units are spaced throughout the crystal and bound together by other cations. The charge on any Si_2O_7-group is $-6 (= 2 \times 4-7 \times 2)$, so that three divalent ions are needed to balance it. The mineral melilite, $Ca_2MgSi_2O_7$, common in slags, has a structure of this kind. The mineral vesuvianite (idocrase) has both independent SiO_4-groups and Si_2O_7-groups (see p. 375); such a combination of two types of structure is, on general principles, unlikely and the mineral is in fact rare. The name *sorosilicate* is derived from the Greek for *group*.

(c) *Ring structures, cyclosilicates* (*cyclos*, ring). When each SiO_4-tetrahedron shares two of its oxygens with neighbouring tetrahedra they may be linked into rings, as shown in Fig. 125c; a ring of three tetrahedra has the composition Si_3O_9 and one of six tetrahedra Si_6O_{18} (both formulae being multiples of SiO_3). An example of the former is the mineral benitoite, $BaTi(Si_3O_9)$; and of the latter, beryl, $Be_3Al_2(Si_6O_{18})$. In beryl the Si_6O_{18}-rings are stacked one above another in columns, which are linked laterally by the Be-ions (in 4-coordination) and the Al-ions (in 6-coordination). Running through the stacks of rings there are thus empty 'tunnels', parallel to the *c*-axis of the crystal; gases have been passed through these spaces in beryl. It is thought that the helium which is often found associated with the mineral may be occluded on the surfaces of the 'tunnels'.

(d) *Single chain structures* (Si_2O_6), *inosilicates* (*inos*, fibre). A linkage of tetrahedra similar to that in the rings described above, but forming straight chains, is found in the important group of rock-forming minerals known as the pyroxenes. The chains consist of a large number of linked SiO_4-groups, each sharing two oxygens (Fig. 125d), and have the composition $n(Si_2O_6)$. The apexes of the tetrahedra forming a chain all point in the same direction. The simplest example is the pyroxene diopside, CaMg (Si_2O_6); here the excess negative charge on the Si_2O_6-chain is balanced by the valencies of the Ca and Mg cations (verify by counting the valencies). The chains run parallel to the c-axis of the mineral and are bonded together by the calcium and magnesium ions which lie between them. Each magnesium is coordinated by 6, and each calcium by 8 oxygens.

In this type of structure, silicon may be replaced by aluminium to a limited extent. Aluminium may also occur among the cations lying between the chains. Thus aluminous pyroxenes are formed, for example augite. The more complex composition of these minerals is discussed in the descriptions of particular species in a later section.

(e) *Double chain structures* (Si_4O_{11}), *inosilicates*. When two single chains are placed side by side, with the apexes of tetrahedra all pointing one way, and are linked by sharing oxygen atoms at regular intervals, a *double chain* results, whose composition is Si_4O_{11} (Fig. 125e; verify by counting the atoms). This type of structure is found in the amphiboles, a large group of rock-forming minerals of which tremolite, $Ca_2Mg_5(Si_4O_{11})_2(OH)_2$, may be taken as an example. All the amphiboles contain hydroxyl (OH), as an essential constituent, to the extent of about one (OH)-radicle to eleven oxygens. Some OH may be replaced by F. The (OH)-groups fit into the spaces in the structure shown in the figure. The double chains run parallel to the length of the amphibole crystals and are held together laterally by the bond strength of the cations which lie between them, in this case Ca being in 8-fold and Mg in 6-fold coordination.

When aluminium replaces some of the silicon in the chains, aluminous amphiboles such as hornblende are formed. Thus, Si_4O_{11} may become $(AlSi_3)O_{11}$, in which case the chain acquires an extra negative charge. This is balanced e.g. by substitution of Al^{3+} for Mg^{2+} among the other cations, or by the addition of univalent alkali metal ions such as Na^{1+}. In the latter case the additional sodium ions are accommodated in spaces which are available between the double chains.

In both the pyroxenes and the amphiboles extensive substitution of cations by others of similar size and charge takes place, e.g. Fe or Mn for Mg; Fe, Mg, or 2Na for Ca, and so on, giving rise to a great variety of compositions which can, however, be accounted for on the basis of the atomic structure.

(f) *Sheet structure* (Si_4O_{10}), *phyllosilicates* (*phylon*, leaf). A sheet structure is formed when the SiO_4-tetrahedra are linked by three of their corners, and extend indefinitely in a two-dimensional network or 'sheet' (Fig. 125f), which has a silicon:oxygen ratio of 4:10. This is found in the micas, chlorites, and other flaky minerals. For example, the mineral pyrophyllite (p. 415) has the composition $Al_2Si_4O_{10}(OH)_2$.

In the micas, silicon is replaced by aluminium to the extent of about one atom in four, and this change is balanced by the addition of positive ions (K); thus muscovite mica has the composition $KAl_2(AlSi_3)O_{10}(OH)_2$. The Si_4O_{10}-sheets are arranged in pairs, with the apexes of their linked tetrahedra pointing inwards in each pair (Fig. 126). In muscovite the two sheets of a pair are held together by aluminium ions which lie between them; in other micas, such as biotite, Mg or Fe ions occupy these positions. Hydroxyl is accommodated in the structure as shown in the figure. Between one pair of sheets and the next pair lie the potassium ions, which are in 12-coordination. This is a much weaker bond than the rest, and the perfect cleavage for which mica is noted takes place

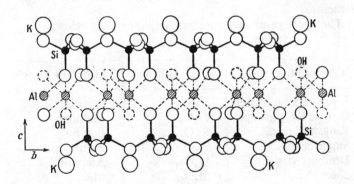

FIG. 126.—The structure of Muscovite, viewed parallel to the Si_4O_{10}-sheets. Each pair of sheets as shown is linked by Al-ions and is separated from the next pair by a layer of potassium ions.

along the layers of potassium ions, parallel to the sheet structure.

In the mineral talc, $Mg_3(Si_4O_{10})(OH)_2$, there is no replacement of Si by Al, and hence no possibility of attaching K-ions between the pairs of sheets. The bonding here is due not to ionic but to residual or stray electrical forces—a weak linkage which is reflected in the characteristic softness of talc.

(g) *Framework structures, SiO_2, tectosilicates* (*tecto*, framework). A three-dimensional framework is formed when each tetrahedron is linked by all four corners, so that every oxygen ion is shared between two tetrahedra. This gives a silicon:oxygen ratio of 1:2, and the charges on such a framework are balanced. Minerals which have this structure include quartz (SiO_2) and other forms of silica, and also the important group of the feldspars, in which aluminium replaces some of the silicon. The chief feldspars are: (i) *Orthoclase*, $KAlSi_3O_8$; here one-quarter of the silicon is substituted by aluminium, and for every Si-ion thus replaced one K-ion is introduced, and is accommodated in large interstices in the framework. By reckoning up the ionic charges, $+$ and $-$, we see that the constitution is neutral. (ii) *Albite*, $NaAlSi_3O_8$, with Na instead of K. (iii) *Anorthite*, $Ca(Al_2Si_2)O_8$; here half the silicon is replaced by aluminium, and a divalent atom (Ca) is needed to restore electrical neutrality.

The feldspathoids and zeolites also possess a framework type of structure.

The above results are summarized in the table below.

Type of Structure	Composition	Mineral Example
Separate SiO_4-groups	SiO_4	Olivine
Double SiO_4-groups	Si_2O_7	Melilite
Ring structures	Si_6O_{18}	Beryl
Single chain	Si_2O_6	Pyroxenes
Double chain	Si_4O_{11}	Amphiboles
Sheet	Si_4O_{10}	Micas
Framework	$\begin{cases} (Al,Si)_nO_{2n} \\ SiO_2 \end{cases}$	Feldspars Quartz

From the examples of ionic substitution (see p. 26) given in the above account of silicate structures, it is evident that such substitutions depend on ionic size rather than on valency. The Table below shows the commoner ions involved in the structure of the silicates arranged in order of increasing ionic radii. From this the possibilities of substitution can be obtained and the Table may be referred to with advantage when the compositions of the silicate minerals are dealt with in later pages.

THE COMMONER IONS IN SILICATE STRUCTURE

(Ionic radii in Å)

Si^{4+}	0·42	Mg^{2+}	0·66				
Al^{3+}	0·51	Ti^{4+}	0·68	Ca^{2+}	0·99	F^{1-}	1·36
Mn^{4+}	0·60	Fe^{2+}	0·74	K^{1+}	1·33	O^{2-}	1·40
Cr^{3+}	0·63	Mn^{2+}	0·80	Ba^{2+}	1·34	OH^{1-}	1·40
Fe^{3+}	0·64	Na^{1+}	0·97				

Classification of the Silicates. The silicates are dealt with here in the order of the structural classes just established. Some silicate minerals have structures belonging to more than one class and are grouped according to their main structural type.

The classification and order of description are as follow:

A. NESOSILICATES

Olivine family: olivine, forsterite, fayalite.
Garnet family.
Sphene.
Topaz.
Aluminium silicates: andalusite, sillimanite, kyanite.
Staurolite.
Chloritoid.

(Zircon, $ZrSiO_4$, is described on p. 332 as an ore of Zirconium.)

B. SOROSILICATES

Epidote family: zoisite, epidote, orthite (allanite).
Idocrase (vesuvianite).
Melilite.

C. CYCLOSILICATES

Tourmaline.
Cordierite.
Axinite.
(Beryl, $Be_3Al_2(Si_6O_{18})$, is described on p. 292 as an ore of Beryllium.)

D. INOSILICATES

Pyroxene family: enstatite, hypersthene, diopside, augite, etc.
Pyroxenoid family: wollastonite, pectolite.
(Spodumene, $LiAlSi_2O_6$, is described on p. 227 as an ore of Lithium (see also under Pyroxenes, p. 380).)
Amphibole family: hornblende, tremolite, actinolite, etc.

E. PHYLLOSILICATES

Mica family: muscovite, biotite, glauconite, etc.
Brittle Micas: margarite.
Chlorite family.
Hydrous magnesium silicates: talc, serpentine, meerschaum, vermiculite.
Hydrous aluminium silicates: the clay-minerals, kaolinite, montmorillonite, etc.
Apophyllite.
Prehnite.

F. TECTOSILICATES

Feldspar family: orthoclase, microcline, plagioclase, etc.
Feldspathoid family: nepheline, leucite, sodalite, etc.
Quartz and other forms of silica (chalcedony, etc.).
Scapolite family.
Zeolite family.

The silicates of certain elements, such as those of the heavy metals, copper, nickel, zinc, iron, etc., are described under their respective metallic components since they are ores of these metals; further, as already mentioned, certain other silicates of rock-forming type have also been described under metallic elements, and for the same reason.

Tests for Silicates. Some silicates gelatinize when boiled with hydrochloric acid; the mineral should first be roasted with sodium carbonate on charcoal. In some cases the silica separates in the form of indistinct flakes. Silica is insoluble in the microcosmic salt bead, and after the fusion of silicates in that bead the silica remains behind as a whitish framework or ghost. Individual silicates are recognized by physical properties such as colour, hardness, specific gravity, crystal form and the like and especially by their optical properties in thin slices under the microscope.

Uses. Many of the silicon minerals are of considerable economic value; rock-forming silicates are exploited mainly on account of certain physical properties that they possess. Examples of such minerals and of their uses are:

(a) Quartz is used for oscillator plates, optical work, etc.; quartz sands are used for glass-making and for various building purposes.

(b) Opal, zircon, sphene, moonstone, topaz, tourmaline, peridot, amethyst, garnet, etc., are used as gemstones.

(c) Tripoli and garnet are used as abrasives.

(d) Feldspars are used in the manufacture of porcelain.

(e) Kaolin is employed in pottery making; clays, used for brick-making, etc., are largely hydrated silicates of aluminium.

(f) Slates and sandstones, into the composition of which quartz and silicates largely enter, are used as building materials.

(g) Mica is of great importance in the electrical industry and for insulating purposes.

(h) Talc is important as a filler for paper, rubber, etc.

(i) Asbestiform minerals are used for the manufacture of non-combustible materials.

(j) The aluminium silicates are employed in the manufacture of sparking plugs and as refractories.

A. NESOSILICATES

The Olivine Family

The olivine family consists of an isomorphous series of neso-silicates with the general formula $R^{2+}_2[SiO_4]$, in which R = magnesium or iron. The structure is described on p. 349. The magnesium end-member is *forsterite*, magnesium silicate, Mg_2SiO_4, the iron end-member is *fayalite*, Fe_2SiO_4; intermediate between these two is *olivine*, magnesium iron silicate, $(Mg,Fe)_2SiO_4$.

OLIVINE, Peridot, Chrysolite

Comp. Magnesium iron silicate $(Mg,Fe)_2SiO_4$, with Mg in excess of Fe in most varieties.

Cryst. Syst. Orthorhombic. Com. Form. Prismatic crystals usually modified by domes and pyramids; also as grains massive or compact. Colour. Shades of green, pale green, olive-green, greyish-green, brownish, rarely yellow; white or yellow in forsterite; brown or black in fayalite. Streak. Colourless. Lustre. Vitreous; transparent to translucent. Fract. Conchoidal. H. 6–7. Sp. Gr. Forsterite 3·2; fayalite 4·3; olivine intermediate.

Opt. Props. In rock-slices, olivine occurs as somewhat rounded elongated sections, traversed by cracks along which the mineral is usually altered into greenish serpentine—the alteration being marked by a network of iron-oxide, as shown in Fig. 127; cleavage is not usually seen; colourless or faint greenish; refractive indices high, increasing with the iron content—forsterite α = 1·640, β = 1·661, γ = 1·680; olivine (optically positive) β = 1·681, olivine (optically negative) β = 1·706; fayalite β = 1·864; birefringence strong, polarization colours being bright colours of Second Order; biaxial, optically positive in varieties with less than about 11 per cent FeO, optically negative in those with more than 11 per cent FeO.

Tests. Most varieties are infusible before the blowpipe; recognized by colour and physical characteristics; decomposed by hydrochloric acid with gelatinization.

FIG. 127.—Olivine altering into Serpentine.

VARIETIES. *Forsterite* is the magnesium olivine, Mg_2SiO_4, in whitish or light green or yellow grains in crystalline limestones; *Fayalite* is the iron olivine, Fe_2SiO_4, brown to black in colour; it is easily fusible before the blowpipe; it occurs in slags, and in cavities, etc., in rhyolites as in the Yellowstone Park, and in granitic pegmatites as in the Mourne Mountains; *Peridot* is a gem variety of olivine, transparent and pale green, found in Egypt, Burma and Brazil.

OCCURRENCE. Olivine is the essential mineral of the igneous rocks known as peridotites, dunite being an almost pure olivine rock; it also occurs in basic igneous rocks such as the olivine-gabbros, basalts and dolerites. The variety forsterite is formed by the dedolomitization of an impure dolomite on metamorphism, as in the forsterite-marbles; the forsterite readily changes into serpentine, thereby giving rise to ophicalcite (see p. 410).

The Garnet Family

In composition the garnets are essentially silicates of various divalent and trivalent metals, their general formula being

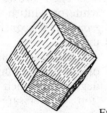
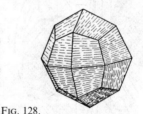

FIG. 128.

Rhombdodecahedron of garnet

Trapezohedron of garnet

$R^{2+}_3R^{3+}_2[SiO_4]_3$, where R^{2+} is calcium, magnesium, iron or manganese, and R^{3+} is iron, aluminium, chromium or titanium. The atomic structure of garnet has separate SiO_4-groups which are bonded together by the metal ions which lie between them.

The following are the principal members of this family:

Grossular. Calcium-aluminium garnet, $Ca_3Al_2(SiO_4)_3$.

Pyrope. Magnesium-aluminium garnet, $Mg_3Al_2(SiO_4)_3$.

Almandine. Iron-aluminium garnet, $Fe_3Al_2(SiO_4)_3$.

Spessartite. Manganese-aluminium garnet, $Mn_3Al_2(SiO_4)_3$.

Andradite. Calcium-iron garnet, $Ca_3Fe_2(SiO_4)_3$.

Uvarovite. Calcium-chromium garnet, $Ca_3Cr_2(SiO_4)_3$.

In *crystallography*, the garnets all crystallize in the cubic system, and occur as rhombdodecahedra (110) or trapezohedra (211), or as combinations of these two forms, as shown in Fig. 128 and Fig. 40B, p. 103. Garnets occur very rarely in one or two other cubic forms.

With regard to its *characters under the microscope*, garnet occurs in thin sections as somewhat rounded crystals, traversed by branching cracks, and having no cleavage. The colour most usually seen is a slight tinge of pink. The refractive index is very high, varying between 1·74 and 1·94, and so the crystals have well-marked borders and a pitted surface (see Fig. 129). Between crossed nicols, the mineral is normally isotropic, though sometimes strain-polarization or abnormal twinning is shown.

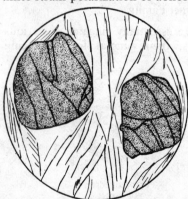

In *hardness*, the garnets range from 6·5 to 7·5, and in *specific gravity* from 3·5 to 4·3. Their *streak* is always white or whitish, and their *fracture* subconchoidal or uneven.

Garnets *occur* in a variety of ways, depending to some degree upon their composition. Garnets are common minerals in metamorphic rocks, such as the gneisses and schists of argillaceous sedimentary parentage, in crystalline limestones, and in metamorphosed basic and other igneous rocks; some varieties occur as primary minerals in igneous rocks, especially of the syenite family; garnets are common minerals in the heavy detrital residues in sediments.

FIG. 129.—Garnet in thin section

The *uses* of garnets are two, as an abrasive and as a gemstone. Garnet is mined from metamorphic rocks in New York and Idaho, U.S.A., and from alluvial deposits in Spain, and elsewhere, and is used as an abrasive, especially in the polishing of wood. Many varieties are cut for gems, some of the names of such gems being Bohemian Garnet, Cape Ruby, carbuncle, cinnamon-stone, etc., in addition to the more scientific names of the garnet varieties. Garnet gems come mostly from Czechoslovakia, India, Ceylon, and South Africa.

GROSSULAR, Grossularite

Comp. Calcium aluminium silicate, $Ca_3Al_2(SiO_4)_3$.

Cryst. Syst. Cubic. Com. Form. Rhombdodecahedron. Colour. Pale olive-green, or greenish-white; occasionally yellow or pink. Lustre. Vitreous; translucent. Sp. Gr. 3·5.

Variety. *Cinnamon-stone* is a calcium-aluminium garnet of a light cinnamon colour, sometimes yellowish; it has a vitreous resinous or dull lustre, and flat conchoidal fracture; it scratches quartz with difficulty; it occurs in Aberdeenshire, Wicklow, etc., and is used in jewellery.

Tests. Fuses easily; after ignition, soluble in hydrochloric acid, yielding gelatinous silica on concentration.

Occurrence. Grossular is characteristic of metamorphosed impure limestones, in which it is associated with other lime-silicates, such as idocrase, wollastonite, etc.

PYROPE, Precious Garnet (in part)

Comp. Magnesium aluminium silicate, $Mg_3Al_2(SiO_4)_3$.

Cryst. Syst. Cubic. Com. Form. Rarely crystallized; usually, in rounded or angular fragments. Colour. Deep crimson or mulberry colour. Lustre. Vitreous; transparent to translucent. Fract. Conchoidal. H. 7·5. Sp. Gr. 3·7.

Tests. Fuses easily; after ignition yields gelatinous silica with hydrochloric acid.

Occurrence. Pyrope occurs in ultrabasic igneous rocks, such

as peridotites, serpentine, etc., where it is associated with olivine, serpentine, chromite, etc., as in Saxony, Czechoslovakia; also occurs in detrital deposits as in Ceylon.

ALMANDINE, ALMANDITE, Precious Garnet (in part); Common Garnet (in part)

COMP. Iron aluminium silicate, $Fe_3Al_2(SiO_4)_3$.

CRYST. SYST. Cubic. COM. FORM. Combination of rhomb-dodecahedron and trapezohedron (see Fig. 40B, p. 103). COLOUR. Deep red. LUSTRE. Vitreous. FRACT. Subconchoidal or uneven. H. 6·5–7·5. SP. GR. 3·9–4·2.

TEST. Fuses to magnetic globule.

VARIETIES. *Common Garnet* is the brownish-red, translucent, subtranslucent or opaque variety; *Precious Garnet* is the deep-red transparent variety.

OCCURRENCE. Almandine is an extremely frequent mineral in metamorphic rocks, mica-schists and gneisses; also in granites; in detrital deposits, as in Ceylon.

SPESSARTITE, Spessartine

COMP. Manganese aluminium silicate, $Mn_3Al_2(SiO_4)_3$.

CRYST. SYST. Cubic. COM. FORM. Rhombdodecahedron. COLOUR. Deep hyacinth, or brownish-red. LUSTRE. Vitreous; slightly translucent on edges. FRACT. Imperfectly conchoidal. H. 7–7·5. SP. GR. 4·15–4·27.

TESTS. Gives manganese reactions before the blowpipe.

OCCURRENCE. A rather rare garnet occurring occasionally in acid igneous rocks such as granites and rhyolites, and in lowly metamorphosed sedimentary rocks such as those used as whet-stones.

ANDRADITE, Common Garnet (in part)

COMP. Calcium iron silicate, $Ca_3Fe_2(SiO_4)_3$.

CRYST. SYST. Cubic. COM. FORM. Rhombdodecahedron and trapezohedron (see Fig. 40, p. 103). COLOUR. Dark brown, yellowish-green, or brownish-green. LUSTRE. Vitreous; opaque. H. Over 7. SP. GR. 3·75–3·78.

TEST. Fuses to magnetic globule.

VARIETIES. *Colophonite* is a coarse granular variety, with a resinous lustre, and of a dark reddish or brownish colour; *Pyreneite* is a black or greyish-black variety, generally occurring in small opaque rhombdodecahedra, as in limestone in the Pyrenees; *Melanite* is a black variety, either dull or with a vitreous lustre, occurring in alkaline igneous rocks such as nepheline-syenites; *Topazolite* is a transparent yellow or green topaz-like variety; *Demantoid* is a bright green variety used as a gem.

OCCURRENCE. Andradite occurs in igneous and metamorphic rocks of many kinds.

UVAROVITE

COMP. Calcium chromium silicate, $Ca_3Cr_2(SiO_4)_3$.

CRYST. SYST. Cubic. COM. FORM. Rhombdodecahedron. COLOUR. Emerald-green. STREAK. Greenish-white. LUSTRE. Vitreous; translucent at the edges. H. 7·5. SP. GR. 3·42.

TESTS. Heated alone before the blowpipe, infusible; gives a clear chrome-green borax bead.

OCCURRENCE. In serpentines rich in chromite, as in the Urals, and Unst in the Shetlands.

SPHENE, Titanite

COMP. Calcium titanosilicate, $CaTiO(SiO_4)$. The composition is variable, Ca being partly replaced by Na, Ti by Al, F, Nb, etc. and O by OH and F.

CRYST. STRUCT. Independent SiO_4-tetrahedra with groups of (CaO_7) and (TiO_6). CRYST. SYST. Monoclinic. COM. FORM. Crystals usually wedge- or lozenge-shaped in habit; also massive. TWINNING. Rather common with twin-plane the orthopinacoid (100), and also on other laws, not so common. CLEAV. Fairly good prismatic. COLOUR. Brown, green, grey, yellow or black. STREAK. White. LUSTRE. Adamantine, or resinous; transparent to opaque. FRACT. Imperfect conchoidal; brittle. H. 5–5.5. SP. GR. 3·54.

OPT. PROPS. In thin sections, sphene occurs as four-sided lozenge-shaped forms, as shown in Fig. 136 on p. 377, or irregular grains; colour in general greyish-purple in soda-rich rocks, and brown in soda-poor rocks, but may be colourless, or yellow; pleochroism marked in many varieties, and may be colourless to plum-red; refractive index very high, β averaging about 1·9; birefringence usually very strong, but the faint polarization colours are usually masked by the body-colour or lost by total reflection; biaxial, optically positive.

TESTS. Heated before the blowpipe, the yellow varieties remain unaltered in colour, but the dark kinds become yellow; partly soluble in hot hydrochloric acid—on the addition of tin, the solution becomes violet when concentrated.

VARIETIES. *Greenovite* is a sphene containing a little manganese oxide, and is red or pinkish in colour; *Leucoxene* is an alteration-product of ilmenite and other titaniferous minerals—the product has been differently determined as sphene, anatase, etc., but is most probably an amorphous hydrated titanium dioxide; most sections of ilmenite show this alteration especially well in reflected light as three sets of lines or stripes making equal angles with one another.

OCCURRENCE. Occurs as an original mineral in acid igneous rocks, generally in accessory amount, but more abundant in rocks rich in lime, as for example the contact-metamorphosed limestones. *Leucoxene* occurs as an alteration-product of ilmenite and other titaniferous minerals.

TOPAZ

COMP. Aluminium fluosilicate, $Al_2F_2SiO_4$; part of the fluorine may be replaced by hydroxyl, OH.

CRYST. SYST. Orthorhombic. COM. FORM. Prismatic crystals made up of prism (110), pinacoid (010), basal pinacoid (001), domes (041) and (201) and bipyramid (111), as shown in Fig. 130; another topaz crystal is figured and described on p. 142; also columnar or granular. CLEAV. Perfect parallel to the basal pinacoid. COLOUR. Wine-yellow, straw-yellow, white, greyish and sometimes blue or pink; the pink colour of much of the topaz sold by jewellers is, however, produced by artificial heat, the stone being wrapped in amadou (a kind of tinder), which is ignited and allowed to smoulder away. STREAK. Colourless. LUSTRE. Vitreous; transparent to subtranslucent. FRACT. Subconchoidal to uneven. H. 8. SP. GR. 3·5–3·6.

OPT. PROPS. Under the microscope, topaz is colourless; the basal cleavage is usually well-marked; refractive index much

FIG. 130. Topaz

higher than that of quartz, e.g. $\alpha = 1·619$, $\beta = 1·620$, $\gamma = 1·627$, so that the relief is distinct; birefringence low—polarization colours like those given by quartz; biaxial, optically positive; distinguished from quartz by biaxial character, cleavage and higher refractive index.

TESTS. Heated alone, infusible; the powdered mineral, when heated, moistened with cobalt nitrate and reheated, assumes a blue colour, due to aluminium; when fused with microcosmic salt gives off silicon fluoride which etches the glass.

OCCURRENCE. Topaz occurs in acid igneous rocks, granites, rhyolites, etc., and good crystals are found projecting into druses as in the Mourne Mountains granite, and in the lithophysal cavities of rhyolites, as in Colorado. It is also a mineral which occurs in tin-bearing pegmatites and in tin veins generally, associated with other pneumatolytic minerals such as fluorspar, cassiterite and tourmaline. It is found also in the zone of contact-alteration adjacent to granite margins.

USE. As a gemstone; the chief sources are the Urals, Brazil, Japan and Rhodesia.

The Aluminium Silicate Family

There are three aluminium silicates of the composition Al_2SiO_5, namely:

Andalusite, orthorhombic.
Sillimanite, orthorhombic.
Kyanite, triclinic.

The crystal-structures of all three minerals are closely related, and contain independent SiO_4-tetrahedra and chains of aluminium-oxygen groups.

The aluminium silicates occur in metamorphic rocks mostly of argillaceous, or clayey, composition. Which of the three forms shall be produced by the metamorphism depends commonly upon the physical factors in action during it. Thus, andalusite is formed under fairly high temperatures and low stress, so that it is typical of normal thermal aureoles around intrusive igneous rocks and of rocks produced by regional metamorphism under similar conditions. Sillimanite is produced at a higher temperature and is stable under a certain amount of stress, so that it is found in rocks of the innermost zone of thermal metamorphism or in regional metamorphic rocks of highest grade. Kyanite is the form stable under stress, so that it characterizes regionally metamorphic rocks of intermediate grade and is absent from normal contact aureoles. The form arising under one condition of metamorphism may be unstable under later conditions, e.g. the andalusite-hornfelses of the Inchbae aureole in Ross-shire, Scotland, are converted into kyanite-schists by the later regional metamorphism; conversely, in Aberdeenshire, kyanite-schists are changed into andalusite-hornfelses on thermal metamorphism.

Andalusite is distinguished from sillimanite in thin section by its negative elongation (see p. 179); kyanite shows oblique extinction in certain sections in the prismatic zone.

The naturally occurring aluminium silicates are industrially important in the manufacture of refractory materials, as the porcelain of sparking-plugs, etc. Deposits of various types occur in the crystalline schists and have been worked in California and the Eastern United States, and in India (Singbhum and Assam), and South Africa.

In addition to the aluminium silicates already mentioned, there is also a compound, of rare occurrence in nature, known as *mullite*, $Al_6Si_2O_{13}$.

ANDALUSITE

COMP. Aluminium silicate, Al_2SiO_5.

CRYST. SYST. Orthorhombic. COM. FORM. Crystals common (see Fig. 131), made up of a nearly square prism (110) with basal plane (001) and sometimes a small dome (011): also granular or massive. CLEAV. Poor, parallel to the prism. COLOUR. Pearl-grey, purplish-red, flesh-red; often altered on the outside to silvery mica. LUSTRE. Vitreous; translucent to opaque. FRACT. Uneven, tough. H. 7·5. Sp. Gr. 3·1–3·3.

OPT. PROPS. In thin section, appears as grains, aggregates or cross-sections of crystals; colourless but sometimes patchily pleochroic in pink to colourless; refractive indices moderately high, e.g. $\alpha = 1·634, \beta = 1·639, \gamma = 1·643$; birefringence weak—polarization colours being little higher than those of quartz; negative elongation, the fast vibration-direction coinciding with

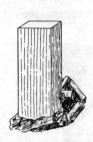
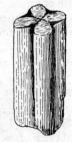

FIG. 131. Andalusite FIG. 132. Chiastolite

the length of crystals; biaxial, optically negative, a basal section yielding an acute bisectrix figure, with large optic axial angle; the variety *chiastolite* (see Fig. 133, p. 368) shows in section black inclusions arranged in a cross-shaped pattern.

TESTS. Heated before the blowpipe, infusible; not acted upon by acids; gives a blue colour when heated with cobalt nitrate solution on charcoal.

VARIETIES. *Chiastolite* or *Macle* is a variety found in some metamorphic rocks, and in certain slates, such as the Skiddaw Slate of Cumberland, and the Killas of Cornwall, resulting from the contact-metamorphism of argillaceous sediments. The crystals when cut or broken across exhibit definite cruciform lozenge-shaped or tessellated markings which are due to impurities enclosed in the crystals during their formation. The British examples are small, but crystals from foreign localities attain considerably larger dimensions. The corners of the crystals may wear away, thus producing a form simulating that of a twin crystal as shown in Fig. 132; *Manganandalusite* or *Viridine* is a manganiferous variety of andalusite.

OCCURRENCE. Andalusite occurs in metamorphosed rocks of clayey composition, as in the andalusite-hornfelses in thermal aureoles, formed under conditions of high temperatures and low stress, and in regional metamorphic rocks, such as the andalusite-schists, which are mostly unconnected with definite igneous intrusions but were presumably formed also at high temperatures and under low stress; andalusite occurs also as an accessory mineral in certain granites, e.g. those of Cornwall, and its presence in such rocks is often connected with the presence of argillaceous inclusions in the igneous rock.

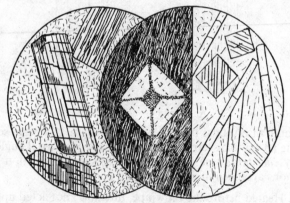

FIG. 133. Aluminium silicates in thin section: left, kyanite; centre, andalusite (chiastolite); right, sillimanite

SILLIMANITE, Fibrolite

COMP. Aluminium silicate, Al_2SiO_5.

CRYST. SYST. Orthorhombic. COM. FORM. Usually occurs as long needle-shaped crystals and in wisp-like aggregates. CLEAV. Perfect parallel to the side pinacoid (010). COLOUR. Shades of brown, grey and green. STREAK. Colourless. LUSTRE. Vitreous; transparent to translucent. FRACT. Uneven. H. 6–7. SP. GR. 3·23.

OPT. PROPS. In thin section appears as colourless long slender crystals, with a cross-fracture, or else as diamond-shaped cross-sections (see Fig. 133), or as mats of fine fibres; refractive indices high, higher than those of andalusite, e.g. $\alpha = 1\cdot659$, $\beta = 1\cdot660$, $\gamma = 1\cdot680$; birefringence also stronger than with andalusite, giving usually Second Order polarization colours; positive elongation (slow along length of crystals)—a distinction from andalusite; biaxial, optically positive, the diamond-shaped basal section yields an acute bisectrix figure with small optic axial angle, the optic axial plane being parallel with the cleavage; sillimanite is distinguished from andalusite by positive elongation, positive sign, smaller optic axial angle, higher refractive index and birefringence.

OCCURRENCE. Occurs in the inner zone of hornfelses resulting from the contact-metamorphism of argillaceous rocks, and in high-grade regionally metamorphosed rocks of similar composition which have been metamorphosed under high temperatures and moderate stress.

KYANITE, Cyanite, Disthene

COMP. Aluminium silicate, Al_2SiO_5.

CRYST. SYST. Triclinic. COM. FORM. Usually occurs in long, thin, blade-like crystals, embedded in schists and gneisses; sometimes in radiating rosettes embedded in quartz. CLEAV. Parallel to the pinacoids, that parallel to (100) being the best. COLOUR. Light blue, sometimes white, sometimes with the middle of the crystal blue and the margins colourless; also grey-green, and rarely black. STREAK. White. LUSTRE. Of cleavage-faces, rather pearly; transparent to subtranslucent. H. Varies on different faces, 4–7. SP. GR. 3·6–3·7.

OPT. PROPS. In thin section, usually colourless, rarely pale blue; if coloured, is weakly pleochroic; sections usually elongated, and show good cleavage-cracks (see Fig. 133); refractive indices high, e.g. $\alpha = 1\cdot712$, $\beta = 1\cdot720$, $\gamma = 1\cdot728$, higher than for andalusite or sillimanite; birefringence lowish, First to low Second Order polarization colours being given; certain sections in the prism zone show oblique extinction, up to 30°—a distinction from anda-lusite and sillimanite; biaxial, optically negative, large optic axial angle.

OCCURRENCE. Characteristic of argillaceous rocks metamor-phosed under high stress and moderate temperature—as in the kyanite-gneisses and schists; also in eclogites—i.e. basic igneous rocks metamorphosed under high-grade conditions.

MULLITE

COMP. Aluminium silicate, $Al_6Si_2O_{13}$.

CHARACTERS AND OCCURRENCE. Occurs as orthorhombic prisms, with the appearance of sillimanite; first found in nature in shales fused by immersion in basic magma in Mull, Scotland; formed when other aluminium silicates are heated, and produced commercially in the electric furnace.

STAUROLITE

COMP. Silicate of iron, magnesium and aluminium, probably $(MgFe^{2+})_2(AlFe^{3+})_9O_6[SiO_4]_4(O.OH)_2$. The structure is com-posed of alternate layers of kyanite and iron hydroxide; mag-nesium and manganese are usually present.

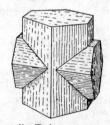

FIG. 134. Staurolite Twins

CRYST. SYST. Orthorhombic. COM. FORM. Occurs in prismatic crystals made up of prism (110), side pinacoid (010), basal pinacoid (001) and sometimes a dome (101). TWINNING. Twinning common on two laws; the first on a pyramid (232) as shown in Fig. 134 left, and the second on a dome (032) as shown in Fig. 134

right (see also p. 155). CLEAV. Interrupted, parallel to the pinacoid (010). COLOUR. Reddish-brown, brownish-black, sometimes yellowish-brown. STREAK. Colourless or greyish. LUSTRE. Sub-vitreous to resinous; crystals usually have dull rough surface; translucent to opaque, usually opaque. FRACT. Conchoidal. H. 7–7·5. SP. GR. 3·7.

OPT. PROPS. In thin sections, appears as yellow to red-brown crystal-sections, markedly pleochroic in yellow and brownish-yellow shades; twinning often seen, cleavage not usually observed; refractive index high, e.g. $\alpha = 1\cdot736$, $\beta = 1\cdot741$, $\gamma = 1\cdot746$; bire-fringence low—polarization colours slightly higher than those of quartz; biaxial, optically positive, large optic axial angle.

TESTS. Varieties containing manganese fuse easily to a black magnetic glass, but the other varieties are infusible.

OCCURRENCE. In metamorphic rocks resulting from the meta-morphism of argillaceous and related rocks—staurolite-schists and gneisses, in which it is associated often with garnet and kya-nite.

CHLORITOID, Ottrelite

COMP. Silicate of iron, magnesium and aluminium, $(Fe^{2+},Mg, Mn)_2(Fe^{3+},Al)Al_3O_2[SiO_4](OH)_4$.

CRYST. STRUCT. Separate SiO_4-tetrahedra form layers parallel to the basal plane (001).

CRYST. SYST. Monoclinic(?) or triclinic. COM. FORM. Crystals tabular, also in laths and plates. CLEAV. Basal good, the lami-nae being brittle; chloritoid is often grouped with the Brittle Micas (p. 405). COLOUR. Dark-green and greenish-black. LUSTRE. Pearly on cleavage-surfaces. H. 6·5. SP. GR. 3·5.

OPT. PROPS. In thin section gives plates or lath-shaped outlines; markedly pleochroic in shades of green, greenish-blue and yellowish; lamellar twinning often seen; refractive indices high, $\beta = 1\cdot72$; birefringence low, about that of quartz; optically posi-tive; the combination of high refractive index and low birefingence distinguishes chloritoid from the micas and chlorites.

OCCURRENCE. Occurs only in metamorphic rocks of sedimentary origin, especially those of argillaceous composition metamorphosed under strong stress. Ottrelite, the manganese-rich variety, characterizes the ottrelite-phyllites of the Ardennes.

B. SOROSILICATES

The Epidote Family

Introduction. In chemical composition the group of minerals comprising the Epidote Family are complex silicates with the general formula $R^{2+}_2R^{3+}_3O[SiO_4][Si_2O_7]OH$, where R^{2+} = calcium and iron, R^{3+} = aluminium, iron, manganese, cerium, etc. The epidotes are all similar in their atomic structure, a mixed type containing both $[SiO_4]$ and $[Si_2O_7]$ groups.

They may be classified by their crystal systems into:

ORTHORHOMBIC EPIDOTE.
  Zoisite, $Ca_2Al_3(SiO_4)_3(OH)$.

MONOCLINIC EPIDOTE.
  Clinozoisite, $Ca_2Al_3(SiO_4)_3(OH)$.
  Epidote, $Ca_2(Al,Fe)_3(SiO_4)_3(OH)$.
  Piedmontite, $Ca_2(Al,Fe,Mn)_3(SiO_4)_3(OH)$.
  Allanite, *Orthite*, $(Ca,Fe)_2(Al,Fe,Ce)_3(SiO_4)_3(OH)$.

Orthorhombic Epidote

ZOISITE

COMP. Complex silicate of calcium and aluminium, $Ca_2Al_3(SiO_4)_3(OH)$ or $Ca_2Al.Al_2O.[SiO_4].[Si_2O_7]OH$; replacement of aluminium by iron provides a transition to epidote.

CRYST. SYST. Orthorhombic. COM. FORM. Crystals elongated along the *b*-axis often striated longitudinally; commonly columnar or massive. CLEAV. Perfect parallel to (001). COLOUR. White, grey, greenish, sometimes rose-red (*thulite*). LUSTRE. Vitreous; transparent to translucent. H. 6–6·5. SP. GR. 3·2–3·37.

OPT. PROPS. In rock-sections, zoisite appears usually as shapeless grains or prismatic crystal sections; usually colourless, but pleochroic in pinks in thulite; refractive indices high, e.g. α =

$1\cdot700$, $\beta = 1\cdot703$, $\lambda = 1\cdot718$; abnormal polarization colours— inky-blue usually—in certain sections; optically positive, variable optic axial angle.

VARIETY. *Thulite* is the rose-pink variety already mentioned.

OCCURRENCE. Occurs in metamorphic rocks, especially those formed from igneous rocks rich in lime-feldspars, gabbros, etc., and where it is often accompanied by metamorphic amphiboles— as in the greenstones, amphibolites and the like; also in metamorphosed impure limestones; occurs commonly as a constituent of *saussurite*, a mixture of epidotes, albite, etc., resulting from the alteration of lime-plagioclases as in the saussurite-gabbros.

Monoclinic Epidote

CLINOZOISITE

COMP. Like zoisite, silicate of calcium and aluminium, $Ca_2Al_3(SiO_4)_3(OH)$; iron usually present, providing a transition to epidote.

CRYST. SYST. Monoclinic. COM. FORM. Crystals like those of epidote. COLOUR. Grey or greyish-white. H. 6–7. SP. GR. $3\cdot2$–$3\cdot4$.

OPT. PROPS. In thin section, colourless or faint yellow-green; refractive index high, $\beta = 1\cdot720$; birefringence low—polarization colours of low order, but the abnormal inky-blue colour of zoisite not shown; optically positive, distinguishing it from epidote.

OCCURRENCE. As a secondary mineral in igneous rocks where it is produced by the alteration of ferromagnesian minerals and of calcic plagioclase, and as a constituent of metamorphic rocks derived from basic igneous, and impure calcareous, rocks.

EPIDOTE, Pistacite

COMP. Silicate of calcium, aluminium and iron, $Ca_2(Al,Fe)_3(SiO_4)_3(OH)$ or $Ca_2Fe^{3+}Al_2O.[SiO_4][Si_2O_7]OH$.

CRYST. SYST. Monoclinic. COM. FORM. Crystals elongated along the b-axis, as shown in Fig. 135; in divergent aggregates and granular masses. CLEAV. Perfect parallel to the basal plane.

COLOUR. Shades of green—pistachio-green, blackish-green, dark oil-green: red in *withamite*. LUSTRE. Vitreous; transparent to opaque. FRACT. Uneven. H. 6–7. SP. GR. 3·25–3·5.

OPT. PROPS. The shapes of sections are either elongated or rounded—rare sections parallel to the clinopinacoid are six-sided, and show the basal cleavage parallel to a pair of the edges; faintly or markedly pleochroic in yellows and yellow-greens; refractive indices high, e.g. $\alpha = 1·733$, $\beta = 1·755$, $\gamma = 1·768$; birefringence

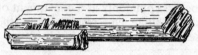

FIG. 135.   Epidote

strong—the polarization colours being bright Second and Third Order colours; optically negative, large optic axial angle.

TESTS. Heated before the blowpipe gives reactions for iron.

VARIETIES. *Pistacite* is a pistachio-green variety; *Arendalite* is a variety from Arendal in Norway, occurring in very fine crystals, externally blackish-green, and of a dark oil-green on fractured surfaces; *Withamite* is a red variety occurring in andesites in Glencoe, Scotland, and in thin section strongly pleochroic in red and yellow.

OCCURRENCE. Commonly in metamorphic rocks, chiefly those of two derivations—from impure calcareous rocks or from igneous rocks rich in lime-feldspar; *epidosite* is a metamorphic rock formed almost entirely of granular epidote; as an alteration-product of many ferromagnesian rock-forming silicates; believed to occur as a primary igneous mineral in certain granitic rocks.

PIEDMONTITE

COMP. Silicate of calcium, aluminium, iron and manganese, $Ca_2(Al,Fe,Mn)_3(SiO_4)_3(OH)$.

CHARACTERS AND OCCURRENCE. A manganiferous epidote, of a dark reddish colour, showing in thin sections a strong pleochroism in yellow, violet and red tones, $\beta = 1·78$, birefringence strong, optically positive; crystal characters like those of epidote; heated before the blowpipe, fuses readily to a black glass, a character distinguishing it from pistacite and zoisite, which fuse only on thin edges; occurs as a constituent of certain metamorphic rocks—piedmontite-mica-schists, etc., and as a secondary mineral in some porphyries, as in the classical *porfido antico rosso* of Egypt.

ALLANITE, Orthite

COMP. Silicate of calcium, aluminium, iron and the cerium metals, $(Ca,Ce)_2(Al,Fe)_3(SiO_4)_3(OH)$ or $(Ca,Ce)_2(Fe^{2+}Fe^{3+})$ $Al_2O.[SiO_4][Si_2O_7].OH$.

CHARACTERS AND OCCURRENCE. A cerium-bearing epidote, brown to black in colour, occurring in tabular or prismatic crystals, or in grains; in thin section, shows strong pleochroism in shades of brown; refractive index high, polarization colours of high order but usually masked by body-colour; allanite often forms a core to epidote crystals; occurs as an accessory mineral in granites, syenites and diorites, or in their metamorphic derivatives.

IDOCRASE, Vesuvianite

COMP. Silicate of calcium, aluminium and magnesium, with some replacement by iron. X-ray analysis has shown that both SiO_4 and Si_2O_7 groups are present, giving the formula: $Ca_{10}Al_4$ $(Mg,Fe)_2(Si_2O_7)_2(SiO_4)_5(OH)_4$.

CRYST. SYST. Tetragonal. COM. FORM. Crystals prismatic, usually with bipyramids and prisms of both orders and basal pinacoid as shown in Fig. 57, p. 119; also massive. CLEAV. Not very distinct parallel to faces of prism of first order (110); basal cleavage more imperfect still. COLOUR. Brown, green, yellowish. STREAK. White. LUSTRE. Vitreous or vitro-resinous; subtransparent to subtranslucent. FRACT. Subconchoidal or uneven. H. 6·5. SP. GR. 3·35–3·45.

OPT. PROPS. In thin sections, colour is pale yellowish-brown or colourless, beautiful zoning sometimes seen; pleochroic only in thick plates; refractive index high, e.g. about 1·73; birefringence weak, and abnormal ultra-blue polarization colours usually shown and often patchily distributed; optically negative, but sometimes optically positive, but optical character often difficult to determine because of the low birefringence.

Tests. Before the blowpipe, fusible with intumescence, forming a greenish or brownish glass.

Occurrence. Occurs, associated with grossular, wollastonite, scapolite, etc., in impure limestones which have been subjected to contact-metamorphism, as in the limestone blocks of Monte Somma, Vesuvius; also in rocks of similar composition in areas of regional metamorphism as in the Loch Tay Limestone of Scotland.

MELILITE

Comp. Essentially $Ca_2MgSi_2O_7$, with some substitution of Na for Ca, and Al for Si or Mg.

Cryst. Struct. See p. 351.

Characters and Occurrence. A tetragonal mineral occurring in tabular crystals or grains. of white, yellowish or greenish colour; in thin sections yellowish or colourless, with moderately high refractive index, e.g. $\omega = 1\cdot634$, $\varepsilon = 1\cdot629$, and very weak birefringence—often inclusions shaped like pegs are seen, running parallel to the c-axis and thus across tabular crystals; decomposed by hydrochloric acid with gelatinization; melilite occurs in lavas of basic type, low in silica and without feldspar—melilite-basalts, and nepheline- and leucite-basalts; a common constituent of slags.

C. CYCLOSILICATES

TOURMALINE

Comp. Complex borosilicate of aluminium, together with alkali metals or iron and magnesium. The general formula can be written $XY_3B_3(Al,Fe^{3+})_6O_{27}(OH,F)_4$ where $X = Na,Ca$, and $Y = Mg,Fe^{2+},Al,Li$. The structure has been shown to contain Si_6O_{18}-rings, see p. 351. Varieties arise according to which metals predominate, e.g. common black tourmaline contains Na,Ca,Mg, Fe^{2+} and Fe^{3+}.

Cryst. Syst. Hexagonal, ditrigonal-pyramidal, hemimorphic, Tourmaline type, see p. 129. Com. Form. Prismatic crystals three-sided in cross-section (Fig. 69, p. 132), with hexagonal prism and trigonal prism, terminated by trigonal pyramid; hemimorphic (see p. 129); needle-like crystals common, often in radiating groups; also massive, compact or columnar. CLEAV. Rhombohedral; difficult. COLOUR. Commonly black or bluish-black; more rarely blue, green or red, and very rarely colourless; colours sometimes arranged in zones about the vertical axes of crystals. STREAK. Colourless. LUSTRE. Vitreous; transparent to opaque. FRACT. Subconchoidal or uneven; brittle. H. 7–7·5. SP. GR. 2·98–3·2.

OPT. PROPS. Under the microscope, tourmaline occurs either as three- or six-sided cross-sections, or as badly terminated elongated longitudinal sections (see Fig. 136), or else in groups of radiating needles; colour is usually dark brown, green or yellow, with strong pleochroism in these colours—greatest absorption occurs when the polarized light vibrates across the length of the tourmaline prism; pleochroic halos common; refractive index high, e.g. $\omega = 1·642$, $\varepsilon = 1·622$; birefringence moderate, but polarization colours often masked by the strong body-colour; elongation negative, length-fast; uniaxial, optically negative— basal sections give a negative uniaxial interference-figure.

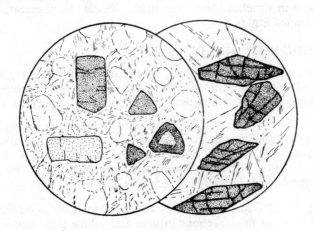

FIG. 136.—Tourmaline (left), sphene (right)

TESTS. Heated before the blowpipe, dark varieties intumesce and fuse with difficulty; red and green varieties only become milk-white and fuse slightly on the edges.

VARIETIES. *Rubellite* is a red or pink variety, transparent and sometimes cut as a gem; *Indicolite* is an indigo-blue variety; *Brazilian Sapphire* is a transparent Berlin-blue variety, cut as a gemstone; *Brazilian Emerald* is a transparent green type; *Peridot of Ceylon* is a honey-yellow variety; *Schorl* is the black opaque variety occurring as aggregates of needle-like or columnar crystals as in the granites of Cornwall.

OCCURRENCE. Occurs as an accessory mineral in many granites, syenites and the more acid rocks generally; commonly in peg-matites, and as a product of pneumatolytic action, both in the parent igneous rock and in the adjacent country-rock; it may also occur associated in veins with lead and cobalt minerals; tourma-line is common as an accessory component also of various meta-morphic rocks, mica-schists, gneisses, crystalline limestones, etc., and in some cases is there produced by vapours traversing the rocks during the metamorphic period; finally it occurs as a com-mon 'heavy residue' in many sedimentary rocks.

USES. Some varieties, as already noted, are used as gemstones; these gem varieties come from Brazil, Russia, Madagascar, and the United States.

CORDIERITE, Iolite, Dichroite

COMP. Silicate of aluminium, magnesium and iron, $(Mg,Fe)_2$ $Al_3[AlSi_5O_{18}]$. The ratio of $Mg:Fe$ varies, and iron-rich cordierites are known. The atomic structure is related to that of beryl (p. 351).

CRYST. SYST. Orthorhombic. COM. FORM. Short pseudo-hexagonal crystals; usually granular or massive. CLEAV. Poor parallel to the pinacoids. COLOUR. Blue of various shades. STREAK. Colour-less. LUSTRE. Vitreous; transparent to translucent. FRACT. Subconchoidal; brittle. H. 7–7·5. SP. GR. 2·6–2·7.

OPT. PROPS. Usually colourless in thin sections, but is markedly pleochroic in thick sections in blue and yellow tints: refractive indices near those of quartz and balsam, e.g. $\alpha = 1.535$, $\beta =$

1·540, $\gamma = 1\cdot544$; birefringence near that of quartz but usually slightly higher—in thin sections in which the highest polarization colour of quartz is First Order grey, cordierite often gives First Order yellow; twinning in sectors often shown; pleochroic halos around zircon inclusions often seen; alters to a yellowish micaceous product, *pinite*, often at the margin and along cracks; biaxial, usually optically negative; distinguished from quartz by its biaxial character and alteration.

TESTS. Heated before the blowpipe, cordierite loses transparency, and fuses with difficulty on the edges; the glassy or resinous appearance is characteristic, and fusibility on the edges distinguishes cordierite from quartz; it is much softer than sapphire.

OCCURRENCE. Cordierite occurs chiefly in metamorphic rocks both of regional metamorphic origin as in the cordierite-gneisses of high grade, and of contact-metamorphic origin as in the cordierite-hornfels of argillaceous composition; also occurs as a magmatic mineral in norites due to contamination of gabbro magma by argillaceous sediments, and as pinite pseudomorphs in granites, and in other igneous rocks.

USES. Sometimes used as a gemstone.

AXINITE

COMP. Boro-silicate of aluminium and calcium, with varying amounts of iron and manganese, approximately $(Ca,Fe^{2+})_3$. $Al_2BO_3[Si_4O_{12}]OH$.

CRYST. SYST. Triclinic. COM. FORM. Usually in thin and very sharp-edged crystals, as shown in Fig. 84, p. 149; sometimes, but rarely, massive or lamellar. CLEAV. Distinct, parallel to the pinacoid (010). COLOUR. Clove-brown, plum-blue, and pearly grey. STREAK. Colourless. LUSTRE. Highly vitreous; transparent to subtranslucent. FRACT. Conchoidal; brittle. H. 6·5–7. SP. GR. 3·27.

OPT. PROPS. In thin sections, colourless or yellowish or violet; refractive index moderate ($\beta = 1\cdot68$), birefringence weak, same as that of quartz; biaxial, optically negative.

TESTS. Heated before the blowpipe, fuses readily with intumescence, and colours the outer flame a pale green; not acted upon by acids unless previously heated, and then it gelatinizes.

OCCURRENCE. Occurs as a mineral of contact-metamorphism, pyrometasomatism or pneumatolysis where lime-rich rocks are in contact with certain igneous rocks, as the Cornish granites (Botallac, Lostwithiel, etc.).

D. INOSILICATES

The Pyroxene Family

Introduction. The Pyroxenes are a group of allied minerals which have certain physical and chemical characters in common, and possess the Si_2O_6 chain structure (p. 351).

In *chemical composition* the pyroxenes are silicates of iron, magnesium and calcium, sometimes with aluminium; some varieties contain sodium or lithium. Thus, several of the pyroxenes have a silicate composition of $R_2[Si_2O_6]$, where R is magnesium, iron, or calcium, or more rarely manganese or zinc; in other pyroxenes there is a substitution of aluminium for part of the silicon, giving a formula of the type $R_2[(Si,Al)_2O_6]$ where R is Ca, Mg, Fe^{2+}, Al or Fe^{3+}. The alkali-pyroxenes have a composition represented by $R^{1+}R^{3+}[Si_2O_6]$ where R^{1+} is sodium or lithium, and R^{3+} is iron or aluminium (see examples given below). There are many varieties of pyroxene, due to the replacement of atoms of one metal by another within the limits of the formulae: that is, the total number of atoms in any group (such as the R group) is always the same. Only aluminium can replace silicon.

The pyroxenes *crystallize* in the orthorhombic, monoclinic and triclinic crystal systems, the prismatic angles of the crystals being nearly right angles; most of the pyroxenes are characterized by a good *prismatic cleavage* (Fig. 138). The orthorhombic and monoclinic members have closely similar atomic structures.

The chief, or most typical, pyroxene is augite, and the names augite and pyroxene are often applied to the same mineral. The differences between augite and hornblende—the chief member of the amphibole family which is in many respects closely analogous with the pyroxene family—are indicated in the introduction to the amphiboles on p. 351.

The Pyroxenes may be divided for our purposes into the groups given below; varieties based mainly on physical characters are considered under their appropriate chemical groups.

I. ORTHORHOMBIC PYROXENES.

Enstatite, $MgSiO_3$ (or $Mg_2Si_2O_6$).
Hypersthene, $(Mg,Fe^{2+})SiO_3$.

II. MONOCLINIC PYROXENES.

(a) DIOPSIDE-HEDENBERGITE SERIES.
Diopside, $CaMgSi_2O_6$
Hedenbergite, $CaFe^{2+}Si_2O_6$.
(b) PIGEONITE, $(Ca,Mg)(Mg,Fe)Si_2O_6$.
(c) AUGITE SERIES (aluminous).
Augite, $(Ca,Mg,Fe,Al)_2(Al,Si)_2O_6$.
(d) ALKALI PYROXENE SERIES.
Acmite, Aegirite, $NaFe^{3+}Si_2O_6$.
Aegirite-augite, transitional between Augite and Aegirite.
Jadeite, $NaAlSi_2O_6$.
(*Spodumene*, $LiAlSi_2O_6$, described on p. 227 under Lithium.)

I. Orthorhombic Pyroxenes

Introduction. The orthorhombic pyroxenes are characterized by two chief minerals, enstatite, $MgSiO_3$, and hypersthene, $(Fe,Mg)SiO_3$, and a third species bronzite, a variety based on physical properties. There is probably a complete isomorphous series between the magnesium silicate, $MgSiO_3$, and iron silicate, $FeSiO_3$ now named *ferrosilite*—though the latter has not yet been distinguished in nature. Hypersthenes with over 80 per cent $FeSiO_3$ are known. There is a gradation in physical properties—specific gravity, refractive index, birefringence, etc.—from enstatite to the most iron-rich hypersthene yet discovered. Orthorhombic pyroxenes with less than about 15 per cent $FeSiO_3$ are optically positive and are enstatite; those with more than 15 per cent $FeSiO_3$ are optically negative and are hypersthene.

ENSTATITE

COMP. Magnesium silicate, $MgSiO_3$, with up to 15 per cent of iron silicate, $FeSiO_3$, giving the general composition $(Mg,Fe)SiO_3$ or $(Mg,Fe)_2Si_2O_6$.

CRYST. SYST. Orthorhombic. COM. FORM. Stout prismatic crystals, showing prism and pinacoids (100) and (010); usually occurs massive and lamellar. CLEAV. Well-developed parallel to the prism (110), giving two sets of cleavage-planes which intersect at nearly 90°; parting parallel to (010) often good. COLOUR. Grey, green, brown, yellow, colourless. LUSTRE. Vitreous, pearly, sometimes fibrous-looking on cleavage-planes; subtranslucent to opaque. H. 5·5. SP. GR. 3·1–3·3, increasing with increase of iron content.

OPT. PROPS. Occurs as colourless to pale-greenish grains and crystal sections in rock-slices; transverse sections of crystals show two sets of cleavage-cracks at nearly 90°; longitudinal sections show one set only; the transverse sections usually 8-sided, squarish, with larger prism edges cut off by small pinacoidal edges; refractive indices, $\alpha = 1·656$, $\beta = 1·659$, $\lambda = 1·665$, and increasing with iron content; birefringence is low, the polarization colours being not much higher than those given by quartz; extinction parallel to the prismatic cleavages in longitudinal sections; since the body-colour is usually weak, the pleochroism is not well-marked; biaxial, with large optic axial angle, and optically positive; transverse sections showing two sets of well-marked prismatic cleavages at about 90° give in convergent light a bisectrix— a feature distinguishing the orthorhombic pyroxenes from the monoclinic, from which they are also separated by lower birefringence; enstatite and hypersthene sometimes alter into a fibrous aggregate called *bastite*, or *schillerspar* which has the composition of serpentine—the bastite fibres are arranged parallel to the *c*-axis of the pyroxene (see also p. 409).

VARIETY. *Bronzite* is an iron-bearing variety of enstatite; it usually has a bronze-like or pearly metallic lustre and is foliaceous; it is the most infusible standard, 6, in Von Kobel's Scale of Fusibility and can only be rounded on the edges of fine splinters in the blowpipe flame.

OCCURRENCE. Enstatite occurs as a primary constituent of the intermediate and basic igneous rocks, such as gabbros and diorites and their dyke and effusive equivalents; it also occurs in some peridotites and serpentines, as in the Kimberley diamond-bearing blue-ground; occurs also in meteorites.

HYPERSTHENE

COMP. Iron magnesium silicate, $(Mg,Fe)SiO_3$, with more than 15 per cent $FeSiO_3$; hypersthenes with over 80 per cent $FeSiO_3$ are known.

CRYST. SYST. Orthorhombic. COM. FORM. Crystals rare, prismatic; usually foliaceous or massive. CLEAV. Prismatic (110) good, also a parting parallel to the side pinacoid (010) good, giving three sets of cleavage or parting lines in transverse sections. COLOUR. Brownish-green, greyish or greenish-black, brown, sometimes almost black. LUSTRE. Sub-metallic; schillerization is very characteristic of hypersthene, and is due to the presence of minute scales possibly of brookite or iron oxide arranged in parallel planes; translucent to opaque. FRACT. Uneven; brittle. H. 5–6. SP. GR. 3·4–3·5, increasing with iron content.

OPT. PROPS. In thin sections, hypersthene shows the same habits as those mentioned for enstatite above; body-colour is more marked in hypersthene, and often a well-marked pleochroism, X = pink, Y = yellow, Z = green, is seen; refractive indices higher than those of enstatite, from $\alpha = 1·673$, $\beta = 1·678$, $\gamma = 1·683$ to $\alpha = 1·715$, $\beta = 1·728$, $\gamma = 1·731$, increasing with the iron-content; birefringence similarly increasing, the polarization colours being of First Order; extinction in longitudinal sections parallel to prismatic cleavage; biaxial, varying but usually large optic angle, and optically negative; inclusions common, consisting of tiny plates, rods or blades, arranged in various sets of parallel planes—schiller-plates; sometimes hypersthene shows a lamellar intergrowth with monoclinic pyroxene.

TESTS. Heated before the blowpipe, fuses to a black enamel, and on charcoal to a magnetic mass.

OCCURRENCE. Occurs in basic igneous rocks, such as norite (hypersthene-labradorite), gabbros, etc., in intermediate rocks also, especially andesite; in the series of metamorphosed igneous rocks known as charnockites, in various types of crystalline schists derived from igneous rocks, and in contact-metamorphosed dominantly argillaceous rocks—the hypersthene-hornfelses.

II. Monoclinic Pyroxenes

Introduction. As indicated on p. 381, there are several series amongst the monoclinic pyroxenes, some of which are dealt with here. The first group of monoclinic pyroxenes, the Diopside-hedenbergite Series, is non-aluminous, whilst the Augite Series contains aluminium; the alkali-pyroxenes are characterized by containing soda. Only the most important series are considered here in detail, less important groups being mentioned briefly.

(a) Diopside-Hedenbergite Series

This series has as its two end-members *diopside*, $CaMgSi_2O_6$, and *hedenbergite*, $CaFeSi_2O_6$. As the proportion of the hedenbergite increases (by substitution of Fe^{2+} for Mg), the refractive index increases, the extinction-angle on the clinopinacoid increases, the birefringence decreases and the size of the optic axial angle remains at about 60°.

DIOPSIDE

COMP. Calcium magnesium silicate, $CaMg(Si_2O_6)$.

CRYST. SYST. Monoclinic. COM. FORM. Prismatic crystals; usually granular. CLEAV. Good parallel to prism (110); a marked parting or cleavage is seen in the variety *diallage* parallel to the pinacoid (100); a parting or striation parallel to the basal pinacoid (001) is often seen in the variety *sahlite*. COLOUR. White, green, darkish green; sometimes colourless. LUSTRE. Vitreous; transparent to opaque. H. 5–6. SP. GR. 3·2–3·38.

OPT. PROPS. In thin section, diopside is colourless or faint greenish, shows the prismatic cleavage as two sets of cracks intersecting at nearly 90° in transverse sections; refractive index and birefringence fairly high, e.g. $\alpha = 1\cdot673$, $\beta = 1\cdot680$, $\gamma = 1\cdot703$—polarization colours, pretty colours of Second and Third Orders;

extinction-angle on clinopinacoid (i.e. angle between cleavage and the slow vibration-direction) is 38–40°; biaxial, optic axial angle moderate (60°), optically positive.

VARIETIES. *Malacolite* is a translucent white, yellow, pale green or colourless variety; *Sahlite* has a more dingy green colour, less lustre and a coarser structure than normal diopside; it often shows a striation parallel to the basal pinacoid—the sahlitic striation. *Coccolite* is a granular variety, white or green in colour; *Chrome-diopside* is a bright green variety, containing a few per cent of Cr_2O_3; *Mansjöite* is a fluorine-bearing diopside-hedenbergite; *Diallage*, see under augite.

OCCURRENCE. Diopside occurs in various igneous rocks—lime-enriched pegmatites and granites, in basic rocks, etc.; extensively in metamorphosed impure calcareous rocks, such as the crystal-line limestones of regional metamorphic origin, and the calc-silicate-hornfelses of contact metamorphic origin.

HEDENBERGITE

COMP. Calcium iron silicate, $CaFe(Si_2O_6)$, with varieties transitional towards diopside.

CRYST. SYST. Monoclinic. COM. FORM. Crystals and lamellar masses. COLOUR. Black. H. 6. SP. GR. 3·7.

OPT. PROPS. General characters as for diopside; refractive indices higher than for diopside, e.g. $\alpha = 1·739$, $\beta = 1·745$, $\gamma = 1·757$, varying with the iron content; extinction-angle on the clino-pinacoid, up to 48°.

OCCURRENCE. The most important type of occurrence of heden-bergite is as a skarn-mineral (see p. 214) at the contacts of granitic rocks with limestones, being associated there with iron-garnet and iron-ores.

(b) Pigeonite

Pigeonite is intermediate in composition between diopside-hedenbergite and *clinoenstatite*, a monoclinic pyroxene with the composition $MgSiO_3$ and rare in nature. The diagnostic character of pigeonite is the small optic axial angle, some samples being uniaxial. Pigeonite occurs in quickly-cooled volcanic rocks, especially those of andesitic composition.

OTHER NON-ALUMINOUS NON-SODIC MONOCLINIC PYROXENES

Schefferite. An iron-manganese pyroxene, black in colour, occurring as a skarn-mineral.

Jeffersonite. A manganese-zinc pyroxene occurring at Franklin Furnace, New Jersey (see pp. 300, 302).

(c) Augite Series

AUGITE

COMP. Silicate of calcium, magnesium, iron and aluminium, $(Ca,Mg,Fe,Al)_2(Al,Si)_2O_6$; composition variable and transitions occur to other types of monoclinic pyroxene.

CRYST. SYST. Monoclinic. COM. FORM. Crystals common—combinations of prism (110), clinopinacoid (010), orthopinacoid (100) and negative hemipyramid ($\bar{1}$11) as shown in Fig. 137; the prism angle is nearly 90°; augite also occurs massive and coarsely lamellar, sometimes granular or fibrous. TWINNING. Crystals often twinned on the orthopinacoid, giving a marked re-entrant angle. CLEAV. Prismatic cleavage good, giving two sets of cleavage-planes meeting at an angle of nearly 90°; a cleavage or parting parallel to the orthopinacoid (100) characterizes the variety *diallage*; parting or striation parallel to (001) seen in *sahlite* types. COLOUR. Black and greenish-black. LUSTRE. Vitreous, inclining to resinous; opaque. H. 5–6. SP. GR. 3·2–3·5.

OPT. PROPS. In thin sections, crystals give eight-sided cross-sections and four-sided longitudinal sections, the former showing the prismatic cleavages, two sets nearly at right angles, and the latter one set only (see Fig. 138); colour is colourless to yellowish-brown, and some titaniferous varieties are purplish; the more strongly coloured varieties are feebly pleochroic; body-colour is sometimes arranged in zoned or hour-glass fashions; refractive index much higher than that of balsam, for augite from Renfrew, Canada, $\alpha = 1\cdot698$, $\beta = 1\cdot704$, $\gamma = 1\cdot723$, and $\gamma - \alpha = \cdot025$—

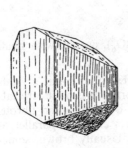

FIG. 137. Augite

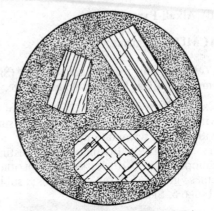

FIG. 138. Sections of augite, showing cleavages: above prismatic sections, below transverse section.

polarization colours are nice bright colours of Second and Third Orders; extinction-angle on the clinopinacoid (i.e. angle between cleavage and slow vibration-direction) is 45–50°; biaxial, large optic axial angle, optically positive; transverse sections in convergent light do not yield a bisectrix, a distinction from orthorhombic pyroxenes; for distinction from hornblende, see p. 391.

VARIETY. *Diallage* is a variety of diopside or augite which in hand-specimen appears as lamellar or foliaceous masses, sometimes fibrous, and usually having a metallic or brassy lustre; colour grass-green, brown or grey, and the mineral is translucent. The lamellar structure is due to a parting parallel to the orthopinacoid (100), and other partings occur. Diallage is characteristic of gabbros and the parting appears under the microscope as a series of fine parallel lines; often schiller-plates are seen; in general optical properties, it resembles diopside or augite.

OCCURRENCE. Augite occurs as short prismatic crystals in many volcanic rocks, andesites, basalts, etc.; and as crystals or plates in dyke and plutonic rocks especially those of basic composition, gabbros, dolerite, and in diorite, etc.; also in ultrabasic rocks such as pyroxenites and peridotites. Augite also occurs in metamorphic rocks such as pyroxene-gneisses, and in pyroxene-granulites and contact-altered igneous rocks.

(d) Alkali-Pyroxenes

ACMITE, Aegirine, Aegirite

COMP. Sodium iron silicate, $NaFe^{3+}(Si_2O_6)$, usually with small amounts of Ca, Mg, and Al in aegirite.

CRYST. SYST. Monoclinic. COM. FORM. Long prismatic crystals with sharp terminations (*acmite*), shorter prismatic crystals with blunt terminations (*aegirine*); also acicular or fibrous crystals. CLEAV. Prismatic (110) distinct; cleavage parallel to clinopinacoid (010) less good. COLOUR. Usually brown, sometimes green. LUSTRE. Vitreous; subtransparent to opaque. H. 6–6·5. SP. GR. 3·5–3·55.

OPT. PROPS. Cross-sections of crystals often six-sided owing to absence of clinopinacoid, and then resemble those of amphibole, but show the prismatic cleavages at nearly 90°; refractive indices for the aegirine of Langesundfiord, Norway, are $\alpha = 1·763$, $\beta = 1·799$, $\gamma = 1·813$—higher than for augite and diopside; birefringence strong; acmite in section is coloured brownish-green and is markedly pleochroic in shades of brown and brownish-green— aegirine is green in section, and pleochroic in grass-green and yellow-green tints; extinction on the clinopinacoid, fast vibration-direction to the cleavage, is low, 2–6°; this low extinction-angle and the strong pleochroism distinguish these soda-pyroxenes from other pyroxenes.

OCCURRENCE. In soda-rich igneous rocks such as nepheline-syenites, phonolites, etc.

Aegirine-Augite, Aegirite-Augite

A series transitional between augite and aegirine, with intermediate characters. In thin sections, shows a marked pleochroism in shades of green and yellow; extinction on the clinopinacoid— fast vibration-direction to the cleavage—varies from a few degrees up to nearly 40°, depending on the proportions of Na and Fe^{3+} present. Aegirine-augite occurs in soda-rich igneous rocks, such as syenites, nepheline-syenites, phonolites, etc.

JADEITE

COMP. Sodium aluminium silicate, $NaAl(Si_2O_6)$.

CRYST. SYST. Monoclinic. COM. FORM. Usually massive. CLEAV. Prismatic, giving two sets nearly at 90°. COLOUR. Shades of green. LUSTRE. Subvitreous; translucent. H. 6·5–7. SP. GR. 3·3–3·35.

OCCURRENCE. Mode of occurrence not well known—possibly of metamorphic origin; localities are Burma, South China, Tibet, New Zealand, Mexico.

USE. As an ornamental stone—constituting one variety of *jade*; another variety of this highly prized material is the amphibole nephrite.

The Pyroxenoid Family

WOLLASTONITE, Tabular Spar

COMP. Calcium silicate, $CaSiO_3$.

CRYST. STRUCT. Chain-structure with chains elongated along the *b*-axis and different in detail from the pyroxene-chains (p. 351). CRYST. SYST. In two modifications, the commoner triclinic, the rarer monoclinic. COM. FORM. Crystals usually tabular, parallel either to the basal pinacoid (001) or the front pinacoid (100), and elongated along the *b*-axis—combination of prisms, pinacoids and domes; also found massive and cleavable, with a long fibrous or columnar structure, the fibres being either parallel or interlaced. CLEAV. Perfect parallel to the front pinacoid. COLOUR. White, grey, sometimes yellowish, reddish or brownish. LUSTRE. Vitreous, rather pearly on cleavage-planes; subtransparent to translucent. H. 4·5–5. SP. GR. 2·8–2·9.

OPT. PROPS. In thin sections, colourless; refractive indices moderately high, e.g. $\alpha = 1·621$, $\beta = 1·633$, $\gamma = 1·635$; birefringence moderate—polarization colours of First and low Second Order colours depending on the section; optically negative, smallish optic axial angle, and the plane of the optic axes usually lies across the elongation-direction of the crystals, a circumstance distinguishing wollastonite from diopside, etc.

TESTS. Heated before the blowpipe, fuses easily on the edges; gelatinizes with hydrochloric acid.

OCCURRENCE. As a product of the contact-metamorphism of impure limestones, in which it is associated with diopside, grossular, etc.—the modification occurring appears to be the triclinic form; the monoclinic modification occurs in limestone blocks ejected from volcanoes; also occurs in certain igneous rocks which have been contaminated with limestone, and in some nepheline-bearing basic igneous rocks.

PECTOLITE

COMP. Hydrous calcium sodium silicate, $Ca_2NaH(SiO_3)_3$

CRYST. SYST. Monoclinic. COM. FORM. Usually in masses composed of divergent or parallel fibres closely compacted.

COLOUR. White or greyish. LUSTRE. Silky or subvitreous when fibrous; dull when massive; subtranslucent to opaque. H. 5. SP. GR. 2·7–2·9.

TESTS. Heated in the closed tube, gives off water; heated before the blowpipe, fuses to a glass; yellow sodium flame; with hydrochloric acid, gelatinizes.

OCCURRENCE. Zeolitic (see p. 446); common in amygdales.

The Amphibole Family

Introduction. The amphibole group, like the pyroxenes, includes a number of important minerals whose physical and chemical characters serve to link them together in one family. X-ray analysis has shown that they all possess the Si_4O_{11} double chain type of structure (p. 352).

In *chemical composition* they are analogous with the pyroxenes and are silicates of magnesium, iron, calcium, sometimes sodium (rarely potassium), with or without aluminium. Their composition is variable, but can be represented by a formula of the type X_{7-8} $(Si_4O_{11})_2(OH)_2$, where X includes mainly Ca, Na, Mg, Fe, Al and Fe^{3+}, with traces of other elements (e.g. Mn). Some part of the Si may be replaced by Al. Hydroxyl (OH) is always present, to the extent of about one (OH)-group to every eleven oxygens.

The amphiboles *crystallize* in the orthorhombic, monoclinic and triclinic crystal systems, the prismatic angle of the crystals being nearly 120°; there is a good cleavage parallel to the prism (110), (Fig. 140).

The elementary distinctions between the pyroxenes and amphiboles—two of the most widespread and abundant rock-forming silicates—are given in the following table:

Amphiboles	*Pyroxenes*
1. Prism-angle c.124°.	1. Prism-angle c. 87°.
2. Cleavages at c. 124°.	2. Cleavages at c. 87°.
3. Bladed forms common.	3. Bladed forms uncommon.
4. Commonly pleochroic.	4. Commonly non-pleochroic.
5. Common crystals terminated by three faces.	5. Common crystals terminated by two faces.
6. Twins with no re-entrant angle.	6. Twins with re-entrant angle.
7. Orthopinacoid not present.	7. Orthopinacoid present.
8. Transverse sections, six-sided.	8. Transverse sections, eight-sided.
9. Extinction-angle on the clinopinacoid, about 16°.	9. Extinction-angle on the clinopinacoid, about 48°.

The Amphiboles may be divided into groups, as shown below, that are in some respects analogous with similar groups in the Pyroxenes:

I. ORTHORHOMBIC AMPHIBOLE.

Anthophyllite, $(Mg,Fe^{2+})_7Si_8O_{22}(OH)_2$.

II. MONOCLINIC AMPHIBOLES.

(a) CUMMINGTONITE-GRUNERITE SERIES.
Cummingtonite, $(Mg,Fe)_7Si_8O_{22}(OH)_2$.
Grunerite, $(Fe,Mg)_7Si_8O_{22}(OH)_2$.

(b) TREMOLITE-ACTINOLITE SERIES.
Tremolite, $Ca_2Mg_5Si_8O_{22}(OH)_2$.
Actinolite, $Ca_2(Mg,Fe)_5Si_8O_{22}(OH)_2$.

(c) HORNBLENDE SERIES.
Hornblende, $(Ca,Na,Mg,Fe,Al)_{7-8}(Al,Si)_8O_{22}(OH)_2$, with varieties arising from substitution among the cations (e.g. Fe^{3+} for Al).

(d) ALKALI AMPHIBOLE SERIES.

Glaucophane, $Na_2(Mg,Fe)_3(Al,Fe^{3+})_2Si_8O_{22}(OH)_2$.

Riebeckite, $Na_2Fe^{2+}_3Fe^{3+}_2Si_8O_{22}(OH)_2$.

Arfvedsonite, $Na_3Mg_4AlSi_8O_{22}(OH)_2$.

III. TRICLINIC AMPHIBOLE.

Cossyrite, an aluminium silicate of Na, Fe, and Ti.

I. Orthorhombic Amphibole

ANTHOPHYLLITE

COMP. Magnesium iron-silicate, $(Mg,Fe^{2+})_7Si_8O_{22}(OH)_2$, with some substitution of Si by Al.

CRYST. SYST. Orthorhombic. CLEAV. Perfect parallel to prism (110). COM. FORM. Usually in aggregates of prismatic needles and in radiating fibres. COLOUR. Shades of brown. LUSTRE. Vitreous; transparent to subtranslucent. H. 5·5–6. SP. GR. 3–3·2, increasing with iron content.

OPT. PROPS. In rock-slices, gives elongated prismatic sections, with positive elongation; perfect prismatic cleavages at 124° in transverse sections; colourless, yellowish or greenish, pleochroic; refractive indices for anthophyllite from Kongsberg, Norway, $\alpha = 1·633$, $\beta = 1·642$, $\gamma = 1·657$; optically negative for most examples, the optically positive variety being called *gedrite*, which contains more iron and also aluminium; straight extinction.

VARIETY. *Amosite* is a variety of anthophyllite-gedrite occurring as long fibres, and so providing one type of asbestos (see p. 395).

OCCURRENCE. As a constituent of certain metamorphic rocks, usually derived from basic or ultrabasic igneous rocks—anthophyllite-schists and gneisses.

II. Monoclinic Amphiboles

(a) Cummingtonite-Grunerite Series

CUMMINGTONITE-GRUNERITE

COMP. Magnesium iron silicate, $(Mg,Fe)_7Si_8O_{22}(OH)_2$, the relative proportions of magnesium and iron varying from about

Mg:Fe = 7:3 to all Fe; *cummingtonite* has an Mg/Fe ratio greater than 1, *grunerite* has more Fe than Mg—and ranges up to all Fe.

CRYST. SYST. Monoclinic. COM. FORM. Usually fibrous or lamellar. CLEAV. Perfect prismatic. COLOUR. Greyish-brown or brown. H. 5–6. SP. GR. 3·2–3·5, increasing with the iron content.

OPT. PROPS. In thin sections, gives elongated crystal sections; lozenge-shaped transverse sections show amphibole cleavages intersecting at about 55°; pale coloured—pleochroic in pale yellow, green or brown; refractive indices fairly high, increasing with iron content, for a cummingtonite, $\alpha = 1·639$, $\beta = 1·647$, $\gamma = 1·664$, and for a grunerite, $\alpha = 1·677$, $\beta = 1·697$, $\gamma = 1·717$; birefringence strong, greater than that of common hornblende; extinction-angle on clinopinacoid—slow vibration-direction over to cleavage—is about 18° for cummingtonite and decreases to about 10° in grunerite; cummingtonite is optically positive, grunerite negative; lamellar twinning commonly seen.

OCCURRENCE. As constituents of metamorphic rocks.

(b) Tremolite-Actinolite Series

TREMOLITE-ACTINOLITE

COMP. Varying from non-aluminous calcium magnesium silicate, $Ca_2Mg_5Si_8O_{22}(OH)_2$, (*tremolite*), to calcium magnesium iron silicate, $Ca_2(Mg,Fe)_5Si_8O_{22}(OH)_2$, (*actinolite*). X-ray analysis shows that the (OH) group enters into the composition to the extent of about one (OH) to every eleven oxygens.

CRYST. SYST. Monoclinic. COM. FORM. Crystals long, slender or blade-like prisms; also columnar, fibrous, radiating, compact or granular. CLEAV. Perfect parallel to the prism (110), the cleavage-angle being about 56°. COLOUR. Tremolite is white or dark grey, actinolite green. LUSTRE. Vitreous, especially in actinolite; transparent to translucent. H. 5–6. SP. GR. 2·9–3·2, increasing with iron content.

OPT. PROPS. In thin sections, give long bladed prismatic sections, and diamond-shaped cross-sections, the latter showing the excellent prismatic cleavage at 56°; tremolite is colourless, actinolite yellowish-green, with pleochroism in shades of yellow and green; refractive index increases with iron content, for tremolite, $\alpha = 1·599$, $\beta = 1·613$, $\gamma = 1·625$, for actinolite $\alpha = 1·628$, $\beta = 1·644$, $\gamma = 1·655$; birefringence fairly strong, polarization colours of Second Order; extinction-angle on the clinopinacoid is 15–18°; biaxial, optically negative.

VARIETIES. *Asbestos*, see immediately below; *Nephrite* is one of the varieties of *Jade* (see p. 389), and consists of very compact minute fibres of tremolite or actinolite, the former giving whitish varieties, the latter green; *Uralite* is an actinolitic amphibole which has replaced a pyroxene crystal, retaining the pyroxene shape but showing amphibole cleavages.

OCCURRENCE. Tremolite and actinolite occur in metamorphic rocks of diverse kinds; *Tremolite* occurs in impure crystalline limestones, and in calc-silicate-hornfelses, *i.e.* contact-metamorphosed impure calcareous rocks, and also in metamorphosed ultrabasic and basic rocks, such as serpentines and greenstones; *Actinolite* is a common metamorphic mineral arising by the metamorphism of pyroxenes and hornblendes and so occurring in the derivatives of basic and ultrabasic igneous rocks such as actinolite-schists and greenstones.

ASBESTOS

Mineralogically, asbestos includes the fibrous forms of amphibole; the fibres are generally very long, fine, flexible, and easily separated by the fingers; the colour varies from white to greenish and brownish. The ancients called similar material *amianthus*, undefiled, in allusion to the ease with which cloth woven from it was cleaned by throwing into the fire—the name amianthus is now restricted to the more silky kinds. *Mountain Cork*, *Mountain Leather* and *Mountain Wood* are varieties of asbestos which vary in compactness and in the matting of their fibres. The term asbestos in the strictest sense is confined to the fibrous forms of actinolite, but the asbestos of commerce includes fibrous varieties of a number of silicates which are now considered.

COMMERCIAL ASBESTOS

Under the term *commercial asbestos* are included the following fibrous minerals:

Chrysotile, fibrous serpentine (see p. 409).

Actinolite, asbestos proper.

Amosite, fibrous anthophyllite, and *Anthophyllite* (see p. 392).

Crocidolite, a fibrous soda-amphibole (see p. 398).

All these minerals occur in long fibrous crystals. The commercial value of the mineral depends almost wholly on its property of being spun, and good asbestos yields long silky fibres when rubbed between the fingers. The heat-resisting value of all the mineral varieties is about the same; chrysotile is of most general use, though certain other forms, e.g. crocidolite and amosite, are sometimes preferred for their acid-resisting qualities. Chrysotile is decomposed by hydrochloric acid, the other asbestos minerals are not.

The usefulness of asbestos depends upon its resistance to heat and its property of being spun into yarn. The better grades, those with long fibres, are woven into fire-proof fabrics and are also used for brake-linings; shorter fibres are utilized in the manufacture of asbestos sheets, boards, roofing tiles, felt, of boiler-coverings, fire-proof paints, insulating cements, etc.

Asbestos, in the commercial sense, occurs in three chief ways—as 'cross-fibre' when the fibres are at right angles to the vein-walls, as 'slip-fibre' when the fibres are parallel with the walls and are formed along planes of movement, and 'mass-fibre' when the material occurs in confused groupings as in anthophyllite types. The chief producers of commercial asbestos are Canada, Russia, Rhodesia, U.S.A., China and South Africa; the annual production is about $3\frac{1}{2}$ million tons.

Further details of the various minerals comprising commercial asbestos are given under their respective descriptions.

HORNBLENDE

COMP. Silicate of aluminium, calcium, magnesium and iron, with sodium; variable composition represented by the formula $(Ca,Mg,Fe,Na,Al)_{7-8}(Al,Si)_8O_{22}(OH)_2$, in which Al replaces some part of the Si, and the proportions of metal atoms in the first bracket may vary within the limits indicated, i.e. a total number of atoms not exceeding four to every eleven oxygens; (OH) is an essential constituent.

CRYST. SYST. Monoclinic. COM. FORM. Crystals common, prismatic in habit, being combinations of prism (110), clinopinacoid (010), clinodome (011) and hemi-orthodome (101) as sketched in Fig. 139 (in this figure, the clinopinacoid is on the right)—note the three terminal faces; the angle of the prism is

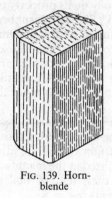

FIG. 139. Hornblende

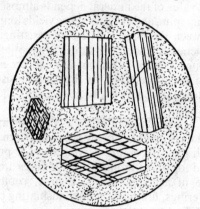

FIG. 140. Sections of hornblende, showing cleavages: above prismatic sections, below transverse section.

55°49′; also occurs in long blade-like forms, or massive or granular. TWINNING. Twin-plane the orthopinacoid (100), usually simple; this twinning causes, as it were, the four clinodome faces to come to the top and the two hemi-orthodome faces to be at the bottom of the crystal and hence these twins, though showing no re-entrant angle, can be readily recognized. CLEAV. Perfect parallel to the prism (110) producing two sets of cleavage-planes meeting at nearly 120°. COLOUR. Black, or greenish-black. LUSTRE. Vitreous; transparent rarely; translucent to opaque. FRACT. Uneven. H. 5–6. SP. GR. 3–3·47.

OPT. PROPS. In thin sections, gives six-sided transverse sections, showing two sets of cleavages meeting at nearly 120° and longitudinal sections showing one set of cleavage-lines (see Fig. 140);

colour in thin section, shades of yellow, green or brown; markedly pleochroic in these shades—the variety *basaltic hornblende* usually shows pleochroism all in brown tints; refractive indices fairly high, e.g. for pargasite from Pargas, $\alpha = 1\cdot616$, $\beta = 1\cdot620$, $\gamma = 1\cdot635$, and common hornblende from Kragerö, $\alpha = 1\cdot629$, $\beta = 1\cdot642$, $\gamma = 1\cdot653$; birefringence fairly strong, giving Second Order colours; extinction on the clinopinacoid—slow vibration-direction over to cleavage—is 18° to 20°; usually optically negative.

TESTS. Physical properties distinctive; heated before the blow-pipe, fuses easily, forming a magnetic globule.

VARIETIES. *Edenite* is a light-coloured hornblende poor in iron; *Pargasite* is a dark-green or bluish-green variety; *Basaltic Hornblende* is a brown or black variety, containing titanium and sodium, occurring in basic igneous rocks rich in iron; it is pleochroic in shades of brown.

OCCURRENCE. Hornblende occurs as a primary mineral in acid and intermediate igneous rocks—granites, syenites, diorites, etc. —more rarely in basic rocks, and sometimes in ultrabasic rocks— hornblendite; common in metamorphic rocks derived from igneous rocks—as in the hornblende-gneisses, hornblende-schists and amphibolites.

(d) Alkali Amphiboles

GLAUCOPHANE

COMP. Silicate of sodium, magnesium, iron, and aluminium, $Na_2(Mg,Fe)_3(Al,Fe^{3+})_2Si_8O_{22}(OH)_2$.

CRYST. SYST. Monoclinic. COM. FORM. Prismatic crystals; usually in fibrous, massive or granular forms. CLEAV. Perfect prismatic. COLOUR. Blue, bluish-black, bluish-grey. LUSTRE. Vitreous; translucent. H. 6–6·5. SP. GR. 3–3·1.

OPT. PROPS. In thin section, strongly pleochroic in blue, violet, yellow-green tints; extinction on the clinopinacoid low, about 5°; optically negative.

OCCURRENCE. In metamorphic rocks, e.g. glaucophane-schists, produced by the metamorphism of soda-rich igneous rocks, and as a constituent of metamorphic rocks of sedimentary origin also.

RIEBECKITE

COMP. Essentially a sodium iron silicate, with a formula of the type $Na_2Fe^{2+}{}_3Fe^{3+}{}_2Si_8O_{22}(OH)_2$.

CRYST. SYST. Monoclinic. COM. FORM. Prismatic crystals, radiating tufts and rather shapeless aggregates. CLEAV. Perfect prismatic. COLOUR. Blue or nearly black. LUSTRE. Vitreous. H. 4. SP. GR. 3·43.

OPT. PROPS. Usually appears as small shapeless aggregates of needles, etc.; strongly pleochroic in deep blue to green—the body-colour often exceedingly strong, the mineral being almost opaque in some sections; refractive index high, birefringence low, but polarization colours are usually masked by the body-colour; extinction on clinopinacoid very low, only a few degrees.

VARIETIES. *Crocidolite* is probably a variety of riebeckite, indigo-blue in colour and fibrous in structure, and forming one of the varieties of commercial asbestos (see above, p. 395); found in Griqualand, South Africa, and elsewhere; by alteration it assumes a golden-yellowish brown colour, and, when infiltrated with silica, constitutes the *Cat's Eye* or *Tiger's Eye* used for ornaments; *Crossite* is a soda-amphibole intermediate between glaucophane and riebeckite.

OCCURRENCE. As a constituent of acid igneous rocks rich in soda, riebeckite-granite, riebeckite-granophyre, etc.

ARFVEDSONITE

COMP. Sodium magnesium aluminium silicate, $Na_3Mg_4AlSi_8O_{22}(OH)_2$.

CRYST. SYST. Monoclinic. COM. FORM. Elongated prismatic crystals, often in aggregates. CLEAV. Perfect prismatic. COLOUR. Black. LUSTRE. Vitreous; opaque. H. 6. SP. GR. 3·45.

OPT. PROPS. In thin section, strongly pleochroic in blue and yellow.

VARIETIES. *Kataphorite* and *Barkevikite* are varieties of arfvedsonite with brownish tints among their pleochroic schemes.

OCCURRENCE. As constituents of igneous rocks rich in soda, such as nepheline-syenites and related pegmatites.

III. Triclinic Amphibole

Cossyrite, Aenigmatite, is a silicate of iron, titanium and sodium with aluminium also; it occurs as black, triclinic crystals in the soda-rich trachytes (pantellarites) of Pantellaria in the Mediterranean, and in similar rocks in East Africa; under the microscope it appears as small brown intensely pleochroic sections; though usually considered with the amphiboles, cossyrite shows important structural differences from this group and pyroxene-type chains may be present.

E. PHYLLOSILICATES

The Mica Family

Micas are distinguished by a perfect basal cleavage, which causes them to split up into thin elastic plates, and by their splendent pearly, somewhat metallic, lustre. They all crystallize in the monoclinic system, but the forms approximate to those of the hexagonal system. They possess the Si_4O_{10}-sheet structure (see pp. 353-4, and Fig. 126), and the perfect cleavage takes place parallel to the Si_4O_{10}-sheets.

In composition the micas are silicates of aluminium and potassium, together with magnesium and iron in the dark varieties such as biotite; some varieties contain sodium, lithium, or titanium. Hydroxyl is always present, and is commonly replaced in part by fluorine.

When a blunt steel punch is placed on a cleaved plate of mica and lightly struck, a small six-rayed star, the *percussion-figure*, is produced. The three cracks which constitute these stars have a constant relation to the form of the crystal from which the plate is cleaved, and one of the cracks is always in the direction of the plane of symmetry. Cleavage-plates of micas give a biaxial interference figure in convergent polarized light, and, by the orientation of this figure with regard to the plane of symmetry, as revealed by the percussion-figure, micas may be divided into two groups. In the first of these groups, the '*Muscovites*', a straight line joining the eyes of the interference figure (that is, the trace of the optic axial plane, see p. 189) is perpendicular to the plane of symmetry, and in the second group, the '*Biotites*', this line lies in the plane of symmetry.

The members of the family dealt with here are:

Muscovite—potassium mica, white mica.
Paragonite—sodium mica.
Lepidolite—lithium potassium mica.
Biotite—iron magnesium mica, black mica.
Phlogopite—magnesium mica.
Zinnwaldite—lithium biotite.
Glauconite.

The specific gravity of the micas ranges from 2·7 to 3·1, and the average hardness is 2·5.

The micas differ from the chlorites and other micaceous minerals in (1) their content of alkalies, (2) the elasticity of their cleavage-flakes, and (3) certain optical properties mentioned below.

Muscovite and phlogopite are of considerable industrial importance, especially in the electrical industry; lepidolite is an ore of lithium.

MUSCOVITE, Common Mica, Potash Mica, Muscovy Glass

COMP. Silicate of aluminium and potassium, with hydroxyl and fluorine, $KAl_2(AlSi_3)O_{10}(OH,F)_2$. Al is substituted for Si in the (Si_4O_{10})-group to the extent of about one atom in four.

CRYST. SYST. Monoclinic, pseudo-hexagonal. COM. FORM. Six-sided tabular crystals, large plates, massive, or in disseminated scales. CLEAV. Perfect parallel to the basal pinacoid, large and very thin laminae being easily separated; these laminae are flexible and elastic (*cp.* chlorite, p. 406); when held up to a bright light these laminae may exhibit *asterism*, starlike rays of light being transmitted. COLOUR. White, black, brown, yellow or green. LUSTRE. More or less pearly; transparent to translucent. H. 2–2·5. SP. GR. 2·76–3.OPT. PROPS. In thin

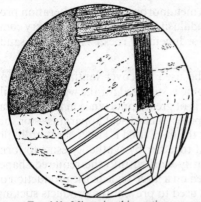

FIG. 141. Micas in thin section:
above *biotite*, below *muscovite*

sections, muscovite appears as shapeless plates or elongated laths, the vertical sections showing the perfect basal cleavage (see Fig. 141, lower part); colourless; the refractive indices average about $\alpha = 1\cdot560$, $\beta = 1\cdot593$, $\gamma = 1\cdot600$—note that the lowest index is not much higher than that of balsam whereas the other two are considerably higher; birefringence strong, the polarization colours for most sections being of high orders, giving delicate pinks and greens—note, however, that basal sections, containing β and γ, show very low polarization colours, as may be seen in flakes of detrital muscovite occurring in sediments; cleavage-flakes, or sections showing no cleavage and giving very low polarization colours, yield a biaxial interference figure, with moderate-sized optic axial angle, optically negative.

TESTS. Before the blowpipe, muscovite whitens and fuses only on thin edges; not decomposed by acids; yields water when heated in the closed tube.

VARIETIES. *Sericite*, *Damourite*, *Gilbertite* are secondary micas resulting in many cases from the alteration of numerous rock-forming minerals, such as feldspar, andalusite, etc.; these varieties occur as fine scales or fibres.

OCCURRENCE. Muscovite occurs as an original constituent of acid igneous rocks, such as granite and pegmatite. Workable quantities of muscovite occur in large plates in pegmatites, the chief producers being India, Russia, the United States and Canada—but most of the mica produced in Canada is phlogopite, for which see below, p. 404. Muscovite is a common constituent of metamorphic rocks—gneisses and mica-schists. The secondary micas occur in igneous and metamorphic rocks as alteration products of feldspars, topaz, andalusite, etc. Muscovite is a very common constituent of detrital sedimentary rocks—micaceous sandstones, clays, etc.

USES. Muscovite used to be used to cover lanthorns, and for lamp chimneys, etc., and some is still used for the windows of oil-stoves, etc. Mica in general is of great importance in the electrical industry for insulating purposes in electrical apparatus—for this purpose it is indispensable; the mica used is either sheet or 'built-up' that is, sheets cemented together by synthetic resins to form blocks, etc., that can be easily worked into the required shape. Ground-up mica is employed on a large scale in the production of roofing material where it is used to prevent the products sticking to one another, and it is also used in the manufacture of lubricants, wall-finishes, artificial stone, rubber tyres, and to give a gloss to wallpaper; powdered mica is used to give the 'frost' effect on Christmas cards and for Christmas-tree 'snow'.

PARAGONITE, Sodium Mica

COMP. Silicate of aluminium and sodium with hydroxyl, $NaAl_2(AlSi_3)O_{10}(OH)_2$, corresponding to muscovite.

CHARACTERS AND OCCURRENCE. A yellowish or greenish mica resembling muscovite in general properties, and occurring in scales in certain metamorphic rocks associated with garnet, kyanite and staurolite.

LEPIDOLITE, Lithium Mica

COMP. A silicate of aluminium, potassium and lithium, with hydroxyl and fluorine, $K(Li,Al)_3(Si,Al)_4O_{10}(OH,F)_2$.

CRYST. SYST. Monoclinic. COM. FORM. Occurs in forms like those of muscovite, but mostly in masses composed of small scales or granules. CLEAV. Perfect basal, as in muscovite. COLOUR. Rose-red, lilac, violet-grey, sometimes white. LUSTRE. Pearly; translucent. H. 2·5–4. SP. GR. 2·8–2·9.

TESTS. Heated before the blowpipe gives red lithium flame; usually gives reaction for fluorine.

OCCURRENCE. Occurs in pegmatites, associated with tourma-line, topaz, and other minerals of pneumatolytic origin, as in the Eastern United States, Elba, Madagascar, etc.; it occurs as a gangue-mineral of tin veins at Zinnwald, Saxony.

BIOTITE

COMP. Silicate of magnesium, iron, aluminium and potassium, with hydroxyl and fluorine, $K(Mg,Fe)_3(AlSi_3)O_{10}(OH,F)_2$. Iron replaces magnesium to a variable extent.

CRYST. SYST. Monoclinic, pseudo-hexagonal. COM. FORM. Six-sided prismatic crystals usually broad and tabular parallel to the basal pinacoid. CLEAV. Perfect basal affording extremely thin flexible and elastic laminae. COLOUR. Black or dark green in thick crystals, while by transmitted light thin laminae appear brown-green or blood-red. LUSTRE. Splendent, and on the cleavage more or less pearly; transparent to opaque. H. 2·5–3. SP. GR. 2·7–3·1.

OPT. PROPS. In thin sections, biotite appears as plates or laths, the perfect basal cleavage being shown in suitable sections (see Fig. 141, upper part, p. 401); coloured, and strongly pleochroic in brown, reddish-brown and yellow, the maximum absorption occurring when the cleavages are parallel with the short diagonal of the polarizer; refractive indices higher than that of balsam, e.g.—$\alpha = 1·584, \beta = 1·648, \gamma = 1·648$; birefringence for vertical sections strong, but polarization colours often masked by the strong body-colour; basal sections show little pleochroism, are almost isotropic and yield a negative almost uniaxial interference figure; often alters into chlorite.

TESTS. Heated with fluxes, gives a strong iron reaction; decomposes in strong sulphuric acid, leaving a residue of siliceous scales.

VARIETIES. *Haughtonite* and *Lepidomelane* are varieties of biotite rich in iron.

OCCURRENCE. Occurs as an original constituent of igneous rocks of all kinds—granites, diorites, gabbros, etc., and their dyke and volcanic representatives; it occurs abundantly as a mineral of metamorphic origin in biotite-gneisses and biotite-schists and in contact-altered clayey rocks, biotite-hornfelses.

PHLOGOPITE

Comp. Silicate of aluminium, magnesium and potassium with hydroxyl and fluorine, $KMg_3(AlSi_3)O_{10}(OH,F)_2$.

Cryst. Syst. Monoclinic. Com. Form. Six-sided prismatic crystals common; also in scales. Cleav.—Perfect parallel to the basal pinacoid, giving thin tough and elastic laminae; asterism is often shown. Colour. White, colourless, brown, copper-red. Lustre. Pearly, often submetallic on cleavage-planes. H. 2·5–3. Sp. Gr. 2·78–2·85.

Opt. Props. In thin sections appears as flakes or laths, with perfect basal cleavage in suitable sections; usually not deeply coloured, and pleochroic from colourless to pale yellows or browns; refractive indices and birefringence as for other micas; in convergent light, cleavage-flakes or basal sections give an interference figure with varying optic axial angle, sometimes as much as 50°.

Occurrence. Phlogopite is a mineral resulting from the dedolomitization of an impure dolomite and is therefore found embedded in crystalline limestones where it is associated with forsterite, diopside, etc.; it is found also in igneous rocks rich in magnesia, but is not common in this association. Phlogopite is extensively worked in Ontario from pockety deposits occurring in crystalline limestone which is invaded by intrusions of pyroxenite.

Uses. See under muscovite.

ZINNWALDITE

Zinnwaldite is a mica whose composition is like that of biotite but which contains lithium in addition. It is violet, pale yellow or brown in colour, and occurs in modifications of the Zinnwald, Erzgebirge, granite, associated with tinstone, etc., and in Cornwall and elsewhere, and is of pneumatolytic origin.

GLAUCONITE

Comp. Essentially a hydrous silicate of iron and potassium, though aluminium, magnesium and calcium are often present, approximately $K(Fe^{3+},Al)_2(Si,Al)_4O_{10}(OH)_2$. Glauconite has been shown to have a similar structure to that of mica.

CHARACTERS AND OCCURRENCE. In form glauconite is amorphous granular or earthy, and in colour, olive-green, yellowish, greyish and blackish-green; it has a dull or glistening lustre, and is opaque; its hardness is 2, and its specific gravity, 2·2–2·4; heated before the blowpipe, it fuses easily to a dark magnetic glass, and gives off water; it occurs extensively disseminated in small grains in the chalk marl, chloritic marl and greensands of the Cretaceous of England, and also in rocks much older than these, as for example, the Cambrian Comley Sandstone; besides such occurrences, it is found in oceanic sediments now in actual process of formation; may arise by the alteration of ferromagnesian silicates, especially biotite. Its presence in a sediment indicates a shallow-water, marine origin for the deposit.

Brittle Micas

MARGARITE

COMP. Silicate of calcium and aluminium with hydroxyl, $CaAl_2(Si_2Al_2O_{10})(OH)_2$.

CRYST. STRUCT. Calcium ions take the place of the potassium ions of muscovite (see p. 353) and in consequence the ratio of Al is increased in the $(Si,Al)_4O_{10}$ group. CRYST. SYST. Monoclinic.

COM. FORM. Micaceous aggregates. CLEAV. Perfect basal, like muscovite, but yielding non-elastic plates hence the name brittle mica. COLOUR. Pink, white or grey. H. 3·5–5, brittle. SP. GR. 3.

OCCURRENCE. Associated with corundum, and in mica-schists.

RELATED MINERALS. Other brittle micas are *clintonite* and *chloritoid* (see p. 371).

The Chlorite Family

Introduction. Under the general name of 'chlorite' are included many allied minerals which are related in composition to the micas, but which contain no alkalies. They have the Si_4O_{10}-sheet structure (p. 353), and in general may be considered as hydrous silicates of aluminium, iron and magnesium. Varieties depend on the proportions of iron and magnesium.

The chlorites are monoclinic, and some members are pseudo-hexagonal. They are green in colour, from which fact they derive their name. The chlorites have a perfect basal cleavage, giving flakes which are flexible but not elastic (*cf.* mica). They are very soft, their hardness averaging about 2.

Numerous species of chlorite have been described, but few are sufficiently well-defined to be described here. Among the varieties are:

> *Clinochlore*, a chlorite occurring in monoclinic tabular crystals distinctly biaxial, and optically positive.
> *Penninite*, a chlorite occurring in pseudo-hexagonal crystals, but really monoclinic; gives an almost uniaxial inter-ference-figure, optically positive or negative.
> *Ripidolite*, occurs in tubular, radiating and granular forms.

CHLORITE

COMP. Hydrous silicate of aluminium, iron and magnesium, approximately $(Mg,Fe)_5Al(AlSi_3)O_{10}(OH)_9$.

CRYST. SYST. Monoclinic; forms are at times pseudo-hexagonal.
COM. FORM. In tabular crystals; commonest in granular masses, disseminated scales and foliae in metamorphic rocks; frequently encrusting; also in forms with a compact radiating structure.
CLEAV. Perfect parallel to the basal pinacoid, giving flexible, but not elastic, flakes. COLOUR. Green of various shades. LUSTRE. Rather pearly; subtransparent to opaque. FEEL. Feels very slightly greasy when granular or in scales. H. 1·5–2·5; scratched by fingernail. SP. GR. 2·65–2·94.

OPT. PROPS. In thin section, chlorite appears either as radiating aggregates filling cavities, as an alteration product of minerals such as biotite, hornblende, etc., or as small irregular flakes and laths; cleavage usually seen; faintly to moderately pleochroic in shades of green and yellow; lamellar twinning frequent; refractive indices moderate, e.g. $\beta = 1·56–1·60$; birefringence very low—polarization colours usually abnormal, ultra-blues and ultra-browns (see p. 183); biaxial positive (*clinochlore*), uniaxial positive or negative (*penninite*).

TESTS. Distinguished by colour and physical properties; heated in closed tube gives water; fuses with difficulty on thin edges.

OCCURRENCE. In igneous rocks, chlorite is a secondary mineral resulting from the alteration of biotite, hornblende, etc.; as a filling in amygdales; abundant as a mineral of metamorphic origin in rocks both of sedimentary and igneous parentage—chlorite-phyllites, chlorite-schists, etc.

Hydrous Magnesium Silicates

This division includes talc, serpentine and meerschaum, which are all hydrous silicates of magnesium. The chemically more complex magnesium silicate vermiculite, in which aluminium and iron may be present, is for convenience placed in this group.

Talc. $Mg_3[Si_4O_{10}](OH)_2$.
Serpentine. $Mg_6[Si_4O_{10}](OH)_8$.
Meerschaum. $Mg_4[Si_6O_{15}](OH)_2.6H_2O$.
Vermiculite. $Mg_3[Si_4O_{10}](OH)_2.nH_2O$.

TALC

COMP. Hydrous magnesium silicate, $Mg_3[Si_4O_{10}](OH)_2$.

CRYST. STRUCT. Sheet-structure, see p. 354. CRYST. SYST. Monoclinic. COM. FORM. Crystals rare, tabular; often massive with foliaceous structure; also granular-massive, compact and cryptocrystalline. CLEAV. Perfect basal, giving thin plates which are flexible but not elastic (*cp.* mica). COLOUR. White, silvery white, apple-green, greenish-grey, dark green. LUSTRE. Pearly; subtransparent to translucent. FEEL. Greasy. H. 1; softest grade in Mohs' scale; cut easily with a knife, or scratched with a fingernail. SP. GR. 2·7–2·8.

OPT. PROPS. Usually occurs under the microscope as laths, wisps or plates in schistose rocks; colourless; refractive indices mostly higher than that of balsam, $\alpha = 1·539$, $\beta = 1·589$, $\gamma = 1·589$; birefringence strong—polarization colours delicate high order pinks and greens; optically negative, small optic axial angle; talc in section resembles muscovite, but is readily distinguished in the hand-specimen of the rock, and also by its usual associates in rocks—serpentine, chlorite, tremolite, etc.

TESTS. Heated alone before the blowpipe, whitens and exfoliates, fusing to an enamel on the edges only; not decomposed by acids, except the variety rensselaerite.

VARIETIES. *Steatite, Soapstone* is a massive variety of talc, mostly white, or grey of various shades, sometimes greenish or reddish, and having a greasy or soapy feel; *Potstone* is an impure massive talc or soapstone, in colour greyish-green, dark-green, iron-grey or brownish-black; it is easily turned on the lathe and, as it stands the fire well, it is made into vessels for cooking; *Rensselaerite* is a variety of soapstone pseudomorphous after pyroxene and occurring in Jefferson County, New York, and Canada; it takes a high polish and is made into inkstands and ornamental articles; colour is white, yellow or black, and it is harder than normal talc; *French Chalk* is a steatite used by tailors for marking cloth; *Indurated Talc* is an impure slaty variety, somewhat harder than French Chalk.

OCCURRENCE. Talc occurs as a secondary mineral resulting from the hydration of magnesium-bearing rocks, such as peridotites, gabbros, dolomites, etc.; the change may be produced in various ways—by the contact-action of granitic magmas, by the action of stress during regional metamorphism, or by the action of magmatic waters. Talc thus occurs most commonly in the crystalline schists, as talc-schists and steatite. It is produced in the United States, Japan, France, Norway, Italy, India, Austria and elsewhere.

USES. In addition to the uses mentioned in the description of the varieties, talc is employed as a filler for paints, paper, rubber, etc., and in plasters, foundry-facings, and lubricants, for removing grease from cloth, etc., in leather-making, for crayons, toilet powder, etc., and as an absorbent for nitroglycerine; soapstone slabs are employed for hearthstones, switchboards, sinks, laboratory table-tops, acid tanks, etc.; harder varieties are carved into ornaments.

SERPENTINE

COMP. Hydrous magnesium silicate, $Mg_6[Si_4O_{10}](OH)_8$.

CRYST. SYST. Monoclinic. COM. FORM. Crystals not known, the crystal system being proved by optical properties; occurs massive, granular or fibrous, sometimes foliaceous. CLEAV. One distinct parting. COLOUR. Different shades of green to almost black, sometimes red, yellow or brown; massive varieties are often veined and spotted with white, green, red, etc.; the white veins are often steatite which in many cases envelop crushed fragments of the darker serpentine, thus producing very ornamental patches of breccia. LUSTRE. Subresinous, greasy; translucent to opaque. FEEL. Sometimes slightly soapy. FRACT. Conchoidal, tough. H. 3–4; can be cut with a knife; easily turned into vases, chimney-pieces, etc., and is used in internal architectural decoration. SP. GR. 2·5–2·6.

OPT. PROPS. Under the microscope, serpentine appears often as a complete or partial pseudomorph after an olivine-, pyroxene- or amphibole-rock—the alteration from olivine being most common, and in this case serpentine occurs as rounded pseudomorphous crystals showing a network of black particles of magnetite, which has been thrown out of combination by the alteration of the olivine into serpentine; colour greenish-yellow or colourless; refractive indices lowish, $\alpha = 1·560$, $\beta = 1·570$, $\gamma = 1·571$, not much higher than that of balsam; birefringence weak, about that of quartz—polarization colours being low order greys, and in some sections the mineral appears almost isotropic; two varieties may be distinguished, (a) *fibrous*, *chrysotile*, optically positive, elongation positive; and (b) *lamellar*, *antigorite*, optically negative, elongation positive.

TESTS. Yields water on heating; fuses on the edges with difficulty; recognized by colour and softness.

VARIETIES. *Precious* or *Noble Serpentine* is a translucent variety, oil-green in colour; *Antigorite* is a lamellar variety characteristic of serpentines produced under stress in dislocation-metamorphism; optically negative; *Chrysotile* is a fibrous variety occurring as small veins in massive serpentine; it has the physical properties of asbestos and forms part of the asbestos of commerce, of which it furnishes the most important part; it is extensively worked in Eastern Canada (see p. 395); *Picrolite* is a columnar variety occurring especially along planes of slight shear. *Bastite*, *Schillerspar* is a serpentine resulting from the alteration of rhombic pyroxene (see p. 382); it is olive-green, blackish-green or brownish in colour; lustre of cleavage-surfaces (010) metallic to pearly. H. 3·5–4. Sp. Gr. 2·5–2·7; in thin section, bastite appears as a fibrous pseudomorph after a rhombic pyroxene, the fibres lying parallel to the *c*-axis of the pyroxene—the polarization colours are low, pleochroism weak; it is found in serpentine at Baste in the Harz, in the Cornish serpentines and in most occurrences of serpentine, and tends to enhance the beauty of the stone; rhombic pyroxenes are often replaced by bastite in basalts, andesites, etc. *Ophicalcite* is a rock made up of green serpentine in white calcite and results from the dedolomitization on metamorphism of a siliceous dolomite; forsterite (olivine) and calcite are first formed, and then the olivine is altered into serpentine.

OCCURRENCE. Serpentine results from the alteration, either by metamorphism of one kind or another or by end-stage action, of rocks rich in magnesium, and therefore of those rocks containing olivine, pyroxene or amphibole, such as peridotites, picrites, etc., as at the Lizard, Portsoy in Banffshire, Unst and Fetlar in the Shetlands, Galway, Wicklow, Eastern Canada, Eastern United States, etc.; Serpentine is also formed by dedolomitization, as mentioned above under ophicalcite.

USES. As a building stone, and for ornamental work in general, chiefly interior decoration; the fibrous varieties supply a great proportion of the asbestos of commerce (see p. 395).

MEERSCHAUM, Sepiolite

COMP. Hydrous magnesium silicate, $Mg_4[Si_6O_{15}](OH)_2.6H_2O$.

CRYST. SYST. Appears to be a mixture of amorphous material (meerschaum) and fibrous biaxial material (sepiolite). COM. FORM. In compact earthy clay-like amorphous masses, or fibrous. COLOUR. White, greyish-white, sometimes with a faint yellowish or pink tint. LUSTRE. Dull and earthy; opaque. H. 2–2·5; easily scratched with the fingernail. SP. GR. 2; when dry, floats on water.

TESTS. Decomposed in hydrochloric acid with gelatinization; when heated gives off water; heated on charcoal with cobalt nitrate gives a pink mass.

OCCURRENCE. In beds or irregular masses in alluvial deposits derived from serpentine masses, as in Asia Minor; in veins with silica, possibly derived from dolomitic rocks, as in New Mexico; and in small quantity in many serpentine deposits. It is procured chiefly from Turkey, Samos and Negropont in the Grecian Archipelago, Morocco and Spain, and from Kenya and Tanzania.

USES. Its chief use is for making pipes and pipe-bowls, for which purpose it is admirably adapted by reason of its absorbent nature and its lightness; moreover, when well coloured by long smoking, it displays additional beauty in the eyes of connoisseurs. Before the meerschaum is made into pipes it is soaked first in tallow and afterwards in wax, and it then takes a good polish. It was formerly used in North Africa as a substitute for soap. Occasionally it is used as a building-stone.

VERMICULITE

COMP. Hydrous magnesium silicate, $Mg_3[Si_4O_{10}](OH)_2.nH_2O$, but with Al replacing Si with consequent adjustment among the cations by introduction of Mg, Fe and Al.

CRYST. STRUCT. Sheet structure resembling that of talc (p. 354) but with water molecules present in some of the sheets. CRYST. SYST. Monoclinic, usually pseudomorphous after biotite or phlogopite. CLEAV. Parallel to (001). H. 1·5. SP. GR. 2·4. COLOUR. Yellow, brown, green.

TESTS. Expands on heating with the escape of steam and the forcing of the sheets apart. OCCURRENCE. As a hydrothermal modification of mica.

USES. In the building industry for the production of insulating materials, lightweight cements and plasters and for refractory purposes, these uses depending on its expansion on heating.

PRODUCTION. From Montana, U.S.A. and Palabora, South Africa.

Hydrous Aluminium Silicates, Clay Minerals

In this group is considered a number of hydrous aluminium silicates which are important constituents of various types of clays. Clays result from the alteration or weathering of rocks chiefly of igneous origin. The weathered material either remains where it is formed and so gives rise to residual clays, or is transported by various agencies—water, glacier ice, wind—and deposited as beds in the sea or in lakes, or as a blanket of boulder-clay, or as loess or adobe deposits. Clays have the properties of becoming plastic when wet, of then being easily moulded, and of becoming hard and rock-like when heated to a suitable temperature. Here are also considered briefly certain other clay-like substances, such as fuller's earth and bentonite which do not become plastic when wetted.

The clay-minerals occur as very minute flaky crystals which have the Si_4O_{10}-sheet structure (p. 353). Those dealt with here are as follow:

> *Kaolinite, Nacrite, Dickite*, $Al_4Si_4O_{10}(OH)_8$.
> *Halloysite*, $Al_4Si_4O_{10}(OH)_8.8H_2O$.
> *Montmorillonite*, $Al_4Si_8O_{20}(OH)_4.nH_2O$, with $Mg \rightarrow Al$.
> *Beidellite*, similar to Montmorillonite, but with $Al \rightarrow Si$.
> *Illite*, $K_yAl_4(Si_{8-y}Al_y)O_{20}(OH)_4$.
> *Pyrophyllite*, $Al_2Si_4O_{10}(OH)_2$.
> *Allophane*, amorphous silica-alumina gel.

With these and other related minerals are associated in clays such minerals as quartz, feldspar, micas, iron-oxides, etc. Individual species of clay-minerals cannot be distinguished from one another except by elaborate chemical or X-ray investigations.

KAOLINITE, Nacrite, Dickite, China Clay, Kaolin

COMP. Hydrous aluminium silicate, $Al_4Si_4O_{10}(OH)_8$.

CRYST. SYST. Triclinic. COM. FORM. Crystals, small pseudo-hexagonal plates; usually a very soft, fine, clayey material, crumbling to powder when pressed between the fingers. CLEAV. Basal perfect. COLOUR. White when pure; grey and yellowish. LUSTRE. Dull and earthy. FEEL AND SMELL. Greasy feel and argillaceous smell. H. 2–2·5 of crystals. SP. GR. 2·6.

OPT. PROPS. The refractive indices of the kaolin minerals are slightly higher than balsam, the birefringence is weak; three species, kaolinite, nacrite and dickite have been established on the basis of their detailed optical properties, but a discussion of these is beyond the scope of this book.

TESTS. Heated in charcoal with cobalt nitrate gives a blue mass due to aluminium; yields water on heating in closed tube; insoluble in acids.

VARIETIES. *Kaolinite*, *Nacrite* and *Dickite*, as already mentioned, are crystallized varieties founded on optical properties; *Kaolin* or *China Clay* consists of partly crystalline and partly amorphous material; *Lithomarge* is a white, yellow or reddish clay, consisting of kaolinite and halloysite, and often speckled and mottled, adhering strongly to the tongue, and having a greasy feel; it yields to the fingernail, affording a shining streak; it is infusible before the blowpipe; occurs in Cornwall, Saxony and elsewhere.

OCCURRENCE. Nacrite and dickite are relatively rare minerals occurring for the most part in association with metallic ores; kaolinite and kaolin result from the alteration of the feldspars of granites. This alteration may be caused by two processes: (1) the ordinary weathering of the feldspar, first into a clay-mineral allied to kaolinite, but with less water, and then into kaolinite, or (2) by the action of gases on the feldspar (pneumatolysis, see p. 213). The second mode of origin has been held to account for the Cornish occurrences at least, since the kaolin is there associated with cassiterite, tourmaline and other minerals of undoubted pneumatolytic origin. Workable deposits considered to have been formed by one or other of the two processes occur, and have been worked, in Cornwall, United States, France, China and Malaysia.

USES. For the manufacture of fine porcelain and china, porcelain fittings, etc., and as fillers in paper, rubber and paint manufacture.

HALLOYSITE

COMP. Hydrous aluminium silicate, $Al_4Si_4O_{10}(OH)_8.8H_2O$, like kaolinite but containing water molecules between its sheets.

CHARACTERS AND OCCURRENCE. A kaolin-like mineral, occurring with kaolinite in kaolin deposits. Reacts as for kaolinite. The brownish-yellow or red clay called *bole* includes types that are impure halloysite; bole can be scratched with the fingernail and then gives a shining streak; it breaks with a somewhat conchoidal fracture, and falls to pieces with a crackling noise when placed in water; heated before the blowpipe it fuses easily to a yellow or green enamel; it is found chiefly in Italy, Silesia and Asia Minor.

MONTMORILLONITE

COMP. Essentially $Al_4Si_8O_{20}(OH)_4.nH_2O$, with substitution of Mg for part of the Al. As a result of this substitution, positive ions such as Na^{1+} or Ca^{2+} are attached to the surfaces or edges of the minute crystals, thus balancing the negative charges which are left when Mg^{2+} takes the place of Al^{3+} (see p. 26). In this way, varieties known as sodium-montmorillonite and calcium-montmorillonite are formed. The positive ions (Na^{1+}, Ca^{2+}) are exchangeable bases, and their presence accounts for the high base-exchange capacity of the mineral.

Layers of molecular water may also occur between the aluminium-silicon sheet units in montmorillonite. This water may be driven off on heating; the mineral then swells greatly on absorbing water again—a property which gives it important uses in oilfield work (see under *Bentonite*, below).

CHARACTERS AND OCCURRENCE. A soft, white, greyish or greenish clay-mineral, occurring as an alteration-product of aluminium silicates. Montmorillonite, together with beidellite (see below), constitutes a large proportion of the clay *bentonite* which occurs as thin beds in the Cretaceous and Tertiary rocks of Western United States; bentonite is believed to result from the decomposition of volcanic ash, and is employed for various purposes such as the decolorizing of oils, water-softening, as a filler, for the thickening of drilling muds in sinking oil-wells, and as an absorbent in many processes.

BEIDELLITE

COMP. Hydrous aluminium silicate, essentially $Al_4Si_8O_{20}(OH)_4.nH_2O$ with some replacement of Si by Al.

CHARACTERS AND OCCURRENCE. Occurs in thin plates, white or reddish in colour, and usually results from the alteration of micas, etc.

ILLITE

COMP. Hydrous potassium aluminium silicate, $K_yAl_4(Si_{8-y}Al_y)O_{20}(OH)_4$.

STRUCT. Sheet structure intermediate between that of montmorillonite and that of muscovite, the three minerals probably forming a reversible genetic sequence.

OCCURRENCE. Formed by the decay of muscovite or feldspar, either by weathering or hydrothermal processes. Illite is a dominant mineral in many shales and may thus be original or changed from some other clay-mineral.

PYROPHYLLITE

COMP. Hydrous aluminium silicate, $Al_2Si_4O_{10}(OH)_2$.

CHARACTERS AND OCCURRENCE. A mineral with many of the physical properties of talc, but gives the blue aluminium reaction when heated with cobalt nitrate on charcoal; it occurs chiefly as foliated masses in crystalline schists. It is used for the same purposes as talc and is extensively quarried in North Carolina, U.S.A. and South Korea.

ALLOPHANE

COMP. Hydrous aluminium silicate. The name is now used to denote the non-crystalline part of a clay that is soluble in dilute hydrochloric acid (Grim).

Fuller's Earth is a greenish-brown, greenish-grey, bluish or yellowish material, soft and earthy in texture, with a soapy feel; it yields to the fingernail with a shining streak, and adheres to the tongue; when placed in water it falls to powder, but does not form a paste; heated before the blowpipe, it fuses to a porous slag and ultimately to a white blebby glass; it has been suggested that fuller's earth is composed dominantly of montmorillonite, but it is probable that other hydrated aluminium silicates are also present together with a variety of other minerals such as quartz, mica, glauconite, etc. Fuller's earth was formerly used for 'fulling' or cleaning woollen fabrics and cloth, its absorbent properties causing it to remove greasy and oily matters; its modern use is in the refining of oils and fats, a use depending also upon the absorbent qualities of the hydrated aluminium silicates contained in it. In England, fuller's earth is found at Nutfield, near Reigate, Detling near Maidstone, Bletchingley in Surrey, Woburn in Bedfordshire, Bath, etc.; it is worked extensively also in Florida, Georgia, Arkansas, etc., in the United States, and in Germany. It occurs chiefly as beds derived from weathered basic volcanic or pyroclastic rocks.

APOPHYLLITE

COMP. Hydrous silicate of calcium and potassium, sometimes also a little fluorine, $KFCa_4[Si_8O_{20}].8H_2O$.

CRYST. SYST. Tetragonal. COM. FORM. Crystals of two habits, as shown in Fig. 30, p. 88; first combinations of prism (100), basal pinacoid (001), and minor forms, giving crystals with a cube-like aspect; and, second, prismatic crystals, made up of prism (100) and bipyramid (111), or of bipyramid alone; also occurs massive and foliaceous. CLEAV. Perfect basal. COLOUR. Milky-white to colourless; greyish, sometimes greenish, yellowish or reddish. STREAK. Colourless. LUSTRE. Vitreous; that of basal pinacoid, pearly; translucent to transparent, rarely opaque. FRACT. Uneven; brittle. H. 4·5–5. SP. GR. 2·3–2·4.

TESTS. Heated in closed tube gives off water, exfoliates and whitens; heated before the blowpipe, exfoliates and fuses to a white vesicular enamel; colours the flame violet, due to potassium; soluble in hydrochloric acid with separation of silica.

OCCURRENCE. In amygdales in basic lavas, often accompanied by zeolites (p.446).

PREHNITE

COMP. Hydrous silicate of calcium and aluminium, $Ca_2Al[AlSi_3O_{10}](OH)_2$.

CRYST. SYST. Orthorhombic, hemimorphic. COM. FORM. Usually in botryoidal masses, with a radiating crystalline structure. COLOUR. Pale green, sometimes colourless. STREAK. Colourless. LUSTRE. Vitreous; subtransparent to translucent. FRACT. Uneven; rather brittle. H. 6–6·5. SP. GR. 2·8–2·9.

TESTS. Heated in closed tube, gives off a little water; heated before the blowpipe, fuses with intumescence to a bubbly enamel-like glass; after fusion, gelatinizes with hydrochloric acid; water is given off by prehnite only at a red heat, and accordingly this mineral is often excluded from the zeolite family.

OCCURRENCE. Occurs in certain amphibolites, crystalline limestones, etc., and with zeolites in the amygdales of basalts.

F. TECTOSILICATES

The Feldspar Family

The feldspars are the most important group of the rock-forming minerals. They are the dominant or important components of most igneous rocks, and they are employed in the classification of such rocks. They are aluminous silicates of potassium, sodium, calcium or barium, and may be considered as isomorphous compounds of the four substances listed below, the first three of these being of the greatest importance, whilst the fourth is rare. These are:

Orthoclase (Or); potassium aluminium silicate, $K[AlSi_3O_8]$.
Albite (Ab); sodium aluminium silicate, $Na[AlSi_3O_8]$.
Anorthite (An); calcium aluminium silicate, $Ca[Al_2Si_2O_8]$.
Celsian (Ce); barium aluminium silicate, $Ba[Al_2Si_2O_8]$.

The atomic structure of the feldspars is of the 3-dimensional framework type (see p. 354). Replacement of part of the silicon by aluminium occurs in all the feldspars, and the alkali metal ions (Na, K, etc.) are held in spaces in the frameworks. Isomorphous substitution of one kind of alkali metal for another gives rise to the varieties of feldspar. Thus, when Na replaces some of the K in orthoclase, the variety *soda-orthoclase* is formed; replacement of some K by Ba (an ion of similar size to that of K) gives *hyalophane*, intermediate between orthoclase and celsian. The *plagioclase series* consists of solid solutions of albite and anorthite in all

proportions; expressed in another way, the replacement NaSi →
CaAl results in a complete gradation between the two end members of the plagioclases. For the purposes of description the feldspars as a whole may be divided into the following groups:

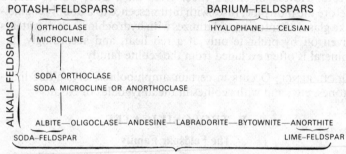

THE PLAGIOCLASE FELDSPARS

The following general characters of the Feldspars may be
noted. Their *colour* is whitish, greyish or pale shades of red. Their
hardness is about 6 and their *specific gravity* ranges from 2·5 to
over 3. With regard to their *crystallization*, orthoclase, hyalophane, celsian and soda-orthoclase crystallize in the monoclinic
system, and microcline, soda-microcline and the albite-anorthite
series in the triclinic system. The crystals of the different varieties
are like one another in *habit*—prism, side-pinacoid and basal pinacoid faces being dominant. Feldspars have two principal *cleavages*,
in the monoclinic system parallel to the basal pinacoid and the
clinopinacoid, in the triclinic system parallel to the basal pinacoid
and the side pinacoid; that is, there are two cleavages at right
angles, or nearly at right angles. *Twinning* is in general simple in
orthoclase, and repeated in the plagioclases. The alkali-feldspars
alter commonly into sericitic mica and, under certain circumstances, into kaolin; the lime-feldspars alter into saussurite, a
mixture of albite, epidote, etc. With regard to their *occurrence in
igneous rocks*, the alkali-feldspars characterize the more acid rocks,
the lime-feldspars the more basic rocks.

CELSIAN

COMP. Barium aluminium silicate, $BaAl_2Si_2O_8$; compare the composition of anorthite.

CRYST. SYST. Monoclinic. COM. FORM. In crystals; usually massive. TWINNING. As for orthoclase. CLEAV. As for orthoclase. COLOUR. Usually colourless. H. 6–6·5. SP. GR. 3·37.

OCCURRENCE. Celsian is not a common feldspar; it occurs in dolomitic limestone at Jakobsberg, Sweden, and elsewhere.

HYALOPHANE

COMP. Silicate of aluminium, barium and potassium, intermediate between orthoclase, $KAlSi_3O_8$ and celsian, $BaAl_2Si_2O_8$, with K in excess of Ba.

CHARACTERS AND OCCURRENCE. Hyalophane is a rare mineral; it is like orthoclase, and especially the adularia variety of this, in physical properties; it occurs in dolomite in the Binnental, Switzerland, and at Jakobsberg, Sweden.

ORTHOCLASE

COMP. Potassium aluminium silicate, $KAlSi_3O_8$; in some varieties there is a small amount of substitution of Na for K, and these provide a transition towards soda-orthoclase, mentioned below.

CRYST. SYST. Monoclinic. COM. FORM. Crystals are common, prismatic in habit, and made up of the prism (110), clinopinacoid (010), basal plane (001) and often hemi-orthodomes such as (20$\overline{1}$) and (10$\overline{1}$), as shown in Fig. 82, p. 147; crystals are sometimes flattened parallel to the clinopinacoid; in the variety *adularia* the clinopinacoid is usually lacking; orthoclase also occurs massive, or with roughly lamellar or granular structure. TWINNING. Twinning is common on three laws, (1) *Carlsbad Law*, the commonest, with twin-plane the orthopinacoid (100) and composition-plane the clinopinacoid (010), (2) *Baveno Law*, with the twin-plane and composition-plane the clinodome (021), and (3) *Manebach Law*, with the twin-plane and composition-plane the basal pinacoid (001)—these three types of twins are figured and described in Fig. 88 and pp. 155–156. CLEAV. Perfect parallel to the basal pinacoid (001), and slightly less good parallel to the clinopinacoid (010), thus giving in some sections two sets of cleavage-lines at right angles. COLOUR. White, reddish-white, red, flesh-coloured, various shades of grey and greenish-grey; also colourless. LUSTRE. Vitreous to pearly on cleavage; semitransparent to translucent, or opaque. FRACT. Conchoidal to uneven or splintery. H. 6. SP. GR. 2·57.

OPT. PROPS. Orthoclase occurs in rectangular sections, or as shapeless grains in rock-slices; it is usually colourless, but often cloudy owing to alteration; the rectangular cleavages are not often seen, except possibly at the edges of the slice; the refractive indices are $\alpha = 1\cdot518$, $\beta = 1\cdot524$, $\gamma = 1\cdot526$, so that all are lower than those of Canada balsam or quartz; the birefringence is low, slightly lower than that of quartz, so that polarization colours are usually greys of the First Order; twinning is always simple, the Carlsbad twins being most common, and the shape of the two halves of the twins depending on the law of twinning followed, as can be deduced by studying Fig. 88, p. 155, and Fig. 144, left, p. 428; orthoclase is biaxial, optically negative; in thin section it is distinguished from quartz by its lower refractive index, type of twinning and negative sign; albite is positive. The maximum extinction on the side pinacoid is 21°.

TESTS. Orthoclase is distinguished from plagioclase in the hand-specimen by the absence of striations due to lamellar twinning, which are seen on the basal plane in plagioclase; orthoclase fuses only on the edges of thin splinters; heated with borax, forms a transparent glass; insoluble in the microcosmic salt bead; unaffected by acids; gives the potassium flame only with difficulty.

VARIETIES. *Common Orthoclase* includes the subtranslucent or dull varieties; *Adularia* is the low-temperature variety, often colourless, occurring in crystals made up of the prism, basal plane and hemi-orthodome; *Moonstone* is an opalescent to pearly variety—the name is also applied to some varieties of albite; *Sanidine* is the high-temperature variety forming transparent glassy tabular crystals and occurring in the more acid volcanic rocks; *Sunstone* and *Aventurine Feldspar* are adularia spangled with minute crystals and plates of hematite, ilmenite, limonite, etc.; *Murchisonite* is a red feldspar with a peculiar yellowish-golden lustre.

OCCURRENCE. Orthoclase occurs as an essential constituent of the more acid igneous rocks, such as granite, syenite, felsite, rhyolite and trachyte; it is found in large crystals as a constituent of the pegmatites, and occurrences of this type are worked in the eastern United States, W. Germany, Norway, Sweden and elsewhere; it is a common mineral in metamorphic rocks; it occurs also as a veinstone, and the variety adularia lines cavities in metamorphic rocks; orthoclase also occurs as grains in feldspathic sandstones or arkoses, these resulting from the incomplete alteration and sorting of weathered granitic rocks.

USES. The alkali-feldspars are used in the manufacture of porcelain and pottery, for the production of glazes on earthenware, sanitary ware and enamelled brick, etc., in the manufacture of opalescent glass, as a binder for abrasive wheels, as a mild abrasive, and in the facing of artificial building material.

MICROCLINE

COMP. Potassium aluminium silicate, $KAlSi_3O_8$; like orthoclase.

CRYST. SYST. Triclinic; the angle between the basal pinacoid and the brachypinacoid is nearly $90°$, being $89° 30'$. COM. FORM. Crystals common, like those of orthoclase in habit, being combinations of hemiprisms, side pinacoid, and basal pinacoid; also massive and granular. TWINNING. Simple twins like those of orthoclase, and, in addition, repeated twinning on two laws, (1) the *Albite Law*, twin-plane the pinacoid (010), and (2) the *Pericline Law*, twin-axis the *b*-axis (see further under Plagioclase, p. 424); the directions of these two twinnings are nearly at right angles to one another, and show on the basal pinacoid as two sets of fine striations and cause the 'cross-hatched' appearance between crossed nicols as described below. CLEAV. The usual feldspar cleavage; perfect parallel to the basal pinacoid (001), slightly less perfect parallel to the pinacoid (010). COLOUR. Greyish-white, pinkish, or flesh-red; also bright-green as in *Amazonstone*. LUSTRE. Vitreous; transparent to translucent. H. 6–6·5. SP. GR. 2·56.

OPT. PROPS. Microcline appears in rock-slices as small shapeless grains and plates; like orthoclase in ordinary light, but between crossed nicols shows the characteristic 'cross-hatched' appearance due to the repeated wedge-shaped twinning on the Albite and Pericline laws, as shown in Fig. 142; Refractive indices α = 1·522, β = 1·526, γ = 1·530; optically negative; microcline is distinguished from other feldspars by the cross-hatched twinning, and by the positions of the twinnings in cleavage-flakes—a cleavage-flake parallel to the *basal pinacoid* (001) shows two sets of twins at right angles, both sets being sharp and distinct and not moving much when the focus is altered, and a cleavage-flake

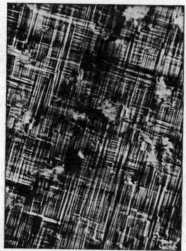

FIG. 142.   Microcline.
Between crossed nicols.

parallel to the *pinacoid* (010) shows pericline twinning making an angle of 80° with the basal cleavage; further, extinction on the basal pinacoid is 15°.

TESTS. In hand-specimen, microcline is distinguished from orthoclase by the presence of striations on the basal pinacoid due to twinning.

OCCURRENCE. Microcline is the lowest-temperature form of potassium feldspar and occurs in acid igneous rocks, especially granite and pegmatites; also in feldspathic sandstones and arkoses derived from such rocks, as in the Torridon Sandstone of the North-West Highlands of Scotland.

SODA-ORTHOCLASE

Soda-orthoclase is a link between orthoclase and albite, consisting of orthoclase in which part of the K is substituted by Na, its composition being expressed as $(K,Na)AlSi_3O_8$; it is like orthoclase in general properties.

ANORTHOCLASE or SODA-MICROCLINE

COMP. An alkali feldspar in which Na is in excess of K, with the formula $(Na,K)AlSi_3O_8$.

CRYST. SYST. Triclinic. COM. FORM. Crystals often rhomb-shaped, since the side pinacoid is usually absent; also granular. TWINNING. As in microcline; the cross-hatched twinning as seen in slices is extremely minute, a feature characterizing this feldspar.

OPT. PROPS. Like microcline, except for the extreme fineness of the cross twinning; the optic axial angle of anorthoclase is small, about 45°, and extinction on the basal pinacoid is 2°—these characters serving to distinguish this species from other feldspars.

OCCURRENCE. As phenocrysts in acid and sub-acid soda-rich lavas such as those of Pantellaria in the Mediterranean, the *Rhombenporphyries* of the Oslo district, the kenyite of Kenya, etc.

PERTHITE

Perthite consists of laminar intergrowths of albite or oligoclase in orthoclase or microcline, the soda-feldspar being arranged in wisps parallel to the orthopinacoid; *microperthite* is a finer intergrowth recognizable only under the microscope and *cryptoperthite* is a still finer intergrowth. Perthitic intergrowths most likely result from the un-mixing at lower temperatures of potash-feldspars with a small proportion of soda-feldspars which together formed homogeneous crystals at higher temperatures.

The Plagioclase Feldspars

The plagioclases, or soda-lime-feldspars, form a solid-solution series between *albite*, $NaAlSi_3O_8$ and *anorthite*, $CaAl_2Si_2O_8$. If Ab stands for albite and An for anorthite, the limits of the various members of the plagioclase series may be defined as below:

Albite, $Ab_{100}An_0$ to $Ab_{90}An_{10}$, i.e. with less than 10 per cent An.

Oligoclase. $Ab_{90}An_{10}$ to $Ab_{70}An_{30}$, i.e. with 10–30 per cent An.

Andesine, $Ab_{70}An_{30}$ to $Ab_{50}An_{50}$, i.e. with 30–50 per cent An.

Labradorite, $Ab_{50}An_{50}$ to $Ab_{30}An_{70}$, i.e. with 50–70 per cent An.

Bytownite, $Ab_{30}An_{70}$ to $Ab_{10}An_{90}$, i.e. with 70–90 per cent An.

Anorthite, $Ab_{10}An_{90}$ to Ab_0An_{100}, i.e. with more than 90 per cent An.

The plagioclases show a continuous gradation in their physical properties—specific gravity, crystal form, refractive index, etc.—from albite to anorthite, as shown below.

SPECIFIC GRAVITY. The specific gravity increases with the content of anorthite, thus, for artificial plagioclases—albite, An_0, 2·605; oligoclase, An_{25}, 2·649; andesine, An_{33}, 2·660; andesine-labradorite, An_{50}, 2·679; labradorite, An_{67}, 2·710; bytownite, An_{83}, 2·733; anorthite, An_{100}, 2·765. Natural plagioclases, however, are seldom pure enough for specific gravity to be used in their indentification.

CRYSTAL FORM. The plagioclases are triclinic in crystallization, and their crystals often resemble those of orthoclase in habit, being combinations of hemi-prisms, side pinacoid and basal pinacoid (see Fig. 84, p. 149). The axial ratios and angles between the crystallographic axes show a progressive variation from albite to anorthite, as illustrated by the angles between the basal pinacoid (001) and side pinacoid (010)—i.e. between the cleavages—given now: albite, 86°24′; oligoclase, 86°18′; andesine, 86°14′; labradorite, 86°4′; bytownite, 85°56′; anorthite, 85°50′.

CLEAVAGES. Cleavages in plagioclase are the two usual in the feldspars—perfect parallel to the basal pinacoid (001) and slightly less good parallel to the pinacoid (010), the angle between the cleavages in the various species varying as indicated in the previous paragraph.

TWINNING. The plagioclases show *simple* twinning on the Carlsbad, Baveno and Manebach laws as in orthoclase (see p. 155) and *repeated* twinning on two laws, the *Albite Law* and the *Pericline Law*. In the Albite Law, the twin-plane and composition-plane are the pinacoid (010), as indicated in Fig. 87, and this twinning is often repeated to give a series of very fine lamellae, which appear on the basal pinacoid as fine striations. In the Pericline Law, the twin-axis is the *b*-axis and the composition-plane is the *rhombic section*, which is a plane parallel with the *b*-axis and cutting the prism faces in such a way that the intersections form a rhombus; the position of the rhombic section varies in the different plagioclases, being 21° on one side of the basal cleavage in albite and 18° on the other side in anorthite; note that in the plagioclases, therefore, the pericline composition-plane makes an angle not near a right angle with the basal pinacoid—a distinction from microcline, as mentioned on p. 422.

REFRACTIVE INDEX. The refractive index increases from albite to anorthite; for albite $\alpha = 1 \cdot 525$, $\beta = 1 \cdot 529$, $\gamma = 1 \cdot 536$, and for anorthite, $\alpha = 1 \cdot 576$, $\beta = 1 \cdot 584$, $\gamma = 1 \cdot 588$. The variation of refractive indices with composition is shown in Fig. 143, where also the refractive indices of quartz and Canada balsam are shown for comparison.

EXTINCTION ANGLES. The optical orientation of the plagioclases changes with the composition. One consequence of this is that the extinction-angles of cleavage-fragments are different in the different species, and curves showing the variation are given in Fig. 143. This variation provides a method of distinguishing the various plagioclases by measuring the extinction-angle of the cleavage-fragments. The basal cleavage-fragment (001) can be recognized in a crush of the feldspar by its showing twin-lamellae of the (010) twinning, whereas the cleavage-fragments parallel to (010) show no twinning.

The maximum extinction-angles of sections cut perpendicular to the pinacoid (010), i.e. perpendicular to the albite-twinning lamellae, are important in the determination of the composition of plagioclases in thin section. Such sections can be recognized by the fact that when the twin-lamellae are parallel to a cross-wire and also when they are in the 45° positions—that is, in eight positions in a complete revolution of the stage—very few signs of albite-twinning are seen; further, there is very little shift of the composition-plane when the focus is changed. The two sets of albite twin-lamellae extinguish at the same angles, one set on either side of the cross-wire. The rock-section is searched till the plagioclase section showing the maximum extinction-angle is found, and this angle, measured from the fast vibration-direction to the

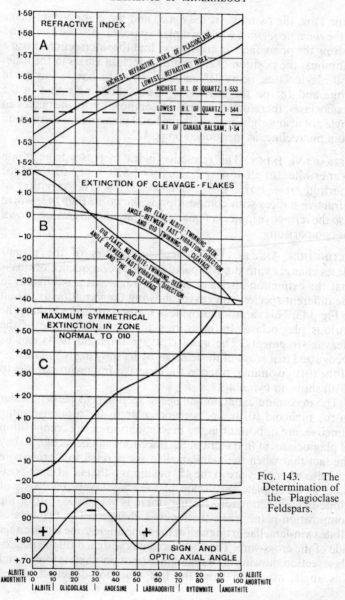

FIG. 143. The Determination of the Plagioclase Feldspars.

twinning-line, is determined. The relation between composition and extinction-angle is shown in Fig. 143, where it will be seen that further observations are necessary to distinguish between certain varieties giving the same extinction-angles.

SIGN OF PLAGIOCLASES. As a consequence of the change in the optical orientation already mentioned, the optic sign of the plagioclases depends on their composition, the relationship being shown in Fig. 143.

THE PLAGIOCLASES IN THIN SECTIONS. Plagioclase feldspars occur often as four-sided, usually elongated, sections in rock-slices, or else as more irregular plates and grains. Cleavage-cracks are rarely seen. The refractive index depends on the composition but, except for the most lime-rich members, their relief is not great. Polarization colours are low—greys of the First Order. Zoning is often seen, and lamellar twinning (see Fig. 144, right) is almost always present, and is usually accompanied by Carlsbad twinning, both types being easily observed between crossed nicols. The symmetrical extinction-angles of albite-twins vary as explained above. The plagioclases are, of course, biaxial, with large optic axial angle, their optic sign depending upon their composition, as shown in Fig. 143. The plagioclases are often cloudy with alteration products, the more sodic types giving sericitic micas and the more basic types a mixture of epidotes, new albite, etc., known as *saussurite*.

DISCRIMINATION OF THE PLAGIOCLASES. The classification of igneous rocks depends upon the nature of the feldspar present in them, so that the determination of a plagioclase in a rock-section is often a matter of importance. The following methods are recommended:

(a) *Refractive Index*. The first observation is the approximate determination of refractive index by the use of Becke's Effect (p. 165) or the Shadow Method (p. 166), a comparison being made with quartz, if it is present in the rock-slice, or with Canada balsam. Inspection of Fig. 143 shows that plagioclases with more than 50 per cent Anorthite have refractive indices all higher than the highest of quartz, and that albite, with less than about 10–12 per cent Anorthite, has refractive indices all lower than the lowest of quartz. Plagioclases with 12 to 40 per cent anorthite have refractive indices not much different from one or the other of quartz.

Most albite sections show refractive indices lower than, oligoclase not much different from, and more lime-rich feldspars higher than balsam.

(b) *Determination of Optical Sign.* If a suitable microscope is available, the determination of the optic sign of the plagioclase is important for the diagnosis of the plagioclase present. A plagioclase section showing lowest polarization colours, or almost isotropic, is chosen for preference, and the optic axis figure obtained as explained on p. 184; the sign is determined with a selenite plate (see p. 191), and the diagram of Fig. 143 consulted. The determination of refractive index and sign goes a long way towards the determination of the anorthite content of the plagioclases present.

Fig. 144.   Feldspars in thin section: Left, *orthoclase* showing Carlsbad twinning (C) and Baveno twinning (B); right, *plagioclase* showing Albite twinning—both figures between crossed nicols.

Even seemingly indeterminate observations on sign, as for instance where there is no curve of the isogyre and the optic axial angle is accordingly about 90°, are of value since, as Fig. 143 shows, only three plagioclase compositions suit such an observation.

(c) *Symmetrical Extinction-angle.* The determination of the maximum extinction-angle in sections perpendicular to the albite-twinning lamellae, as explained above, furnishes a value which can be employed in the determination of the plagioclase present. The values are given in Fig. 143.

ALBITE

COMP. Pure sodium aluminium silicate, $NaAlSi_3O_8$; rock-forming albite may contain up to 10 per cent of anorthite, $CaAl_2Si_2O_8$.

CRYST. SYST. Triclinic. COM. FORM. Often in crystals, tabular parallel to the side pinacoid, and formed of hemiprisms, side pinacoid, basal pinacoid, and hemi-macrodome, as shown in Fig. 84, p. 149; also occurs massive, with a granular or lamellar structure. TWINNING. Twinning on Carlsbad, Albite and Pericline laws very common, as explained above on p. 424, and figured in Fig. 87, p. 154. CLEAV. Perfect parallel to the basal pinacoid, almost as good parallel to the side pinacoid. COLOUR. White, sometimes with a bluish, greyish, reddish or greenish tinge; also colourless. LUSTRE. Vitreous; pearly on basal cleavage-plane; transparent to translucent. FRACT. Uneven. H. 6–6·5. SP. GR. 2·60–2·62.

OPT. PROPS. See above, p. 425; characterized by low refractive index—lower than those of quartz and balsam—positive optic sign, symmetrical extinction-angles of albite-twins 12°–16°.

TESTS. Before the blowpipe fuses with difficulty, colouring the flame yellow.

VARIETIES. *Aventurine* and *Moonstone* are two varieties corresponding to similar varieties of orthoclase; *Pericline* is a white semi-opaque variety occurring as crystals elongated along the *b*-axis; *Peristerite* is a variety with a play of colours like those shown by labradorite (see p. 431), and often containing disseminated grains of quartz; *Cleavelandite* is a lamellar variety of white albite.

OCCURRENCE. Albite occurs in the acid and intermediate igneous rocks—granite, syenite and diorite, and their hypabyssal and volcanic representatives; also as fragmental grains in arkoses and feldspathic sandstones; fairly common in the crystalline schists, especially characteristic of the albite-schists and gneisses, and of some migmatites; a principal component of adinole, a soda-rich rock formed at the contact of certain dolerites with slates, as in Cornwall.

Uses. See under Orthoclase.

OLIGOCLASE

Comp. $Ab_{90}An_{10}$ to $Ab_{70}An_{30}$, where $Ab = NaAlSi_3O_8$, albite, and $An = CaAl_2Si_2O_8$, anorthite.

Cryst. Syst. Triclinic. Com. Form. Crystals not common; usually occurs massive and cleavable. Twinning. As for albite. Cleav. As for albite. Colour. Greyish, greenish, yellowish, white, occasionally with a reddish tinge. Lustre. Resinous on cleavage-planes, vitreous or pearly; translucent on edges; when weathered, dull. Fract. Conchoidal, uneven. H. 6–6·5. Sp. Gr. 2·64.

Opt. Props. See above, p. 425, characterized by refractive index slightly lower than quartz or overlapped by quartz, negative sign, symmetrical extinction angle for albite-twinning only a few degrees.

Tests. Fuses more easily than orthoclase and albite, and forms a clear glass; insoluble in acids.

Occurrence. Oligoclase occurs as an original constituent of sub-acid and intermediate igneous rocks—syenites, diorites and their dyke and volcanic representatives; it accompanies orthoclase in many granites; it occurs in metamorphic rocks of various origins—granite-gneisses, oligoclase-biotite-gneisses of sedimentary parentage, and in many types of migmatitic gneisses.

ANDESINE

Comp. $Ab_{70}An_{30}$ to $Ab_{50}An_{50}$, where $Ab = NaAlSi_3O_8$, albite, and $An = CaAl_2Si_2O_8$, anorthite.

Cryst. Syst. Triclinic. Com. Form. Crystals not common. usually occurs massive and cleavable. Twinning. As for albite. Cleav. As for albite. Colour. White or grey. Lustre. Subvitreous to pearly. H. 6. Sp. Gr. 2·66.

OPT. PROPS. See above, p. 425, characterized by refractive indices being all higher than that of balsam, and overlapping with, or slightly higher than, those of quartz; sodic andesines are optically negative, calcic andesines optically positive, medium andesine showing a straight brush in the optic axis figure; symmetrical extinction in albite-twins is about 20°.

OCCURRENCE. Occurs as a primary mineral in intermediate igneous rocks such as diorites and andesites.

LABRADORITE

COMP. $Ab_{50}An_{50}$ to $Ab_{30}An_{70}$, where $Ab = NaAlSi_3O_8$, albite, and $An = CaAl_2Si_2O_8$, anorthite.

CRYST. SYST. Triclinic. COM. FORM. Crystals not common, thin and tabular, being flattened parallel to the pinacoid (010); usually occurs massive, crystalline and cleavable. TWINNING. See under albite. CLEAV. Usual feldspar cleavage, see under albite. COLOUR. Grey, dark ashy, brown, green, reddish, rarely colourless; often, but not always, shows a rich play of colours, in which blue and green generally predominate. LUSTRE. Vitreous, inclining to pearly on the basal cleavage-planes; subvitreous or subresinous on other surfaces; translucent but only on fine edges. FRACT. Imperfectly conchoidal, uneven, splintery, and brittle. H. 6. SP. GR. 2·67.

OPT. PROPS. See above, p. 425; characterized by all refractive indices being clearly above those of quartz and balsam, by its positive sign, and by its symmetrical extinction-angle in albite-twins of 27–37°.

TESTS. Fuses rather more easily than orthoclase or oligoclase to a colourless glass; powdered labradorite is soluble in hot acid.

OCCURRENCE. Occurs as a primary constituent of basic igneous rocks, such as gabbro, norite, dolerite and basalt.

BYTOWNITE

COMP. $Ab_{30}An_{70}$ to $Ab_{10}An_{90}$, where $Ab = NaAlSi_3O_8$, albite, and $An = CaAl_2Si_2O_8$, anorthite.

CRYST. SYST. Triclinic. COM. FORM. Massive in igneous rocks. TWINNING. See under albite. CLEAV. See under albite. COLOUR. Grey, dark grey, bluish. SP. GR. 2·72. OPT. PROPS. See above, p. 425; characterized by high refractive index for a plagioclase, all refractive indices being well above those of quartz, by negative sign, and by symmetrical extinction angle in albite-twins of more than 38°.

OCCURRENCE. As a primary constituent of basic and ultrabasic igneous rocks, such as olivine-gabbro.

ANORTHITE

COMP. $Ab_{10}An_{90}$ to Ab_0An_{100}, where $Ab = NaAlSi_3O_8$, albite, and $An = CaAl_2Si_2O_8$, anorthite.

CRYST. SYST. Triclinic. COM. FORM. Prismatic crystals elongated parallel to the c-axis, or crystals elongated along b-axis; usually massive and cleavable. TWINNING. See under albite. CLEAV. Usual feldspar cleavage, perfect parallel to the basal pinacoid, not so good parallel to the pinacoid (010). COLOUR. Colourless or white. LUSTRE. Vitreous; pearly on cleavage-planes; transparent to translucent. FRACT. Conchoidal. H. 6–6·5. SP. GR. 2·74.

OPT. PROPS. See above, p. 425; characterized by its very high refractive index for a plagioclase, and by its positive sign.

TESTS. Fuses to a glass; soluble in hydrochloric acid with some separation of gelatinous silica.

OCCURRENCE. Occurs as a primary constituent in basic and ultrabasic igneous rocks, such as gabbros of various types; also as a constituent of the calc-silicate-hornfelses, produced by the thermal metamorphism of impure limestones and marls.

The Feldspathoid Family

In the Feldspathoid Family is grouped a number of minerals which in certain respects show similarities with the feldspars, especially in their types of chemical composition. The members of the Feldspathoid Family considered here are:

Leucite, $K[AlSi_2O_6]$.
Nepheline, $Na[AlSiO_4]$ (artificial).
Naturally occurring nepheline contains some $K[AlSiO_4]$, *Kalsilite*.
Cancrinite, $4(NaAlSiO_4).CaCO_3.H_2O$ (approx).
Sodalite, $3(NaAlSiO_4).NaCl$.
Hauyne, $3(NaAlSiO_4).CaSO_4$.
Nosean, $3(NaAlSiO_4).Na_2SO_4$.
Lazurite, $3(NaAlSiO_4).Na_2S$.

It will be noticed that leucite differs in composition from ortho-clase, $KAlSi_3O_8$, in having a lower ratio of silica to alumina, and that artificial nepheline differs from albite, $NaAlSi_3O_8$, in a similar way. The other feldspathoids show a similar deficiency of silica when compared with albite, and, in addition, contain such molecules as $CaCO_3$, $NaCl$, $CaSO_4$ or Na_2S. The feldspathoids, except lazurite, are primary constituents of igneous rocks, and it follows from this character of their chemical composition that primary quartz and feldspathoid never occur together in the same rock; if free silica had been present in the magma it would have combined with the feldspathoid to form a feldspar. Further, this deficiency of silica restricts the occurrence of feldspathoids (except lazurite again) to igneous rocks low in silica and rich in alkalies—such rocks as nepheline- and leucite-syenites, phono-lites, and nepheline- and leucite-basalts, etc.

LEUCITE, Amphigene

COMP. Potassium aluminium silicate, $KAlSi_2O_6$.

CRYST. SYST. Cubic at 500–600 °C; when leucite is heated to that temperature, its double refraction and other anomalous properties seen at lower temperatures disappear; at ordinary temperatures the crystal angles are approximately those of a tetragonal form. COM. FORM. Crystals are in forms resembling that of the trapez-ohedron (211), rarely with the cube (100) and rhombdodeca-hedron (110) as subordinate forms; also in disseminated grains. CLEAV. Rhombdodecahedral, but very imperfect. COLOUR. White or ashy grey. STREAK. Colourless. LUSTRE. On fractured surfaces, vitreous; translucent to opaque. FRACT. Conchoidal; brittle. H. 5·5–6. SP. GR. 2·5.

OPT. PROPS. Leucite in thin sections of rocks shows rounded water-clear crystals, sometimes containing inclusions symmetrically arranged (see Fig. 145, upper). Its refractive index, 1·508–1·509, is below that of Canada balsam and, although usually isotropic, it may show very low polarization colours or twinning bands, as explained above.

TESTS. Infusible before blowpipe; gives blue aluminium coloration when heated with cobalt nitrate in the oxidizing flame; soluble in hydrochloric acid without gelatinization.

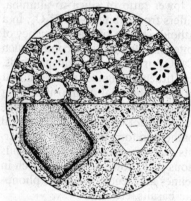

FIG. 145. Feldspathoids in thin section: above, *leucite*; below, left, *nosean*, right, *nepheline*.

OCCURRENCE. Leucite occurs as a primary constituent of volcanic rocks, and usually in fresh or recent types, such as the Vesuvian lavas, and in leucite-basalts, leucite-phonolites, leucite-tephrites, leucitophyres, and leucitites—rocks low in silica and rich in potash.

USES. Leucite has been worked in the Italian leucite-lava fields for the production of potash-fertilizer and of aluminium; in normal years the production is about 40,000 metric tons of leucite rock.

NEPHELINE, NEPHELITE, Eleolite

COMP. Artificial nepheline is sodium aluminium silicate, $NaAlSiO_4$; in natural nepheline potassium is always present, together with silica in excess of the artificial proportions—the composition given as oxides approximating to $K_2O.3Na_2O.4Al_2O_3.9SiO_2$; an analysis of excellent material from Dungannon, Ontario, gave $K_2O.5Na_2O.6Al_2O_3.13SiO_2$; it has been suggested that natural nepheline forms a solid-solution series between nepheline, $NaAlSiO_4$, and kalsilite, $KAlSiO_4$.

CRYST. SYST. Hexagonal. COM. FORM. Crystals with hexagonal prism and basal plane, often modified; also massive. CLEAV. Distinct prismatic, imperfect basal; the cleavages become more distinct as alteration of the mineral progresses. COLOUR. Colourless, white, yellowish, dark green, brownish, etc. LUSTRE. Vitreous or greasy; transparent to opaque. FRACT. Subconchoidal. H. 5·5–6. SP. GR. 2·5–2·6.

OPT. PROPS. In thin sections of rocks nepheline occurs either as hexagonal basal or rectangular longitudinal sections of crystals (see Fig. 145, lower, right) or as more irregular masses; the refractive indices are near that of balsam, $\omega = 1·541, \varepsilon = 1·538$, and the birefringence is weak—the basal sections are isotropic and the longitudinal sections polarize in low greys; uniaxial, negative, but usual sections are often too thin to give a good figure; sometimes shows cleavages emphasized by decomposition products, and then gives straight extinction parallel to these cleavages; nepheline is distinguished from apatite by its lower refractive index, from quartz by its negative sign, lower birefringence, alteration especially along cleavages, from orthoclase by its higher refractive index; if the uncovered rock-section is treated with hydrochloric acid, nepheline is gelatinized and this jelly absorbs stains so that the distribution of the nepheline is shown.

VARIETIES. *Ordinary Nepheline*, the small glassy crystals in lavas; *Eleolite* is a dark-coloured variety, with a greasy lustre, it shows no definite crystalline form and only rough cleavages; it occurs in some syenites.

TESTS. Heated before the blowpipe, nepheline fuses to a colourless glass: gelatinizes with acids.

OCCURRENCE. Nepheline occurs as an original constituent of volcanic rocks such as phonolites, nephelinites, nepheline-basalts, etc., i.e. rocks low in silica, and rich in soda; also plutonic igneous rocks of similar composition, the nepheline-syenites, etc., the term eleolite used to be restricted to the massive non-crystallized varieties occurring in such syenites, but the name should be discarded in petrology.

CANCRINITE

COMP. Hydrated silicate of sodium, calcium and aluminium with carbon dioxide, approximately represented by $4(NaAlSiO_4).CaCO_3.H_2O$.

CRYST. SYST. Hexagonal. COM. FORM. Massive. CLEAV. Perfect prismatic (10$\bar{1}$0). COLOUR. Commonly yellow, but also white, red. H. 5–6. SP. GR. 2·4–2·5.

OPT. PROPS. In thin section colourless, with low refractive indices, $\omega = 1·524$, $\varepsilon = 1·496$, and high double refraction, thus giving between crossed nicols polarization colours of Second and Third Orders; uniaxial negative; recognized by the combination of lowish refractive indices and high polarization colours.

OCCURRENCE. As a constituent of igneous rocks of the nepheline-syenite type.

SODALITE

COMP. Sodium aluminium silicate with sodium chloride represented by $3(NaAlSiO_4).NaCl$; compare the composition of nepheline.

CRYST. SYST. Cubic. FORM. Rhombdodecahedral crystals; also occurs massive. CLEAV. Rhombdodecahedral distinct. COLOUR. Grey, bluish, lavender blue, yellowish. STREAK. Colourless. LUSTRE. Vitreous; subtransparent to translucent. FRACT. Conchoidal or uneven. H. 5·5–6. SP. GR. 2·2.

OPT. PROPS. Under the microscope, sodalite appears as colourless grains, without distinct cleavage; refractive index low, 1·482, much lower than balsam, so that if the diaphragm is inserted, the surface of the sodalite appears pitted due to this difference in refractive index (see p. 164); isotropic; if the uncovered rock-slice is treated with nitric acid which is allowed to evaporate slowly, cubic crystals of sodium chloride are produced.

TESTS. The mineral is soluble in nitric acid, the solution reacting for a chloride; decomposed by hydrochloric acid with separation of gelatinous silica; heated before the blowpipe fuses with intumescence to a colourless glass.

OCCURRENCE. Occurs in soda-rich igneous rocks low in silica, such as nepheline-syenite, etc., where it accompanies nepheline or leucite.

HAUYNE, Hauynite

COMP. Sodium aluminium silicate with calcium sulphate represented by $3(NaAlSiO_4).CaSO_4$; compare the composition of nepheline.

CRYST. SYST. Cubic. COM. FORM. Crystals small octahedra and rhombdodecahedra; usually occurs in rounded or subangular grains. CLEAV. Rhombdodecahedral fairly good. COLOUR. Bright blue or greenish-blue. STREAK. Colourless. H. 5·5–6. SP. GR. 2·4–2·5.

OPT. PROPS. In thin section, hauyne is usually blue, but may be colourless; refractive index low, 1·496, much lower than that of balsam; isotropic between crossed nicols; characterized by an abundance of minute dark inclusions, often arranged in a black border or in some symmetrical pattern or scattered through the crystal; these inclusions together with the low refractive index and isotropism serve to identify the mineral in thin section.

TESTS. Decomposed by hydrochloric acid with the separation of gelatinous silica—the solution gives the reaction for sulphate with barium chloride solution.

OCCURRENCE. As a constituent of igneous rocks, mostly of volcanic origin, that are low in silica and rich in alkalies, as in those of Vesuvius, Eifel, etc.

NOSEAN, Noselite

COMP. Like that of hauyne but contains little or no calcium—sodium aluminium silicate with sodium sulphate represented by $3(NaAlSiO_4).Na_2SO_4$; compare the composition of nepheline.

CHARACTERS AND OCCURRENCE. Like hauyne, described above; cubic; colour greyish or brown; in thin section isotropic, low refractive index 1·495, with inclusions similar to those shown by hauyne (see Fig. 145, lower, left); gelatinizes with acid, the solution giving the reaction for sulphate; occurrence as for hauyne.

LAZURITE, Lapis Lazuli

COMP. Sodium aluminium silicate with sodium sulphide, represented by $3(NaAlSiO_4).Na_2S$; compare the composition of sodalite.

CRYST. SYST. Cubic. COM. FORM. Crystals, cubes and rhombdodecahedra, rare; usually massive and compact. CLEAV. Imperfect rhombdodecahedral. COLOUR. Berlin blue or azure blue. LUSTRE. Vitreous; translucent to opaque. FRACT. Uneven. H. 5·5. SP. GR. 2·38–2·45.

TESTS. Fuses with intumescence to a white glass; decomposed by hydrochloric acid with the evolution of sulphuretted hydrogen, a gelatinous deposit of silica being left; the mineral is often spangled with iron pyrites.

OCCURRENCE. In crystalline limestone near granite contacts and presumably of contact-metamorphic origin; deposits of this type occur at Lake Baikal, and in Iran, Afghanistan, etc.

USES. Lapis Lazuli is cut and polished for ornamental purposes, but is too soft to be much used in jewellery. Ancient Egyptian amulets carved in this material are common. When powdered, lazurite constitutes the blue paint ultramarine; most of the ultramarine, however, is now artificially prepared by heating together clay, sodium carbonate and sulphur, and since the artificial pigment is very much cheaper than the powdered mineral, the latter has been almost entirely superseded.

FORMS OF SILICA

The forms of silica, including the hydrated forms, can be grouped as below:

> *Crystalline.* Quartz, Tridymite, Cristobalite.
> *Cryptocrystalline.* Chalcedony (some), Jasper, Flint, etc.
> *Cryptocrystalline hydrous.* Opal, Chalcedony (some), Sinter, Diatomite, etc.

Crystalline Silica: Quartz, Tridymite, Cristobalite

QUARTZ

COMP. Silicon dioxide, SiO_2.

CRYST. STRUCT. The SiO_4-tetrahedra, linked by all four corners to make three-dimensional frameworks, build up a pattern consisting of a series of linked spirals whose axes lie parallel to the c-axis of the crystal. CRYST. SYST. Hexagonal-trigonal rhombohedral-trapezohedral; Quartz symmetry-class described on pp.132–136. COM. FORM. Crystals usually hexagonal prisms, terminated by positive and negative rhombohedra, these giving a form like a hexagonal bipyramid when equally developed (Fig. 146 left); sometimes the prism is lacking, as shown in Fig. 146 right; for a fuller discussion of the crystallography of quartz, and figures of quartz crystals see pp. 132–136. Quartz crystals often show faces very irregularly developed, and are often distorted. Occasionally crystals contain small cavities partially or wholly filled with liquid; not infrequently acicular crystals of rutile, etc., and scales and grains of chlorite and mica are enveloped. Quartz also occurs massive, granular, and sometimes stalactitic. COLOUR. When pure, colourless; often coloured by impurities, giving rise to certain varieties mentioned below. LUSTRE. Vitreous, occasionally resinous; transparent to opaque. FRACT. Conchoidal; no cleavage. H. 7; cannot be scratched with a knife. SP. GR. 2·65.

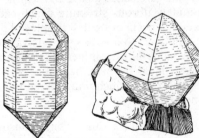

FIG. 146. Quartz

OPT. PROPS. In thin sections of rocks, quartz appears either as grains, or as well-formed crystals, the mode of occurrence depending on the rock; quartz is colourless, transparent and unaltered, shows no cleavage, and an irregular fracture. The refractive indices are slightly above that of balsam, $\omega = 1·544$ and $\varepsilon = 1·553$, so that in thin sections the borders of the grains are not well marked, and the surface of the mineral appears smooth. The birefringence is weak, ·009, and between crossed nicols sections show greys or yellows of the First Order; twinning is not commonly shown; transverse sections are isotropic and give a positive uni-axial interference figure. The elongation is positive.

TESTS. Physical properties distinctive; heated alone before the blowpipe, quartz is unaltered; soluble in the borax and sodium carbonate beads; insoluble in the microcosmic salt bead.

VARIETIES. *Rock Crystal* is the purest and most transparent form of quartz; it is sometimes employed in jewellery, and for making spectacle glasses; *Amethyst* is a purple or violet coloured transparent form of quartz, owing its colour perhaps to manga-nese; it is used in jewellery; *Rose Quartz* is a pale pink or rose-coloured variety of quartz; the colour fades on exposure, but may be restored by moistening; *Cairngorm, Smoky Quartz* are varieties of quartz of a fine smoky-yellow or smoky-brown colour, and *Morion* is a nearly black variety; these types are used in Scottish jewellery; *Milky Quartz* is a common variety, of a milk-white colour—the milkiness being due to the presence of a multitude of very small air-cavities; the milkiness is sometimes merely super-ficial, and such crystals are called *quartz en chemise*; *Cat's Eye* is quartz with a minutely fibrous structure which causes it when suitably cut to exhibit a peculiar opalescent play of light, bearing some fanciful resemblance to a cat's eye; this variety is often a pseudomorph after some fibrous mineral; *Aventurine Quartz* is quartz containing spangles of mica, hematite, etc.; *Ferruginous Quartz* contains iron-oxides, which impart a reddish or brownish colour to the mineral.

OCCURRENCE. Quartz occurs as an original constituent of the more acid igneous rocks—granites, quartz-felsite, rhyolite, etc. It also forms the bulk of the sandstones which result from the breaking-up of igneous rocks containing quartz; sandstones therefore consist of small, usually somewhat angular but occa-sionally well rounded, grains of quartz, which are cemented together by various substances—by quartz as in some types of quartzite, by limonite as in the ferruginous sandstones, by calcite in the calcareous sandstones, by clay as in the argillaceous sand-stones, and so on. Quartz is a common constituent of many meta-morphic rocks. Quartz occurs also as a very prominent veinstone in many mineral veins, and in geodes—potato-shaped stones, with a central hollow into which project quartz crystals.

Uses. Quartz sands, sandstones and quartzites are used in the building trade, and sands are employed in moulding, glass-making, etc. Quartz from various sources is extensively used as an abrasive material, as in scouring soaps, sandpaper, toothpaste, etc. Quartz is employed in the manufacture of pottery, silica-bricks, fused silica-ware, ferro-silicon, and as a flux, as linings for tube-mills, and as a filler.

An important new use of quartz, dependent on its piezo-electric properties, was developed extensively during the war of 1939–45, in connection with short wave radio apparatus. Thin plates cut from certain types of quartz crystal (Dauphiné twins) are used to control the frequency of radio circuits; millions of these oscillator plates have been made, the necessary supplies of suitable quartz coming mainly from Brazil.

TRIDYMITE

COMP. Silica, SiO_2.

CRYST. SYST. Orthorhombic inverting to hexagonal. COM. FORM. Minute, six-sided plates, wedge-shaped twins common. COLOUR. Colourless to white. SP. GR. 2·28–2·33. Refractive index, 1·47 (less than balsam).

OCCURRENCE. Tridymite is the stable form of silica at temperatures between 870° and 1470°C at atmospheric pressure; it occurs in acid volcanic rocks, as in the Drachenfels trachyte, the Tardree (Antrim) rhyolite, the Mount Ranier (Washington) andesite, etc.

CRISTOBALITE

COMP. Silica, SiO_2.

CRYST. SYST. Cubic. Refractive index, 1·48.

CHARACTERS AND OCCURRENCE. The form of silica stable above 1470°C; occurs as minute octahedra or cubes in certain lavas, as in the andesite of San Cristobal Mountains, Mexico, also as small spherulites in acid lavas and glasses.

Cryptocrystalline Silica: Chalcedonic Silica

CHALCEDONIC SILICA

COMP. A mixture of crystalline silica and hydrated silica, that is of quartz and opal; pure chalcedonic silica is silica, SiO_2, and is one end-member of a series of minerals, the other end-member of which is opal, $SiO_2.nH_2O$.

COM. FORM. Possesses an obscure or minutely crystalline (cryptocrystalline) structure; often showing a radiating minutely fibrous structure; usually occurs filling cavities in amygdaloidal rocks, as a veinstone, and as nodules in sedimentary rocks, especially limestone; the surface of these nodules is generally mammillated or botryoidal; sometimes stalactitic. COLOUR. Varied; white, grey, pale blue, bluish-white, brownish, black, etc. LUSTRE. Rather waxy.

OPT. PROPS. In thin sections, the cryptocrystalline varieties of silica are mostly colourless, with refractive index a little lower than that of quartz and of balsam: the micro-fibrous kinds show between crossed nicols a black cross, the arms of which are parallel to the nicol planes, since the fibres have straight extinction; chalcedony gives such a cross, and this fibrous variety has fibres showing negative elongation, the fast direction of vibration being parallel to the length of the fibres.

VARIETIES. Three main varieties of chalcedonic silica can be distinguished, *Chalcedony*, *Flint* and *Jasper*.

(1) *Chalcedony* includes a great number of sub-varieties based mainly on colour; the chief are: *Carnelian* is a well-known translucent variety of a reddish or yellowish-red colour, and *Sard* is a brownish variety, both types being used for signet-rings and similar work; *Prase* is a translucent, dull, leek-green variety of chalcedony, and also of quartz; *Plasma* is a sub-translucent, bright green variety, speckled with white; *Blood Stone* or *Heliotrope* is similar to plasma, but is speckled with red; *Chrysoprase* is an apple-green chalcedony, probably coloured with nickel oxide; *Agate* is a variegated chalcedony, composed of different coloured bands, sometimes with sharp lines of demarkation, sometimes shading off imperceptibly one into another, and affording various

patterns according to the direction in which the stone is cut; the agates are cut and polished for brooches, snuff-boxes and similar articles; they mostly come from Saxony, Bavaria, Arabia, India and Perthshire (*Scotch Pebbles*), and are found filling the vesicles in amygdaloidal volcanic rocks: *Moss Agate* or *Mocha Stone* is a chalcedony containing small dendrites (tree-like growths), which consist of iron-oxide, or of a ferruginous member of the chlorite group of minerals; *Onyx* and *Sardonyx* are flat-banded varieties of chalcedony, onyx having white and grey or brown bands, and sardonyx having white or bluish-white and red or brownish-red bands.

(2) *Flint* is compact cryptocrystalline silica of a black colour, or various shades of grey, and occurs in bands or more usually in irregularly-shaped nodules arranged in layers in the Upper Chalk of England; flint breaks with a well-marked conchoidal fracture, and affords sharp cutting-edges; accordingly the substance was extensively used by prehistoric man for the fabrication of weapons, chisels, hatchets, etc.; before the invention of percussion-locks and matches, flint was employed for gun-flints and for igniting tinder, these uses being based upon the well-known fact of flint generating sparks when struck with steel, small particles of steel being raised to a state of incandescence by the heat produced by the blow; flint is used in tube-mills, and calcined flint in the pottery industry; in Kent, Sussex, etc., flint has been extensively used for road making and building; *Hornstone* and *Chert* are grey to black opaque forms of cryptocrystalline silica, resembling flint, but breaking with a more or less flat fracture, rather different from the conchoidal fracture of flint: nodules and beds occur in limestone formations, as in the Carboniferous Limestone of North Wales.

(3) *Jasper* is an impure opaque form of cryptocrystalline silica, usually of red, brown and yellow colours, rarely green; it is opaque even on the thinnest edges; some varieties, such as *Egyptian* or *Ribbon Jasper*, are beautifully banded with different shades of brown: *Porcelain Jasper* is merely clay or shale altered or baked by contact with a hot igneous rock; it may be distinguished from true jasper by being fusible on the edges before the blowpipe.

Hydrous Silica, Opal

OPAL

COMP. Hydrous silica, $SiO_2.nH_2O$; the percentage of water is usually less than 10 and is contained in sub-microscopic pores in cryptocrystalline cristobalite.

COM. FORM. Compact, reniform or stalactitic. COLOUR. White, grey, yellow, red, brown, etc., the colours often blending and changing according to the direction in which the stone is viewed and displaying beautiful internal reflections and opalescence. LUSTRE. Subvitreous; transparent to nearly opaque. FRACT. Conchoidal. H. 5·5–6·5—softer than quartz. SP. GR. 2·2—lighter than quartz.

OPT. PROPS. Opal is practically isotropic between crossed nicols; in thin section, colourless, with low refractive index, 1·44, considerably lower than that of balsam.

VARIETIES. *Precious Opal* is the gem variety, exhibiting opalescence and a brilliant play of colours; *Hydrophane* is an opaque white or yellowish variety, which when immersed in water becomes translucent and opalescent; *Hyalite* is a transparent colourless glassy variety, occurring in small botryoidal or stalactitic forms; *Menilite* or *Liver Opal* is an opaque liver-coloured variety found in flattened or rounded concretions with pale exteriors, as at Menil Montant, Paris; *Wood Opal* is wood in which the cavities have been filled and the tissues replaced by opal; *Siliceous Sinter* sometimes consists of hydrous, sometimes of anhydrous silica; it has a loose porous texture, and is deposited from the waters of hot springs; it occurs on a grand scale around the hot springs of Taupo, in New Zealand, and is common at the geysers of Iceland and the United States; from its mode of occurrence, it is called *geyserite*; it lines the bores or tubes of geysers, and is deposited in cauliflower-like encrustations on the surface of the neighbouring ground; *Float Stone* is a porous form of silica which floats on water; it is found in the chalk at Menil Montant, Paris. *Diatomaceous Earth, Diatomite, Kieselguhr*, is a deposit of the tests and skeletons of siliceous organisms, such as the algae and diatoms, and consists of opal; it forms beds in ponds and lakes, and thick deposits occur in situations where siliceous volcanic emanations have supplied abundant material for diatom growth, as in the Miocene beds of California; the uses of diatomite depend mainly on the size and shape of the diatoms; it is used as an absorbent, as a polishing powder, as a filtering medium, and especially as an insulator for high temperatures, in cement, glazes, pigments, and for a great variety of minor uses; the chief producers are United States, Denmark, Algeria, Italy, France and West Germany.

OCCURRENCE. The modes of occurrence of many varieties have been given with their descriptions; opal itself is most likely a dried-up gel, and owes its distinctive properties to its structure, closely packed spheres of silica being arranged as a three-dimensional diffraction-grating and giving rise to the play of colours (see p. 444); it occurs filling cracks and cavities in igneous rocks, and also embedded in flint-like nodules in sandstones and shales, etc. The chief producers of precious opal are Australia—New South Wales, South Australia, and Queensland—and Mexico.

The Scapolite Family

The chemical composition of the Scapolite Family can be represented in terms of two end-members:

Marialite, Ma—$3(NaAlSi_3O_8).NaCl$.
Meionite, Me—$3(CaAl_2Si_2O_8).CaCO_3$.

Note that marialite has the composition 3 albite + NaCl, and meionite, 3 anorthite + $CaCO_3$.

Scapolites are tetragonal in crystallization. They show a gradation in physical properties from one end-member to the other—the specific gravity, refractive index and birefringence increasing with the content of meionite.

SCAPOLITE, Wernerite, Dipyre

COMP. Silicate of aluminium, sodium and calcium, with sodium chloride and calcium carbonate radicles present also, being made up of varying amounts of marialite $(3NaAlSi_3O_8.NaCl)$ and meionite $(3CaAl_2Si_2O_8.CaCO_3)$; common scapolite corresponds to $Ma_{20}Me_{80}$ to $Ma_{60}Me_{40}$.

CRYST. SYST. Tetragonal. COM. FORM. Tetragonal and ditetragonal prisms, or first and second order prisms, terminated by bipyramids; also occurs massive or granular and sometimes columnar. CLEAV. Distinct parallel to first and second order prisms. COLOUR. White, or pale shades of blue, green, and red. STREAK. Colourless. LUSTRE. Vitreous to pearly, or rather resinous; transparent to nearly opaque. FRACT. Subconchoidal; brittle. H. 5–6. SP. GR. 2·6–2·75, increasing with amount of meionite molecule.

OPT. PROPS. Colourless in thin sections, prismatic cleavages often seen as interrupted lines; refractive indices moderate, $\omega = 1{\cdot}55{-}1{\cdot}60$, $\varepsilon = 1{\cdot}54{-}1{\cdot}57$, increasing with amount of meionite molecule; birefringence weak in marialite-rich and strong in meionite-rich types, but in usual varieties polarization colours are bright; occasionally swarms of tiny inclusions are seen; uniaxial, optically negative—thus distinguished from feldspars, quartz, brucite, wollastonite, etc.

TESTS. Heated before the blowpipe, fuses with intumescence to a white glass; imperfectly decomposed by hydrochloric acid, meionite being easily decomposed, marialite less easily.

OCCURRENCE. Scapolite occurs either in metamorphic rocks, such as contact-altered impure limestones or regionally metamorphosed gneisses and amphibolites, or else as a secondary mineral in igneous rocks, where it is an alteration-product of lime-rich plagioclase feldspars.

The Zeolite Family

In composition the Zeolites are hydrated silicates of calcium and aluminium, sometimes with sodium and potassium, and in many ways are analogous in composition to the feldspars. They result in general from the alteration of the feldspars and aluminous minerals of igneous rocks and, with the possible exception of analcite, occur only as secondary minerals, filling cavities, joint-spaces, cracks and fissures, in such rocks as basalts, scoriaceous lavas, etc. Their formation represents the final stages in the cooling-down of igneous magmas, and they are therefore linked up in origin with many ore-deposits.

When heated before the blowpipe, the zeolites froth or boil up, a circumstance from which their name is derived, from the Greek, *zein*, to boil, and *lithos*, stone. The zeolites are rather soft minerals, their hardness varying from 3·5 to 5·5. Their specific gravities vary from 2 to 2·4. They are mostly decomposed with acids with the separation of gelatinous silica. The zeolites are readily recognized by their mode of occurrence.

The *fibrous zeolites* have framework structures of linked SiO_4-tetrahedra, which are arranged in groups of five (within the framework); as will be seen from the formulae below, there is a group of Si_5O_{10} or $5(SiO_2)$-type in each case, with Al replacing part of the Si.

In *other zeolites*, different frameworks occur, but all conform to the composition $n(SiO_2)$. Aluminium replaces part of the silicon, and the resulting negative charges on the framework are balanced by cations such as Na and Ca, which lie in open spaces in the frameworks. Water molecules are also accommodated in these spaces. Such a structural arrangement makes possible the well-known base-exchange capacity of many of the zeolites.

In this book, the following arrangement of the zeolites is adopted:

Cubic Zeolite	Analcite, $Na(AlSi_2)O_6.H_2O$.
Fibrous Zeolites	Natrolite, $Na_2[Al_2Si_3O_{10}].2H_2O$.
	Scolecite, $Ca[Al_2Si_3O_{10}].3H_2O$.
	Mesolite, intermediate between natrolite and scolecite.
	Thomsonite, $NaCa_2[Al_5Si_5O_{20}].6H_2O$.
Other Zeolites	Heulandite, $Ca_2(Al_4Si_{14})O_{36}.12H_2O$.
	Phillipsite, $(K, Na, Ca)(Al_2Si_4)O_{12}.4\frac{1}{2}H_2O$.
	Harmotome, $(K,Ba)(Al_2Si_5)O_{14}.5H_2O$.
	Stilbite, $(Na_2,Ca)[Al_2Si_7O_{18}].7H_2O$.
	Chabazite, $(Ca,Na)[Al_2Si_4O_{12}].6H_2O$.
	Laumontite, $(Ca)[Al_2Si_4O_{12}].4H_2O$.

Minerals sometimes classed with the Zeolites

Apophyllite, $KFCa_4Si_8O_{20}.8H_2O$, resembling Zeolites in giving off water when heated (see p. 416).

Pectolite, $Ca_2NaH(SiO_3)_3$ zeolitic in occurrence (see p. 390).

Prehnite, $Ca_2Al[AlSi_3O_{10}](OH)_2$, sometimes zeolitic in occurrence (see p. 416).

ANALCITE Analcime

COMP. Hydrous silicate of sodium and aluminium, $Na(AlSi_2)O_6.H_2O$.

CRYST. SYST. Cubic. COM. FORM. Crystals, the trapezohedron (211) very common; also occurs massive or granular. CLEAV. Cubic, obscure. COLOUR. Milk-white; often colourless, greyish, greenish, reddish-white or pink. STREAK. White. LUSTRE. Vitreous; transparent to nearly opaque. FRACT. Subconchoidal and uneven; brittle. H. 5–5·5. SP. GR. 2·25.

OPT. PROPS. Colourless in section, refractive index much lower than that of balsam, n = 1·487; isotropic between crossed nicols, though strain-polarization and optical anomalies may be shown; by shutting off some of the light passing through the microscope by a diaphragm, the analcite appears mottled and with strong relief.

TESTS. Heated in closed tube yields water; heated on charcoal, fuses to a clear, colourless globule; decomposed by hydrochloric acid with the separation of silica; colours flame yellow.

OCCURRENCE. Typically zeolitic (see p. 446); considered to be a primary mineral in certain dolerites.

NATROLITE

COMP. Hydrous silicate of sodium and aluminium, $Na_2(Al_2Si_3O_{10}).2H_2O$.

CRYST. SYST. Orthorhombic. COM. FORM. Small prismatic crystals known; usually in slender acicular crystals; also massive, compact, granular, fibrous or radiating. CLEAV. Perfect prismatic. COLOUR. White, sometimes yellowish or reddish. STREAK. White. LUSTRE. Vitreous or pearly; transparent to translucent. H. 5–5·5. SP. GR. 2·2–2·25.

TESTS. Fusible in a candle flame; fuses quietly to a clear bead; gelatinizes with acid; yields water in the closed tube.

OCCURRENCE. Zeolitic; found as a secondary mineral in the amygdales in the basalts of Antrim, Scotland, New Jersey, etc.

SCOLECITE

COMP. Hydrous silicate of calcium and aluminium, $Ca(Al_2Si_3O_{10}).3H_2O$.

CRYST. SYST. Monoclinic. COM. FORM. In radiating prismatic crystal groups, fibrous forms, nodular or massive. CLEAV. Good prismatic. COLOUR. White. LUSTRE. Vitreous; of fibrous types, silky; transparent to subtranslucent. H. 5–5·5. SP. GR. 2·2–2·4.

TESTS. Heated in a closed tube, gives water; gelatinizes with acid; heated before the blowpipe, fuses with wormy intumescence, and often forms a frothy mass.

OCCURRENCE. Zeolitic (see p. 446).

MESOLITE

COMP. Hydrous silicate of calcium, sodium and aluminium, intermediate between natrolite and scolecite.

CRYST. SYST. Monoclinic, as shown by X-ray tests. COM. FORM. Usually in tufts composed of very delicate acicular crystals; also sometimes massive. COLOUR. White or greyish. LUSTRE. Vitreous; when fibrous or massive, silky; transparent to translucent; opaque when massive. H. 5. SP. GR. 2–2·4.

TESTS. Gives off water when heated in the closed tube; gelatinizes with hydrochloric acid; heated before the blowpipe, it becomes opaque and fuses with worm-like intumescence to an enamel.

THOMSONITE

COMP. Hydrous silicate of calcium, sodium and aluminium, $NaCa_2[Al_5Si_5O_{20}]. 6H_2O$.

CRYST. SYST. Orthorhombic. COM. FORM. Usually in columnar or radiating crystalline masses. CLEAV. Perfect parallel to the pinacoid (010). COLOUR. Snow-white. LUSTRE. Vitreous to pearly; transparent to translucent. FRACT. Subconchoidal. H. 5–5·5. SP. GR. 2·3–2·4.

TESTS. Yields water when heated in the closed tube; fuses with intumescence to a white blebby enamel; soluble in hydrochloric acid with gelatinization.

OCCURRENCE. Zeolitic (see p. 446).

HEULANDITE

COMP. Hydrous silicate of calcium and aluminium, $Ca_2(Al_4Si_{14})O_{36}.12H_2O$.

CRYST. SYST. Monoclinic. COM. FORM. Crystals with clino-pinacoid (010) and hemi-orthodomes, (201) and ($\bar{2}$01), largely developed; also globular. CLEAV. Perfect parallel to the clino-pinacoid. COLOUR. White, brick-red, brown. STREAK. White. LUSTRE. Vitreous; of clinopinacoid planes, pearly; transparent to subtranslucent. FRACT. Subconchoidal or uneven; brittle. H. 3·5–4. SP. GR. 2·2.

TESTS. Before the blowpipe, intumesces and fuses; decomposed by acids without gelatinization, with separation of silica.

OCCURRENCE. Zeolitic (see p. 446).

PHILLIPSITE

COMP. Hydrous silicate of calcium, potassium, sodium and aluminium, $(K,Na,Ca)(Al_2Si_4)O_{12}.4\frac{1}{2}H_2O$.

CRYST. SYST. Monoclinic. COM. FORM. Crystals penetrating cruciform twins, often grouped in radiating aggregates. COLOUR. White, reddish. STREAK. Colourless. LUSTRE. Vitreous. FRACTURE. Uneven; brittle. H. 4–4·5. SP. GR. 2·2.

TESTS. Heated in closed tube gives water; heated on charcoal, fuses quietly to a bubbly enamel; decomposed by acid with separation of silica.

OCCURRENCE. Zeolitic (see p. 446); also found in the deep-sea deposits.

HARMOTOME, Cross-stone

COMP. Hydrous silicate of potassium, barium and aluminium, $(K,Ba)(Al_2Si_5)O_{14}.5H_2O$.

CRYST. SYST. Monoclinic. COM. FORM. The crystals are always cruciform penetration twins, much resembling those of stilbite; they are either simple twins, or groups consisting of four individuals; these fourlings sometimes show re-entrant angles, whence the name *cross-stone*, or they may present the aspect of a square prism combined with faces of a pyramid, which latter are really prism faces. COLOUR. White, or shades of grey, yellow or brown. STREAK. White. LUSTRE. Vitreous; subtransparent to translucent. FRACT. Uneven; brittle. H. 4–4·5. SP. GR. 2·3–2·5.

TESTS. Heated before the blowpipe, whitens, crumbles, and fuses without intumescence to a white translucent glass; decomposed by hydrochloric acid without gelatinization.

OCCURRENCE. Zeolitic (see p. 446).

STILBITE

COMP. Hydrous silicate of calcium, sodium and aluminium, $(Na_2,Ca)(Al_2Si_7O_{18}).7H_2O$.

CRYST. SYST. Monoclinic. COM. FORM. Crystals usually thin, tabular parallel to the clinopinacoid, and compound, being twinned combinations of the unit prism with clinopinacoid and basal pinacoid; extremely common in sheaf-like aggregates, as shown in Fig. 147, and in divergent and radiating forms. CLEAV. Perfect clinopinacoidal. COLOUR. Usually white, sometimes red, yellow or brown. STREAK. Colourless. LUSTRE. On cleavage-faces, pearly; elsewhere vitreous; subtransparent to translucent. H. 3·5–4. SP. GR. 2·1–2·2.

FIG. 147. Stilbite

TESTS. Heated before the blowpipe, fuses with wormy intumescence to a white enamel; heated in closed tube gives water; decomposed by hydrochloric acid with separation of silica.

OCCURRENCE. Typically zeolitic (see p. 446); especially common filling the steam cavities of lavas, as in the basalts of Antrim, etc.

CHABAZITE

COMP. Hydrous silicate of calcium, sodium and aluminium, approximately $(Ca,Na)(Al_2Si_4O_{12}).6H_2O$.

CRYST. SYST. Hexagonal-trigonal, Calcite Type. COM. FORM. Crystals combinations of positive and negative rhombohedra; also massive. COLOUR. White, yellowish, reddish. LUSTRE. Vitreous; transparent to translucent. H. 4–4·5. SP. GR. 2·1.

TESTS. Heated before the blowpipe, intumesces, whitens and fuses to a glass; decomposed by acid with separation of silica; heated in a closed tube, gives water.

VARIETY. *Phacolite* is a variety occurring in colourless lenticular crystals due to twinning.

OCCURRENCE. Typically zeolitic (see p. 446).

LAUMONTITE

COMP. Hydrous silicate of calcium and aluminium, $Ca(Al_2Si_4O_{12}).4H_2O$.

CRYST. SYST. Monoclinic. COM. FORM. Common in prismatic crystals; also fibrous or columnar. COLOUR. White, greyish, or yellowish. STREAK. Colourless. LUSTRE. Vitreous when unaltered, dull and pulverulent when altered; transparent to translucent when unaltered, but becomes opaque and white on exposure. H. 3·5–4. SP. GR. 2·2–2·3.

TESTS. Heated before the blowpipe fuses with intumescence to a white enamel; soluble in hydrochloric acid with gelatinization.

OCCURRENCE. Zeolitic (see p. 446).

TIN MINERALS

Tin (Sn) is said to have been found native but, if it does occur so, is of very rare occurrence. It is chiefly found in the form of the oxide, *cassiterite* or *tinstone*, which is the main source of the metal, only a small amount being obtained from tin sulphides occurring with cassiterite in Bolivia. Tin is a bright, white metal, malleable and ductile. It has a specific gravity of 7·3 and melts at 232°C. A bar of the metal emits a crackling sound when bent.

Cassiterite is obtained commercially from both lodes and alluvial (placer) deposits. In the former it may be associated with arsenic, copper and iron minerals, wolfram, etc.; in alluvial deposits it is often associated with ilmenite or titaniferous iron ore, monazite, zircon, topaz, tourmaline, etc. The proportion of tin in ores is usually expressed in pounds of *black tin*, that is, cassiterite containing about 70 per cent of the metal, per ton of ore. In the case of alluvial deposits, less than 2 pounds per ton has been profitably worked. The alluvial or mine stuff is concentrated up to as near 70 per cent of metallic tin as is practicable by means of shaking tables and other mechanical contrivances; when the tin-stone is associated with ilmenite, wolfram, or other magnetic minerals, electromagnetic separators are employed. The dressed product or concentrate is reduced in reverberatory furnaces, and the metal further purified by electrolytic processes.

The chief use of tin is in the manufacture of tin-plate, which is sheet-iron coated with a very thin coating of tin, the tin-plate being employed for the production of cans, etc. Another very important use is for the manufacture of a number of important alloys, such as pewter, the various solders, bearing-metals, type-metal, bronze, gun-metal, bell metal, fusible metal, etc. Salts of tin are employed in calico-printing, dyeing, silk-making, in the ceramic industry, etc.

The average mine-production of tin ore annually is well over 160,000 tons in tin content. The leading producers are Malaysia, China, Bolivia, Thailand, Indonesia and U.S.S.R. Other important producers are Congo, Nigeria and Australia, with minor amounts from Cornwall, Portugal and elsewhere. Tin-ore is one of the few important industrial ores not produced in significant amount in the United States. The recovery of scrap tin—*secondary tin*—was important in industry, especially in the United States, where possibly half of the annual consumption of the metal was formerly supplied from this source; the tin-coating on cans is now too thin to repay recovery. Tin-smelting plants are concentrated in Malaysia, Britain, China, U.S.S.R., U.S.A. and Holland, which between them account for a very dominant proportion of annual world output.

TESTS. The following are the chief tests for tin. When heated on charcoal with sodium carbonate and charcoal, tin compounds are reduced to the metal which is soft and malleable. The encrustation given by heating tin compounds alone on charcoal, when moistened with cobalt nitrate and strongly reheated, assumes a blue-green colour. The tin bead when treated with warm nitric acid becomes coated with a white covering of hydroxide.

As already noted, the chief mineral of tin is the oxide *cassiterite*. In addition to this compound, tin occurs in a few other minerals, mostly complex sulphides, only one of which is sufficiently important to be described here, namely, *stannine* or *tin pyrites*, Cu_2SnFeS_4. The important tin minerals are therefore:

Oxide	Cassiterite, Tinstone, SnO_2.
Sulphide	Stannine, Tin Pyrites, Cu_2SnFeS_4.

CASSITERITE, Tinstone

COMP. Tin oxide, SnO_2: tin, 78·6 per cent.

CRYST. SYST. Tetragonal, Zircon Type. COM. FORM. Crystals consist of tetragonal prisms terminated by tetragonal bipyramids; knee-shaped twins often seen (p. 154); also occurs massive, or fibrous, or disseminated in small grains; in alluvial deposits as rolled water-worn grains. COLOUR. Usually black or brown, rarely yellow or colourless. STREAK. White or pale grey to brownish. LUSTRE. Adamantine and, on crystals, usually very brilliant; crystals, when of pale colour, nearly transparent, when dark, opaque. FRACT. Subconchoidal or uneven; brittle. H. 6–7. SP. GR. 6·8–7·1.

TESTS. Heated alone before the blowpipe, infusible; heated with sodium carbonate and charcoal on charcoal, yields a globule of metallic tin; the sublimate resulting from heating on charcoal with sodium carbonate, when moistened with cobalt nitrate and strongly reheated, assumes a blue-green colour.

VARIETIES. *Wood Tin* has a structure which is compact and fibrous internally, and exhibiting concentric bands, thus resembling wood; it occurs in reniform masses; *Toad's Eye Tin* shows the characters of wood tin, but on a smaller scale; *Stream Tin* is rolled and worn cassiterite, resulting from the wearing-away of tin veins or of rocks containing the ore; it occurs in the beds of streams and in the alluvial deposits which border them; much of the tin ore sent into the market is derived from this source.

OCCURRENCE. The most important primary mode of occurrence of cassiterite is in pneumatolytic veins associated with granitic and allied rocks, as in Cornwall, Saxony, Tasmania, etc.; the chief veinstone is quartz, associated with such boron or fluorine minerals as fluorspar, topaz, tourmaline, axinite and apatite; the adjacent country-rock is altered in suitable cases into a mixture of quartz, muscovite, topaz or tourmaline, called *greisen*; tin-silver veins of a rather different type are important in Bolivia where they are associated with porphyries of hypabyssal or volcanic origin; fully one-half of the world's supply of tin is obtained from placer deposits resulting from the degradation of tin veins, this type of deposit supplying the great outputs of Malaysia, Indonesia and Thailand; and an interesting eluvial placer occurs immediately adjacent to the primary tin veins of Mount Bischoff, Tasmania. In addition to the lode and alluvial tin deposits, which are of the greatest economic importance, cassiterite occurs as an original constituent of igneous rocks, such as granites and pegmatites, and as a contact-metamorphic deposit in limestone adjacent to granite contacts.

STANNINE, STANNITE, TIN PYRITES, Bell Metal Ore

COMP. Sulphide of tin, copper and iron, Cu_2SnFeS_4; tin, 27·5 per cent; zinc is usually present in varying quantity.

CRYST. SYST. Tetragonal, appearing cubic (tetrahedral) through twinning. COM. FORM. Crystals rare; commonly occurs massive, granular or disseminated. COLOUR. Steel-grey when pure; iron-black, sometimes bronze or bell-metal colour, occasionally with a bluish tarnish; often yellowish from admixture with copper pyrites. STREAK. Blackish. LUSTRE. Metallic; opaque. FRACT. Uneven; brittle. H. 4. SP. GR. 4·4.

TESTS. Heated in the open tube, gives off fumes of sulphur dioxide, and also forms a sublimate of tin oxide close to the assay; heated on charcoal, it fuses after long roasting to a brittle metallic globule which, heated in the oxidizing flame, gives off sulphur, and coats the support with white tin oxide; the roasted mineral affords, in borax, reactions for iron and copper.

OCCURRENCE. Occurs associated with cassiterite, copper pyrites, blende or galena, as in Cornwall, Saxony, etc. It occurs also in the tin lodes of Bolivia, associated with cassiterite, silver minerals, and sulphides of copper, antimony, lead, zinc, bismuth, etc.

LEAD MINERALS

Lead (Pb) is known native, but is of exceedingly rare occurrence. Lead is a bluish-grey metal, whose freshly cut surface shows a bright metallic lustre which, however, quickly oxidizes on exposure to air. It is soft, may be scratched with the fingernail, and makes a black streak on paper. The specific gravity of the metal is 11·34. It fuses at 327°C, and crystallizes when cooled slowly. It has little tenacity, and cannot be drawn into wire but is, however, readily rolled or pressed into thin sheets, or exuded when in a semi-molten condition through dies to form piping. These properties are a consequence of its close-packed cubic structure (see p. 21); they are materially affected by the presence of small quantities of impurities.

Lead is easily reduced from its compounds. It is readily soluble in nitric acid, but is little affected by hydrochloric or sulphuric acid. It forms a number of compounds of great commercial importance, as noted below under the uses of lead.

The principal ores of lead are the sulphide, *galena*, PbS, and the sulphate, *anglesite*, $PbSO_4$, and carbonate, *cerussite*, $PbCO_3$. For the production of lead, the ore is first partly roasted or calcined, and then smelted in reverberatory or blast furnaces. Most lead ores contain silver, and this metal is obtained from the lead by cupellation, repeated melting and crystallization, alloying with zinc, or by electrolytic processes.

Blende is frequently associated with galena and the presence of zinc causes difficulties in smelting. Mechanical separation (dressing) by jigs, flotation-processes, etc., of the two minerals is resorted to, and, with the improvement of such processes, low-grade mixed ores are now being worked at a profit. When antimony is associated with the galena, the ore may be smelted direct for the production of antimonial lead.

The uses of lead and its compounds are manifold. The metal is employed in the construction of accumulators, as is the oxide, for lead sheeting and piping, cable-covers, ammunition, foil, etc. It is a constituent of many valuable alloys, such as pewter, solder, babbitt-metal, type-metal, bronzes, anti-friction metal and fusible metal. Lead compounds are employed extensively as pigments, such as the oxides, red lead and litharge, and the basic carbonate, white lead. The oxide is used in glass-making, as a flux, and in the rubber industry. The nitrate is employed in calico dyeing and printing processes, the arsenate is used as an insecticide, and the acetate is employed in medicine.

The annual world smelter production of lead is over 3 million tons, the dominant producers being the U.S.A. and U.S.S.R. followed by West Germany, Australia, Mexico, Britain and Canada; other notable producers are Belgium, Yugoslavia, France, Japan, China and Peru. The chief ore-production comes from Australia (mostly from the famous Broken Hill mines), U.S.S.R., U.S.A., Canada, Mexico, Peru and Yugoslavia.

The chief primary ore is galena, PbS; deposits of galena oxidize in their upper parts into oxy-salts, of which the most important economically are cerussite, $PbCO_3$, and anglesite, $PbSO_4$. Lead ores occur in a number of ways, not all of economic importance; the chief modes of occurrence are as lodes or veins, as metasomatic replacements and contact-metamorphic deposits, or as disseminations.

TESTS. The chief tests for lead are as follow: Lead compounds colour the blowpipe flame a pale sky-blue; this is a poor colour and of little value and, further, lead compounds attack the platinum wire. When lead minerals are heated alone on charcoal, they give a sulphur-yellow encrustation. When heated with potassium iodide and sulphur, they give a brilliant yellow encrustation—this being a good test. Roasted with sodium carbonate and charcoal on charcoal, lead minerals are reduced to metallic lead, which shows as a lead-grey bead, bright while hot but dull when cold; the bead is malleable and marks paper.

The minerals of lead considered here are:

Sulphide	Galena, PbS.
Oxide	Minium Pb_3O_4.
Carbonate	Cerussite, $PbCO_3$.
Choro-carbonate	Phosgenite, $PbCO_3.PbCl_2$.

Sulphato- carbonate	Leadhillite, $PbSO_4.2PbCO_3.Pb(OH)_2$.
Sulphate	Anglesite, $PbSO_4$.
Basic Sulphates	Plumbojarosite, $PbFe_6(OH)_{12}(SO_4)_4$.
	Linarite, $(Pb,Cu)SO_4.(Pb,Cu)(OH)_2$.
Chloro-phosphate	Pyromorphite, $3Pb_3P_2O_8.PbCl_2$ or (PbCl) $Pb_4(PO_4)_3$.
Chloro-arsenate	Mimetite, $3Pb_3As_2O_8.PbCl_2$ or (PbCl) $Pb_4(AsO_4)_3$.
Chloro-vanadate	Vanadinite, $3Pb_3V_2O_8.PbCl_2$ or (PbCl) $Pb_4(VO_4)_3$.
Chromate	Crocoisite, $PbCrO_4$.
Molybdate	Wulfenite, $PbMoO_4$.

Note also *Jamesonite*, $Pb_4FeSb_6O_{14}$, described with antimony minerals on p. 482, *Bournonite*, $CuPbSbS_3$, described with copper on p. 250, *Freieslebenite*, $(Pb,Ag)_8Sb_5S_{12}$, described with silver on p. 261, and *Nagyagite*, sulpho-telluride of lead and gold, described with gold on p. 269.

GALENA, Lead Glance, Blue Lead

COMP. Lead sulphide, PbS; silver sulphide is almost always present, and galena is one of the most important sources of silver; when sufficient silver is present to be worth extracting, the ore is called 'argentiferous galena'; zinc, cadmium, iron, copper, antimony and gold have also been detected in analyses of this mineral; there are apparently no external characters which serve to distinguish even the highly argentiferous ores from ordinary galena—the question can only be solved by analysis.

CRYST. STRUCT. Atomic arrangement that of rocksalt (p. 76). CRYST. SYST. Cubic, Galena Type. COM. FORM. Cube, often modified by octahedral and other forms, as shown in Fig. 20, p. 79; also occurs massive, and coarsely or finely granular. CLEAV. Perfect cubic, many specimens crumbling readily into small cubes when rubbed or struck. COLOUR. Lead-grey. STREAK. Lead-grey. LUSTRE. Metallic, but often dull, due to tarnish; opaque. FRACT. Flat, even, or subconchoidal. H. 2·5. SP. GR. 7·4–7·6.

TESTS. Heated in the open tube, galena gives off sulphurous

fumes; heated on charcoal, it emits sulphurous fumes, forms a yellow encrustation of lead oxide and fuses to a malleable metallic globule, which marks paper; heated with potassium iodide and sulphur on charcoal, it forms a brilliant yellow encrustation; galena is decomposed by hydrochloric acid, with evolution of sulphuretted hydrogen—on cooling the solution, white crystals of lead chloride are deposited which are soluble on heating.

OCCURRENCE. Galena often occurs associated with blende, and reference should be made to the description of blende occurrences on p. 299; metasomatic disseminations are exemplified by the lead-zinc deposits of the important Tri-State field in the Mississippi Valley—whether the ores in this deposit were derived from below or above is a matter of discussion (see p. 300); another type of metasomatic replacement, but definitely of hydrothermal origin, is shown by the Leadville, Colorado, field, in which limestone is the country-rock; the famous Broken Hill lode in Australia is of hydrothermal origin and provides argentiferous galena—it is remarkable in that among the gangue minerals is garnet; many other types of lead lodes are known, in several of which the galena is associated with silver minerals; pyrometasomatic or contact-metamorphic deposits of galena are relatively unimportant—one such deposit in which the ore occurs in limestone at the contacts with granite-porphyry is exploited at the Magdalena mines, New Mexico; galena of sedimentary origin is not important—examples of such deposits are found in the Permo-Triassic rocks, as at Aix-la-Chapelle, where the sandstone contains a small proportion of galena and cerussite, most likely leached from lead-ores out-cropping within the denudation-area. In Britain, small galena deposits of metasomatic or hydrothermal origin occur in Derby-shire, Flint, Cumberland, Cardigan, Isle of Man, Cornwall, etc.

USES. Galena is the most important ore of lead, nearly all the metal of commerce being derived from this source.

MINIUM, Red Oxide of Lead

COMP. Pb_3O_4.

COM. FORM. Powdery. COLOUR. Bright red, scarlet, or orange-red. STREAK. Orange-yellow. LUSTRE. Faint, greasy or dull; opaque. H. 2–3. SP. GR. 4–6.

TESTS. Before the blowpipe in the reducing flame, yields globules of metallic lead; other lead tests given; oxygen, tested by glowing splinter, is given off on heating in closed tube.

OCCURRENCE. Occurs associated with galena and sometimes with cerussite, being derived by the alteration of these minerals.

USES. The red lead of commerce, which has the same composition as minium, is artificially prepared by heating lead to form the yellow monoxide, and then subjecting the cooled monoxide to heating again at a lower temperature; red lead is used in the manufacture of glass and as a pigment.

CERUSSITE, Ceruse, White Lead Ore

COMP. Lead carbonate, $PbCO_3$.

CRYST. SYST. Orthorhombic, Barytes Type. COM. FORM. Prismatic crystals, variously modified, often showing faces of the prism (110), brachydome (021) and bipyramid (111); twin crystals common, often in cruciform or radiate arrangements; also occurs granular, massive, compact and sometimes stalactitic. COLOUR.— White or greyish, sometimes tinged blue or green by copper salts. STREAK. Colourless. LUSTRE. Adamantine, inclining to vitreous or resinous; transparent to translucent. FRACT. Conchoidal; very brittle. H. 3–3·5. SP. GR. 6·55.

TESTS. Soluble in hydrochloric acid with effervescence; heated before the blowpipe, decrepitates, and fuses; heated on charcoal with sodium carbonate and charcoal yields the lead bead, malleable and marking paper; heated with potassium iodide and sulphur on charcoal, gives a brilliant yellow encrustation.

OCCURRENCE. Cerussite occurs in the oxidation zone of lead veins, associated with galena and anglesite; it may result from the decomposition of anglesite by water charged with bicarbonates, and since anglesite results from the oxidation of galena, the sulphide passing into the sulphate, cerussite may be regarded as indirectly derived from the decomposition of galena; it occurs at most localities for lead ores, e.g. Cornwall, Derbyshire, Durham, Cardigan, Leadhills, in Britain.

USES. When found in quantity, cerussite is a valuable ore, ranking next to galena. The white lead of commerce is the basic lead carbonate, $2PbCO_3.Pb(OH)_2$, and is artificially prepared by various processes, the most general in Britain being the Old Dutch Process, in which white lead is produced by the action of acetic acid on metallic lead cast in the form of gratings—lead acetate is produced, and this is converted into the carbonate by the carbon-dioxide liberated by fermenting tan-bark, etc. White lead may also be prepared by passing carbon dioxide through a solution of basic lead acetate, but this product is usually considered inferior. White lead is used as a pigment, and is sometimes adulterated with barytes, etc.

PHOSGENITE, Cromfordite, Horn Lead

COMP. Chloro-carbonate of lead, $PbCO_3.PbCl_2$.

CRYST. SYST. Tetragonal. COM. FORM. In prismatic crystals. COLOUR. White, grey or yellow. STREAK. White. LUSTRE. Adamantine; transparent to translucent. H. 3. SP. GR. 6–6·3.

TESTS. Dissolves with effervescence in hydrochloric acid; lead tests, as detailed in the introduction to the lead minerals, given; soluble in nitric acid, giving a solution which reacts for chloride with silver nitrate; heated in closed tube, colourless globules of lead chloride obtained.

OCCURRENCE. A rare mineral formed in the oxidation zone of lead deposits, where it is associated with cerussite.

LEADHILLITE

COMP. Sulphato-carbonate of lead, $PbSO_4.2PbCO_3.Pb(OH)_2$.

CHARACTERS AND OCCURRENCE. A greyish-white mineral, crystallizing in the monoclinic system in tabular crystals, with a pearly or resinous lustre; it has a good basal cleavage, and splits into flexible laminae. H. 2·5. SP. GR. 6·26–6·44; it occurs in lead deposits as an alteration-product of galena or cerussite in the zone of oxidation, as at Leadhills (Scotland), Matlock (Derbyshire), etc.

ANGLESITE, Lead Vitriol

COMP. Lead sulphate, $PbSO_4$.

CRYST. SYST. Orthorhombic, Barytes Type. COM. FORM. Prismatic crystals, or tabular parallel to the basal pinacoid, or occasionally in bipyramids variously modified; the faces of the prism and macropinacoid are often vertically striated; also occurs massive and occasionally stalactitic. CLEAV. Parallel to the three pinacoids, but varying in value. COLOUR. White, sometimes with a blue, grey, green, or yellow tint. LUSTRE.—Usually adamantine, sometimes inclining to resinous or vitreous; transparent to opaque. FRACT. Conchoidal; brittle. H. 2·5–3. SP. GR. 6·3–6·4.

TESTS. Heated before the blowpipe, in the oxidizing flame, decrepitates and fuses to a clear globule, which on cooling becomes milk-white; heated in the reducing flame effervesces and yields metallic lead; dissolves in hydrochloric acid, the solution yielding a dense precipitate of barium sulphate on the addition of barium chloride solution, indicating the presence of a sulphate; sulphate is also detected by the silver coin test (see p. 38).

OCCURRENCE. Anglesite, when found in sufficient quantity, is a valuable lead ore; it is usually associated with galena and results from the decomposition of that mineral in the upper portion of lead veins; localities are Parys Mine in Anglesey, Cornwall, Derbyshire, Cumberland, Leadhills, Broken Hill (N.S.W.), etc.

PLUMBOJAROSITE

COMP. Basic sulphate of lead and iron, $PbFe_6(OH)_{12}(SO_4)_4$.

CHARACTERS AND OCCURRENCE. A dark brown mineral, crystallizing in the hexagonal system, in tiny tabular crystals, with specific gravity of 3·67, and occurring in certain mines in Utah in sufficient quantity to be worth working—otherwise it is rare.

LINARITE

COMP. Basic sulphate of copper and lead, $(Pb,Cu)SO_4.(Pb,Cu)(OH)_2$.

CHARACTERS AND OCCURRENCE. A rare mineral, crystallizing in the monoclinic system; it has a deep azure-blue colour, and a pale blue streak ; it occurs in the zone of oxidation of lead-copper veins, and has been found in Cumberland, Leadhills, etc.

The Pyromorphite Set

In what may be called the *Pyromorphite Set* are included three minerals with related properties. Their compositions are:

Pyromorphite	$3Pb_3P_2O_8.PbCl_2$, or $(PbCl)Pb_4(PO_4)_3$.
Mimetite	$3Pb_3As_2O_8.PbCl_2$, or $(PbCl)Pb_4(AsO_4)_3$.
Vanadinite	$3Pb_3V_2O_8.PbCl_2$, or $(PbCl)Pb_4(VO_4)_3$.

These minerals are related to the apatite family, and certain inter-mediate compounds are known.

The three minerals considered here are isomorphous; they crystallize in the hexagonal system, but in a lower symmetry group than beryl. As a rule they show long prismatic crystals, and their colours are usually vivid. They occur in the zone of oxidation of lead deposits, as at numerous mines in Cornwall, Derbyshire, Cumberland, Flintshire, Leadhills, Saxony, Harz, Mexico, and the United States, and sometimes serve as minor ores of lead.

PYROMORPHITE, Green Lead Ore

COMP. Chloro-phosphate of lead, $3Pb_3P_2O_8.PbCl_2$, or better, $(PbCl)Pb_4(PO_4)_3$; sometimes a small quantity of arsenic or calcium is present.

CRYST. SYST. Hexagonal; bipyramidal symmetry class, like apatite. COM. FORM. Prismatic crystals made up of prism, pyramid and basal pinacoid, crystals are usually aggregated or form crusts; also reniform and botryoidal. CLEAV. Parallel to the faces of the prism, in traces. COLOUR. Green, yellow and brown, of different shades; the colours sometimes very vivid. STREAK. White or yellowish-white. LUSTRE. Resinous; sub-transparent to subtranslucent. FRACT. Subconchoidal or uneven; brittle. H. 3·5–4. SP. GR. 6·5–7·1.

TESTS. Heated in the closed tube, gives a white sublimate of lead chloride; heated before the blowpipe, alone, fuses easily, colouring the flame bluish-green; heated on charcoal fuses to a globule which, when cool, assumes a crystalline angular form, but without being reduced to metallic lead—at the same time, the charcoal becomes coated with white lead chloride and yellow lead oxide; heated with sodium carbonate, yields a metallic lead bead; chloride reaction given by microcosmic salt-copper oxide bead test (p. 41); heated with magnesium in the closed tube, and then moistened, gives a smell of phosphoretted hydrogen; soluble in acids.

OCCURRENCE. Pyromorphite is found in company with other ores of lead in the oxidized zone of lead veins, as at the localities cited above.

MIMETITE, Green Lead Ore

COMP. Chloro-arsenate of lead, $3Pb_3As_2O_8.PbCl_2$, or better, $(PbCl)Pb_4(AsO_4)_3$.

CRYST. SYST. Hexagonal, bipyramidal; isomorphous with pyromorphite. COM. FORM. Crystals like those of pyromorphite; also in botryoidal and crusty forms. COLOUR. Pale yellow, brown, white. STREAK. White, or whitish. LUSTRE. Resinous. H. 3·5. SP. GR. 7–7·5.

TESTS. Heated in the closed tube, mimetite behaves like pyromorphite; heated on charcoal in the reducing flame, it yields metallic lead, gives off an arsenical odour, and coats the charcoal with lead chloride, lead oxide and arsenious oxide; flame coloraration, blue and green; soluble in hydrochloric acid.

VARIETY. *Campylite* is a variety occurring in barrel-shaped crystals of a brown or yellowish colour.

OCCURRENCE. In the oxidation zone of lead deposits; it is a minor ore of lead.

VANADINITE

COMP. Chloro-vanadate of lead, $3Pb_3V_2O_8.PbCl_2$, or better, $(PbCl)Pb_4(VO_4)_3$; phosphorus and arsenic are sometimes present in small amounts.

CRYST. SYST. Hexagonal, bipyramidal, isomorphous with pyromorphite and mimetite. COM. FORM. Prismatic crystals like those of pyromorphite, often in parallel groups; also in crusts. COLOUR. Ruby-red, orange-brown, or yellowish. STREAK. White or yellowish. LUSTRE. Resinous. H. 3. SP. GR. 6·7–7·1.

TESTS. Gives the reactions for lead; soluble in nitric acid, a precipitate of silver chloride being given with silver nitrate solution; vanadium given by the microcosmic salt bead—in the oxidizing flame yellow, in the reducing flame, bright green.

OCCURRENCE. In the zone of oxidation of lead veins, associated with lead oxy-salts and the other members of the pyromorphite set, as at many lead-mining localities; also accompanies other vanadium minerals in sediments, as in the Triassic sandstone of Alderley Edge in Cheshire.

USES. As a source of vanadium (see p. 468), and as a minor ore of lead.

CROCOISITE, Crocoite, Crocoise

COMP. Lead chromate, $PbCrO_4$.

CRYST. SYST. Monoclinic. COM. FORM. Prismatic crystals, modified—the faces of the prism often striated longitudinally; also occurs in an imperfectly columnar or granular condition. CLEAV. Prismatic, tolerably distinct; basal cleavage less so. COLOUR. Hyacinth-red of different shades. STREAK. Orange-yellow. LUSTRE. Adamantine to vitreous; translucent. FRACT. Conchoidal or uneven; sectile. H. 2·5–3. SP. GR. 5·9–6·1.

TESTS. Heated in the closed tube, decrepitates and blackens, but reverts to its original colour on cooling; heated on charcoal with sodium carbonate, is reduced to metallic lead, and coats the charcoal with an encrustation of chromium and lead oxides; microcosmic salt bead, emerald-green in both oxidizing and reducing flames.

OCCURRENCE. Occurs where lead lodes traverse rocks containing chromium, as at Beresof in Siberia, in the Urals, Hungary and the Philippines.

WULFENITE

COMP. Lead molybdate, $PbMoO_4$.

CRYST. SYST. Tetragonal. COM. FORM. Prisms and bipyramids, variously modified crystals often tabular; also occurs massive and granular. CLEAV. Parallel to the faces of the pyramid, smooth; there is also a less distinct cleavage parallel to the basal pinacoid. COLOUR. Wax-yellow, orange-yellow, yellowish-grey, greyish-white, brown, and sometimes shades of orange, red or green. STREAK. White. LUSTRE. Waxy or adamantine; subtransparent to subtranslucent. FRACT. Subconchoidal; brittle. H. 3. SP. GR. 6·3–7.

TESTS. Heated before the blowpipe, decrepitates and fuses; with borax in the oxidizing flame, gives a colourless bead, which, in the reducing flame, becomes opaque, black, or dirty green with black specks; with microcosmic salt it gives in the oxidizing flame a yellowish-green bead, which in the reducing flame becomes dark green; heated with sodium carbonate on charcoal, yields metallic lead; wulfenite is decomposed when heated in hydrochloric acid, the addition of a scrap of zinc to the solution causes it to assume a deep blue colour.

OCCURRENCE. Found in the oxidized portions of lead deposits.

GROUP 5A

VANADIUM, TANTALUM

VANADIUM MINERALS

Vanadium (V) does not occur free in nature. It is a whitish silvery metal, melting at about 1720°C. It has a great affinity for oxygen, a property underlying its use in metallurgy. Vanadium ores are treated in various ways, smelting in the electric furnace, reduction by the Thermit process (see p. 318), etc., to produce ferro-vanadium, with some 30 per cent vanadium.

Vanadium is used chiefly in the manufacture of special steels, such as high-speed tool-steels—the vanadium acting as a scavenger for oxygen, and also imparting special properties of toughness, etc., to the steel. In addition, other alloys are becoming important. Vanadium salts are used for various processes connected with chemical manufacture, printing of fabrics, dyeing, ceramics, etc.

Vanadium minerals are not abundant. The chief from the industrial viewpoint are:

Sulphide

Patronite, possibly VS_4, is an important vanadium ore; it occurs in a lens-shaped deposit at Minasraga, Peru, and the ore is associated with nickel and molybdenum sulphides and asphaltic material, the whole looking like a slaty coal.

Silicate

Roscoelite, the vanadium mica, in which vanadium has replaced to a small extent the aluminium of muscovite; roscoelite occurs in certain gold-quartz veins, and as flakes replacing the cement of certain sandstones.

Vanadate of Uranium and Potassium

Carnotite, $K_2(UO_2)_2(VO_4)_2.3H_2O$, which is one of the sources of radium, and is therefore described with the Uranium Minerals on p. 496; it occurs as seams and pockets in sandstones in Colorado and elsewhere.

Chloro-vanadate of Lead

Vanadinite, $(PbCl)Pb_4(PO_4)_3$, is a member of the Pyromorphite Set of Lead Minerals, and though it is an ore of vanadium, it is so closely related to the other members of the Set that it is described with them on p. 464. It occurs in the zone of oxidation of lead- and lead-zinc deposits.

In addition to these minerals of economic importance, vanadium enters into the composition of a number of still rarer complex minerals, some of which furnish a small part of the vanadium output. *Mottramite*, a hydrous vanadate of lead and copper, occurs as coatings in the Triassic sandstone of Mottram St. Andrew and Alderley Edge in Cheshire.

The principal producers of vanadium ores have been Peru, Colorado and Utah in the United States, Broken Hill in Rhodesia, South-west Africa, South Africa and Finland.

TESTS. Vanadium compounds give characteristic reactions in the beads. The borax bead is, in the oxidizing flame, yellow when hot and yellow-green to colourless when cold; in the reducing flame, it is dirty green when hot, clear green when cold. The microcosmic salt bead is, in the oxidizing flame, yellow to amber coloured, and in the reducing flame, green.

TANTALUM MINERALS

Native Tantalum (Ta) has been recorded but is exceedingly rare; it is produced by means of the electric furnace. It is a hard, white, ductile metal of great tensile strength, having a specific gravity of 16·64 and a melting point of 2,850°C. It is extremely resistant to corrosion, and on this account is of use in certain chemical and electrical processes. Its industrial uses, though not many, are of considerable importance; formerly it was extensively used for electric filaments, but for this purpose its place is now taken by tungsten. Tantalum is used in the production of special steels, especially those used for dental and surgical instruments. For the addition to steel, it is more usual to employ ferro-tantalum alloys, which are made from tantalite (iron-manganese-tantalum oxide) in the electric furnace. It is employed also in electrodes, and a compound, tantalum carbide, one of the hardest materials known, is used in tools. The only sources of any commercial importance are the minerals *tantalite* and *pyrochlore*. Tantalite occurs in association with wolfram and tinstone in granitic pegmatites, and the several small production units operate on deposits of this nature, or else alluvial deposits derived from similar occurrences. A part of the tantalum in tantalite is almost invariably replaced by the closely allied metal, *columbium* or *niobium*, and when the tantalum is subordinate in amount, the mineral is known as *columbite* or *niobite*. The tantalum-niobium minerals considered here are:

> *Oxide of Tantalum-Niobium, Iron and Manganese.* Tantalite-Columbite, $(Fe,Mn)(Ta,Nb)_2O_6$.
>
> *Oxide of Tantalum-Niobium, Sodium and Calcium.* Pyrochlore, $(Na,Ca)(Ta,Nb)_2O_6(O,OH,F)$.

TANTALITE-COLUMBITE, Tantalite-Niobite

COMP. Oxide of tantalum-niobium, iron and manganese $(Fe,Mn)(Ta,Nb)_2O_6$; when tantalum exceeds niobium the mineral is called *Tantalite*, and when niobium is in excess, *Columbite*.

CRYST. SYST. Orthorhombic. COM. FORM. Prismatic or tabular crystals, formed of several prisms, the three pinacoids, and one or more bipyramids; the commonest twin-plane is a face of the dome (021); often occurs massive. CLEAV. Parallel to the macropinacoid and brachypinacoid, the former the better. COLOUR. Grey, black or brown; sometimes iridescent. STREAK. Dark-red to black. LUSTRE. Submetallic to subresinous. FRACT. Subconchoidal to uneven. H. 6. SP. GR. 5·3–7·3, increasing with the increasing Ta_2O_5 content.

TESTS. Distinguished from black tourmaline by its higher specific gravity and by the shape of the crystals, and from wolfram by its less good cleavage.

OCCURRENCE. As a constituent of certain granitic pegmatites, as in the Black Hills, South Dakota, where very large crystals of columbite have been mined, and in similar rocks in Western

Australia and elsewhere; tantalite-columbite occurs also in certain alluvial deposits associated with tinstone and wolfram, and such deposits have been worked on a small scale. The main producers are Western Australia, Rhodesia, Nigeria, South Dakota and Brazil.

USES. As a source of tantalum and niobium metals.

PYROCHLORE

COMP. Oxide of tantalum-niobium, sodium and calcium, $(Na,Ca)(Ta,Nb)_2O_6.(O,OH,F)$; rare earths may replace some Na or Ca.

CRYST. SYST. Cubic. COM. FORM. Octahedral crystals, massive or granular. CLEAV. Parallel to the octahedron sometimes developed. COLOUR. Brown to black, lighter-coloured in niobium-rich types.

OCCURRENCE. In alkaline igneous rocks and in pegmatites derived from them. Worked from such deposits in Canada, Brazil and elsewhere.

GROUP 5B
NITROGEN, PHOSPHORUS, ARSENIC, ANTIMONY, BISMUTH

NITROGEN MINERALS

The gas nitrogen (N) makes up some 78 per cent by volume of the atmosphere. It occurs in combination in two principal types of minerals, the *nitrates* and the *ammonium minerals*.

The *nitrates* are salts of nitric acid, HNO_3. These salts are mostly very soluble in water, so that their occurrence as minerals is restricted. The two chief mineral nitrates are:

Sodium nitrate	Soda nitre, Chile saltpetre, $NaNO_3$, described on p. 231.
Potassium nitrate	Nitre, Saltpetre, KNO_3, described on p. 239.

These two minerals are described with the Sodium and Potassium minerals respectively on the pages cited above, where their properties, formation, modes of occurrence and uses are discussed.

The *ammonium radicle*, NH_4, occurs as the cation portion of several mineral salts, for example:

Ammonium Chloride	Sal Ammoniac, NH_4Cl.
Ammonium Sulphate	Mascagnite, $(NH_4)_2SO_4$.

Several other still rarer ammonium salts are known as minerals as, for instance, *Taylorite*, $(NH_4)_2SO_4.5K_2SO_4$, an *Ammonium alum*, $(NH_4)Al(SO_4)_2.12H_2O$, and *Ammonioborite*, $(NH_4)_2B_{10}O_{16}.5H_2O$, but only sal ammoniac and mascagnite are described here.

TESTS. All the ammonium salts are more or less soluble in water, and are easily and entirely volatilized before the blowpipe; this character suffices to distinguish them from other minerals. They also give the characteristic ammonia odour when heated with quick-lime, or when ground up with lime and moistened at the same time with water.

SAL AMMONIAC

COMP. Ammonium chloride, NH_4Cl.

CRYST. SYST. Cubic. COM. FORM. Octahedron, but generally efflorescent or encrusting. COLOUR. White when pure, often yellowish or grey. LUSTRE. Internally vitreous, externally dull; translucent to opaque. TASTE. Pungent, cool and saline. H. 1·5–2. SP. GR. 1·52.

TESTS. When ground in a mortar with soda-lime gives an ammoniacal odour.

OCCURRENCE. Occurs as a white efflorescence in volcanic districts. All sal ammoniac of commerce is an artificial product.

USES. Sal ammoniac is used in medicine, in dyeing, in soldering, in various metallurgical processes, as a chemical reagent, and also in electric batteries.

MASCAGNITE

COMP. Ammonium sulphate, $(NH_4)_2SO_4$.

CRYST. SYST. Orthorhombic.

CHARACTERS AND OCCURRENCE. Occurs as yellowish-grey, pulverulent, mealy crusts in the neighbourhood of volcanoes; also occurs in guano-deposits where it is accompanied by other ammonium sulphates such as *taylorite* $(NH_4)_2SO_4.5K_2SO_4$; mascagnite has a vitreous lustre, and is easily soluble in water.

PHOSPHORUS MINERALS

Phosphorus (P) forms an acid, phosphoric acid, H_3PO_4, and salts of this acid are fairly common as minerals. Usually, however, mineral phosphates are more complex, being phosphates of two or more metals, basic phosphates of various types, or compounds into which other radicles enter. The phosphates dealt with in this book are described with their most suitable metallic element. The chief mineral phosphates are the following, their descriptions being given at the pages cited in brackets:

Amblygonite, lithium aluminium phosphate, $Li(F,OH)AlPO_4$ (p.227).

Apatite, calcium fluo-phosphate and chloro-phosphate, $Ca_5(F,Cl)(PO_4)_3$ (p. 282).

Autunite, hydrous phosphate of calcium and uranium, $Ca(UO_2)_2(PO_4)_2.10-12H_2O$ (p. 495).

Libethenite, hydrous copper phosphate, $4CuO.P_2O_5.H_2O$ (p. 255).

Monazite, phosphate of the cerium metals, $(Ce,La,Yt)PO_4$ (p. 335).

Phosphochalcite, hydrous copper phosphate, $6CuO.P_2O_5.3H_2O$ (p. 255).

Pyromorphite, chloro-phosphate of lead, $Pb_5Cl(PO_4)_3$ (p. 463).

Torbernite, hydrous phosphate of copper and uranium, $Cu(UO_2)_2(PO_4)_2.12H_2O$ (p. 495).

Turquoise, basic hydrous aluminium copper phosphate, $CuAl_6(PO_4)_4(OH)_8.4H_2O$ (p. 325).

Vivianite, hydrous iron phosphate, $Fe_3(PO_4)_2.8H_2O$ (p. 526).

Wavellite, hydrous aluminium phosphate, $Al_6(PO_4)_4(OH)_6.9H_2O$ (p. 326).

The uses of the various economically important phosphates are detailed under their descriptions.

TESTS. Tests for phosphates are as follow: In the flame-test, many phosphates give a pale blue-green colour, which is increased in many cases if the mineral is moistened with strong sulphuric acid; when phosphates are fused with sodium carbonate on charcoal, and the fused mass removed and transferred to a closed tube with a little powdered magnesium and ignited, the phosphate is reduced to phosphide, which, when moistened, gives the well-known disagreeable smell of phosphoretted hydrogen, PH_3.

ARSENIC MINERALS

Arsenic (As) is found native, usually associated with other metals, but never in sufficient quantity to repay working; in combination it is very widely distributed, and occurs in many sulphide ores.

Arsenic is an extremely brittle, steel-grey metal of a brilliant lustre, and having a specific gravity of 5·7. It is obtained from its ores by heating in retorts, but most of the production is as a by-product in the smelting of arsenical lead, silver, cobalt or copper ores.

The metal arsenic is employed in small quantity in the manufacture of lead-shot and certain alloys. The most important industrial compound of arsenic is *white arsenic*, arsenious oxide, As_2O_3, which is obtained in the form of flue-dust or 'soot' in the smelting of arsenical ores proper and of the numerous arsenical ores of other metals. The great proportion of the arsenic used in industry depends for its employment on the poisonous properties of arsenic compounds, the manufacture of insecticides, weedkillers, sheep-dips, etc., absorbing some 70 per cent of the annual output. Other uses are as a decolorizer of glass, in paint-manufacture, textile-printing, etc. It is probable that the world consumption of arsenic annually is about 50,000 tons of white arsenic, the chief producers being the United States, Sweden, Mexico, France, Belgium and Japan; it is an instructive comment on the arsenic situation to realise that one Swedish gold-copper mine could supply the whole world demand.

The most important minerals of arsenic from the economic view-point are *mispickel* or *arsenopyrite*, FeAsS, with the sulphides *orpiment* As_2S_3, and *realgar*, As_2S_2, of less account. *Native arsenic* is fairly widespread, but not commercially important. Arsenic enters into the composition of a number of complex sulphides, from some of which white arsenic is obtained as a by-product; examples of such minerals are: enargite, Cu_3AsS_4, tennantite, $(Cu,Fe)_{12}As_4S_{13}$, proustite, Ag_3AsS_3, and rarer sulphides of arsenic, copper and lead—as indicated below, these minerals are described under their more important metallic component. In addition to the iron sulpharsenide, mispickel, another sulpharsenide, that of cobalt, *cobaltite*, CoAsS, is important, but it is described on p. 529 as an ore of cobalt. Another group of arsenic minerals are the arsenides of cobalt and nickel described with the minerals of these two metals—these arsenides are *smaltite*, $(Co,Ni)As_{3-n}$, *kupfernickel* or *niccolite*, NiAs, and *chloanthite*, $NiAs_2$. Certain arsenates, such as *mimetite*, $(PbCl)Pb_4(AsO_4)_3$, *olivenite*, $Cu_3As_2O_8.Cu(OH)_2$, occur in the oxidized portion of arsenical lead and copper veins. Other arsenates such as *erythrite*, $Co_3As_2O_8.8H_2O$, and *annabergite*, $Ni_3As_2O_8.8H_2O$, characterize weathered cobalt and nickel arsenide ores respectively. Finally, the oxide, *arsenolite*, As_2O_3, is known in small quantity as a decomposition product of other arsenic ores.

The primary arsenic minerals occur in lodes or veins more or less directly connected with igneous intrusions; the arsenates, arsenolite, realgar and orpiment are characteristic of the oxidized portion of such deposits.

TESTS. The following are the chief tests for arsenic. Arsenic compounds, when heated on charcoal, give a white encrustation far from the assay, and at the same time, fumes having a garlic odour are emitted. Heated in the open tube, arsenic compounds give a white sublimate, which is volatile on heating. Heated in the closed tube some arsenic compounds give a shining black sublimate, the arsenic mirror; most arsenates give a similar mirror when heated with charcoal or sodium carbonate in the closed tube.

From the foregoing, the following list of arsenic minerals may be compiled:

Element	Native Arsenic, As.
Oxide	Arsenolite, White Arsenic, As_2O_3.
Sulphides	Orpiment, As_2S_3.
	Realgar, As_2S_2.
Sulpharsenides	Mispickel, Arsenopyrite, FeAsS.
	Cobaltite, CoAsS, (described on p. 529 with the Cobalt minerals).
Arsenides	Kupfernickel, Niccolite, NiAs, (described on p. 533 with the Nickel minerals).
	Chloanthite, $NiAs_2$, (also described with the Nickel minerals on p. 533).
	Smaltite, $(Co,Ni)As_{3-n}$, (described on p. 529 with the Cobalt minerals).
Arsenates	Mimetite, $(PbCl)Pb_4(AsO_4)_3$, (described on p. 464 with Lead minerals).
	Olivenite, $Cu_3As_2O_8.Cu(OH)_2$, (see p. 255).
	Erythrite, $Co_3As_2O_8.8H_2O$, (see p. 530).

Annabergite, $Ni_3As_2O_8.8H_2O$, (see p. 536).

Complex Sulphides { Enargite, Cu_3AsS_4, (described on p. 250 with the Copper minerals).

Tennantite, $(Cu,Fe)_{12}As_4S_{13}$, (described on p. 249 with the Copper minerals).

Proustite, Ag_3AsS_3, (described on p. 260 with the Silver minerals).

NATIVE ARSENIC

COMP. Arsenic, As; often with some antimony and traces of other metals.

CRYST. SYST. Hexagonal-trigonal. COM. FORM. Often occurs granular, massive or reniform, or sometimes columnar or stalactitic. COLOUR. On recent fractures, tin-white, but quickly tarnishing to a dark grey. STREAK. Tin-white. LUSTRE. Nearly metallic. FRACT. Uneven and granular; brittle. H. 3·5. SP. GR. 5·7.

TESTS. Gives the reactions for arsenic mentioned in the introduction on p. 475.

OCCURRENCE. Native arsenic occurs principally as a minor constituent of certain lead, silver, nickel or cobalt ores, as at Freiberg in Saxony, Joachimsthal in Bohemia, etc.

ARSENOLITE, White Arsenic, Arsenious Acid

COMP. Arsenic trioxide, As_2O_3.

CRYST. SYST. Cubic.

CHARACTERS AND OCCURRENCE. Arsenolite is not a mineral of common occurrence in nature, but is sometimes met with in white fine capillary crystals or crusts, resulting from the decomposition of arsenical ores; it is, however, extensively manufactured and, on account of its very poisonous properties, is of great economic importance in the manufacture of insecticides and the like, as explained in the introduction on p. 474.

ORPIMENT

COMP. Arsenic trisulphide; As_2S_3; arsenic 61·0 per cent.

CRYST. SYST. Monoclinic. COM. FORM. Crystals rare, usually occurs foliaceous or massive. CLEAV. Perfect parallel to the clinopinacoid; the laminae are flexible but not elastic; sectile. COLOUR. Fine lemon-yellow. STREAK. Yellow. LUSTRE. Pearly and brilliant on cleavage-faces, elsewhere resinous or dull; subtransparent to subtranslucent. H. 1·5–2. SP. GR. 3·4–3·5.

TESTS. Heated in closed tube, gives a reddish-yellow sublimate; heated on charcoal, emits sulphurous and garlic fumes and forms a white sublimate far from the assay; heated in the open tube, white volatile sublimate and garlic fumes formed.

OCCURRENCE. Occurs in the oxidized portions of arsenic veins; in veins associated with antimony ores as at Kapnik in Transylvania, Kurdistan in Asiatic Turkey, etc.; it also occurs as a deposit from some hot springs, as at Steamboat Springs, Nevada, and as a sublimate from volcanoes at Naples.

REALGAR

COMP. Arsenic monosulphide, As_2S_2; arsenic 70·1 per cent.

CRYST. SYST. Monoclinic. COM. FORM. Prismatic crystals, rare, usually occurs massive or granular. COLOUR. Fine red, or orange. STREAK. Red or orange. LUSTRE. Resinous; transparent to translucent. FRACT. Conchoidal or uneven. H. 1·5–2. SP. GR. 3·56.

TESTS. As for orpiment (see above).

OCCURRENCE. Occurs associated with orpiment, to which mineral it changes on exposure; occurs as a deposit from hot springs, and as a volcanic sublimate; common in veins, where it may occur as nests or nodules in clay, or associated with cinnabar, as in Tuscany, Galicia and Spain.

MISPICKEL, ARSENOPYRITE, Arsenical Pyrites

COMP. Iron sulpharsenide, FeAsS; arsenic 46·0 per cent.

CRYST. SYST. Orthorhombic. COM. FORM. Prism, mostly terminated by horizontally striated faces of the brachydome; twinned crystals common; also massive. CLEAV. Parallel to the faces of the prism. COLOUR Tin-white, or silver-white, inclined to steel-grey, and tarnishing pale copper-colour on exposure. STREAK. Dark greyish-black. LUSTRE. Metallic. FRACT. Uneven; brittle; gives sparks when struck with steel and then emits a garlic odour. H. 5·5–6. SP. GR. 5·9–6·2.

TESTS. Heated before the blowpipe gives rise to arsenical fumes and fuses to a globule which is attracted by the magnet; heated in the closed tube, gives first the red sublimate of arsenic sulphide, and then the black sublimate of arsenic; heated in the open tube gives sulphurous fumes and a white sublimate of arsenic oxide; heated with hydrochloric acid, gives sulphur.

OCCURRENCE. Occurs in tin, copper, cobalt, nickel and especially lead and silver veins, usually traceable to igneous intrusions; examples of pneumatolytic veins (tin and copper) are those of Cornwall and Devon—Tavistock, Botallock, Dolcoath, of the hydrothermal veins (lead, silver, etc.), Saxony, Leadville, etc.; abundant in quartz-rich veins in Hastings County, Ontario.

USES. As an ore of arsenic.

ANTIMONY MINERALS

Antimony (Sb) in a free state is of extremely rare occurrence. The chief source of the metal is the sulphide, *stibnite* or *antimonite*, Sb_2S_3, which is widely distributed, but found in workable quantities in comparatively few localities. Metallic antimony is a tin-white, very brittle metal, with a crystalline structure. It has a specific gravity of 6·7 and melts at 630°C.

For the production of the metal, the sulphide is freed from its gangue by liquation and reduced in reverberatory furnaces, or the crude ore is volatilized in a blast furnace, and the condensed fumes reduced in reverberatory furnaces. On the market the liquated sulphide is called 'crude antimony', while the metal is called 'regulus of antimony'. The first quality of refined antimony is known as 'star antimony', owing to the fern-like markings on its surface. Antimonial lead ores, free from gold and silver, are commonly smelted direct for 'hard' or 'antimonial' lead.

The chief use of antimony metal is in the production of important alloys, such as type-metal, anti-friction metals, pewter, etc., and for hard lead. Its compounds are used for a variety of purposes, such as for pigments, in medicine, as a mordant, in the manufacture of opaque enamel ware, in glass and pottery manufacture and in flame-proofing.

The world production of antimony amounts to over 62,000 tons per year, the chief producers being China and South Africa, followed by Bolivia, U.S.S.R., Mexico, Yugoslavia, Turkey, Canada and Thailand.

The antimony minerals are as follows. The element occurs as *native antimony* in small amounts, but the most important mineral is the sulphide, *antimonite* or *stibnite*, Sb_2S_3. A series of oxygen-compounds occur as weathered products of the sulphide ores, and among these are the oxides *senarmontite*, Sb_2O_3 (cubic), *valentinite*, Sb_2O_3 (orthorhombic), *cervantite*, $Sb_2O_3.Sb_2O_5$ and the oxy-sulphide, *kermesite*, $2Sb_2S_3.Sb_2O_3$. In addition to these purely antimony minerals, the element also enters into the composition of a large number of complex sulphides, of which the following are considered in this book—*jamesonite*, $Pb_4FeSb_6S_{14}$, dealt with here with the antimony minerals; a group of silver antimony sulphides, such as *stephanite*, Ag_5SbS_4, *pyrargyrite*, Ag_3SbS_3., *polybasite*, $(Ag,Cu)_{16}(Sb,As)_2S_{11}$, *freieslebenite* $(Pb,Ag)_8Sb_5S_{12}$, etc., which are considered with the silver minerals, and a group of copper antimony sulphides, such as *bournonite*, $CuPbSbS_3$, *tetrahedrite*, $(Cu,Fe)_{12}Sb_4S_{13}$, *famatinite*, Cu_3SbS_4, which are considered with the copper minerals.

Antimony ores occur both in deposits associated with volcanic rocks and also more deep-seated veins formed under moderate to high temperatures and pressures. Thus stibnite often occurs with mercury ores, but is more common in veins with a gangue of quartz and but a small proportion of other sulphides. Certain of the replacement galena deposits show antimony minerals such as jamesonite and stibnite. The surface oxidation of these primary antimony ores leads to the formation of the oxides and oxy-sulphide mentioned above.

TESTS. The following are important tests for antimony. When antimony compounds are heated on charcoal, a dense white sublimate is formed as an encrustation near the assay—the nearness to the assay, and the absence of any characteristic fumes distinguish the reaction of antimony from that given by arsenic; heated in the open tube, antimony compounds give a white sublimate of oxide of antimony, which appears as a ring near the assay; heated with sodium carbonate in the closed tube, antimony compounds give a red-brown sublimate, which is black when hot.

The following minerals of antimony are considered in this book:

Element	Native Antimony.	
Oxides	Senarmontite, Sb_2O_3, cubic.	
	Valentinite, Sb_2O_3, orthorhombic.	
	Cervantite, $Sb_2O_3.Sb_2O_5$.	
Sulphide	Antimonite, Stibnite, Sb_2S_3.	
Oxysulphide	Kermesite, $2Sb_2S_3.Sb_2O_3$.	
	Jamesonite, $Pb_4FeSb_6S_{14}$.	
Complex Sulphides	Stephanite, Ag_5SbS_4.	Described with the *Silver* minerals on pp. 259–261.
	Pyrargyrite, Ag_3SbS_3.	
	Polybasite, $(Ag,Cu)_{16}(Sb,As)_2S_{11}$.	
	Freieslebenite, $(Pb,Ag)_8Sb_5S_{12}$.	
	Bournonite, $CuPbSbS_3$.	Described with the *Copper* minerals on pp. 248–251
	Tetrahedrite, $(Cu,Fe)_{12}Sb_4S_{13}$.	
	Famatinite, Cu_3SbS_4.	

NATIVE ANTIMONY

COMP. Antimony, Sb; sometimes with traces of silver, iron or arsenic.

CRYST. SYST. Hexagonal-trigonal. COM. FORM. Usually occurs massive, granular or lamellar. CLEAV. Perfect basal. COLOUR. Tin-white. STREAK. Tin-white. LUSTRE. Metallic; opaque. FRACT. Uneven; very brittle. H. 3–3·5. SP. GR. 6·6–6·7.

TESTS. Heated before the blowpipe, fuses easily giving off white fumes of antimonious oxide, which condense and form a white encrustation on the charcoal near the assay.

OCCURRENCE. In veins associated with stibnite or ores of silver or arsenic.

SENARMONTITE

COMP. Antimony trioxide, Sb_2O_3.

CRYST. SYST. Cubic.

CHARACTERS AND OCCURRENCE. A white or greyish mineral occurring in octahedral crystals or crusts, and arising by the oxidation of primary antimony ores; common at Djebel-Haminate Mine, Algeria.

VALENTINITE

COMP. Antimony trioxide, Sb_2O_3.

CRYST. SYST. Orthorhombic.

CHARACTERS AND OCCURRENCE. A white, greyish or reddish mineral, possessing a perfect cleavage parallel to the brachydome, occurring in prismatic crystals or crusts and resulting from the oxidation of various antimony ores, as at the Djebel-Haminate Mine, Algeria.

CERVANTITE

COMP. Antimony oxide, $Sb_2O_3.Sb_2O_5$.

CRYST. SYST. Orthorhombic.

CHARACTERS AND OCCURRENCE. Occurs as acicular or powdery crusts of a white or yellow colour, and resulting from the oxidation of primary antimony ores.

ANTIMONITE, STIBNITE, Antimony Glance, Grey Antimony

COMP. Antimony trisulphide, Sb_2S_3; antimony 71·7 per cent.

CRYST. SYST. Orthorhombic. COM. FORM. Elongated prisms striated longitudinally (see Fig. 148); commonly found in masses of radiating crystals, or with a columnar or bladed structure; sometimes granular. CLEAV. Perfect parallel to the brachypinacoid. COLOUR. Lead-grey. STREAK. Lead-grey. LUSTRE.

Metallic; liable to tarnish, and sometimes iridescent on the surface. FRACT. Subconchoidal; sectile; brittle, but thin laminae slightly flexible. H. 2. SP. GR. 4·5–4·6.

TESTS. Heated in the open tube, antimonite gives off antimonious and sulphurous fumes, the former condensing as a white non-volatile sublimate, while the latter may be recognized by the odour; heated on charcoal, fuses easily and gives a white encrustation near the assay; fuses easily in the flame of a candle.

OCCURRENCE. The chief mode of occurrence is in quartz-stibnite veins, but it also occurs associated with other antimony minerals, lead-zinc sulphides, and with quartz, dolomite, calcite and barytes as veinstones, as in Cornwall, Westphalia, Saxony, etc. The great Chinese deposits of Hunan occur in brecciated sandstone in which the antimonite occurs as irregular stringers, veins and pockets; antimonite is a primary mineral oxidizing to the antimony oxides described above.

USES. Antimonite is the chief source of the antimony of commerce.

FIG. 148. Antimonite

KERMESITE, Red Antimony

COMP. Antimony oxy-sulphide, $2Sb_2S_3.Sb_2O_3$.

CRYST. SYST. Orthorhombic or monoclinic.

CHARACTERS AND OCCURRENCE. Occurs as red needle-shaped crystals, and results from the alteration of stibnite in the oxidized zones of antimony deposits, being a stage towards the formation of the oxides, senarmontite and valentinite, with which it is often associated.

JAMESONITE

COMP. Lead iron antimony sulphide, $Pb_4FeSb_6S_{14}$; antimony, 29·5 per cent.

CRYST. SYST. Monoclinic. COM. FORM. Acicular crystals, often with feather-like forms, giving the variety, *plumosite* or *feather-ore*; also fibrous and massive. CLEAV. Perfect basal. COLOUR Dark lead-grey. STREAK. Greyish-black. LUSTRE. Metallic. H. 2–3. SP. GR. 5·5–6.

TESTS. Gives reactions for antimony in the open tube; heated on charcoal with sodium carbonate and charcoal, gives metallic bead of lead; lead also given by the potassium-iodide and sulphur test (see p. 42); jamesonite is soluble in hydrochloric acid, a precipitate of lead chloride being formed on cooling.

OCCURRENCE. In veins associated with stibnite, and tetrahedrite and other lead-silver sulpho-salts; some localities are Endellion, Cornwall; Foxdale, Isle of Man; Dumfriesshire, Scotland; etc.

BISMUTH MINERALS

Bismuth (Bi) occurs in a free state in nature often associated with silver, gold, copper, lead and other minerals. It is a greyish-white metal, with a slightly reddish tinge, lustrous and very brittle, and having a specific gravity of 9·8 and melting at 271°C. Bismuth is obtained by smelting the dressed ores in small reverberatory furnaces or crucibles. The main source is native bismuth and sulphide ores, but a considerable portion of the production comes from the anode slimes resulting from the electrolytic refining of copper and lead. Production exceeds 4,000 tons annually. Peru, Bolivia, Mexico and Canada are the chief producers, whilst Korea, Yugoslavia, Japan, Sweden, France and several other countries produce some. Bismuth salts are used in medicine, and to a limited extent in pigments, glass, etc.; alloyed with tin, lead, mercury, etc., bismuth forms a series of alloys with low melting points, the fusible metals, which are important in certain industrial processes, e.g. casting, and are also employed in certain appliances such as automatic sprinklers and similar apparatus.

As already noted, bismuth occurs as the *native element* which, with the sulphide, *bismuthinite*, Bi_2S_3, and various complex lead-bismuth sulphides not dealt with in this book, are of primary origin. The chief oxidized minerals are the oxide, *bismuth ochre* or *bismite*, Bi_2O_3, and the basic carbonate, *bismutite*, $Bi_2CO_5.H_2O$. A telluride, *tetradymite*, $Bi_2(Te,S)_3$, is described with the tellurium minerals on p. 502.

Bismuth ores occur in three main associations—(1) with tin and copper minerals, as in the Bolivian deposits, (2) with cobalt, as at Schneeberg, Saxony, and (3) with gold, as in the Australian deposits.

TESTS. Bismuth compounds react as follows. When heated on charcoal with sodium carbonate and charcoal, they give a brittle, metallic bead, which volatilizes on heating to give a yellow encrustation. Heated with potassium iodide and sulphur, bismuth compounds give an encrustation which is yellow near the assay, and scarlet in the outer parts. Solutions of bismuth salts become milky on the addition of water, owing to the formation of insoluble basic compounds, which are redissolved on the addition of an acid.

The bismuth minerals dealt with in this book are:

Element	Native Bismuth, Bi.
Oxide	Bismuth Ochre, Bismite, Bi_2O_3.
Carbonate	Bismutite, $Bi_2CO_5.H_2O$.
Sulphide	Bismuthinite, Bi_2S_3.
Telluride	Tetradymite, $Bi_2(Te,S)_3$, described with the *Tellurium* minerals on p. 502.

NATIVE BISMUTH

COMP. Bismuth, Bi; sometimes with traces of sulphur, arsenic and tellurium.

CRYST. SYST. Hexagonal-trigonal. COM. FORM. Crystals are rhombohedra, much resembling cubes in form; usually found massive, foliaceous, or granular; also in reticulated or plumose forms. CLEAV. Perfect parallel to the basal pinacoid, less good parallel to a rhombohedron. COLOUR. Silver-white, with a faint tinge of red. STREAK. Silver-white. LUSTRE. Metallic; easily tarnishes; opaque. FRACT. When cold, brittle; when heated it is somewhat malleable; sectile. H. 2–2·5. SP. GR. 9·7–9·8.

TESTS. Heated on charcoal, native bismuth fuses and volatilizes, forming an orange-red encrustation; after fusion in a ladle, bismuth crystallizes readily; dissolves in nitric acid, the solution becoming milky when water is added; heated on charcoal with potassium iodide and sulphur, bismuth gives a brilliantly coloured encrustation, yellow near the assay, scarlet farther away.

OCCURRENCE. Occurs in veins associated with ores of tin, silver, cobalt and nickel, and also with pyrites, chalcopyrite, quartz, etc. as in Bolivia, Schneeberg in Saxony, Australia, etc.

USES. As an ore of bismuth.

BISMUTH OCHRE, Bismite

COMP. Bismuth trioxide, Bi_2O_3.

CHARACTERS AND OCCURRENCE. A yellow earthy pulverulent mineral, usually impure and hydrated, occurring as an alteration-product of bismuth and bismuthinite.

BISMUTITE

COMP. Basic bismuth carbonate, perhaps $Bi_2CO_5.H_2O$.

CHARACTERS AND OCCURRENCE. A white, grey or yellowish mineral, occurring as fibrous or earthy crusts, and resulting from the alteration of native bismuth and bismuthinite.

BISMUTHINITE, Bismuth Glance

COMP. Bismuth sulphide, Bi_2S_3.

CRYST. SYST. Orthorhombic. COM. FORM. Small needle-like crystals; usually massive. COLOUR. Lead-grey, but tarnish common. LUSTRE. Metallic; opaque. H. 2. SP. GR. 6·4–6·5.

TESTS. Heated before the blowpipe, bismuthinite fuses easily; heated on charcoal with potassium iodide and sulphur, it gives a yellow and bright red encrustation; heated in the open tube, gives sulphurous fumes; soluble in nitric acid, a white precipitate being formed on addition of water.

OCCURRENCE. Occurs in veins associated with copper, lead, tin, and other ores, as in Cornwall, Cumberland, Saxony, Bolivia, etc.

USE. As an ore of bismuth, as in certain mines in Bolivia.

GROUP 6A
CHROMIUM, MOLYBDENUM, TUNGSTEN, URANIUM

CHROMIUM MINERALS

Chromium (Cr), never found in nature except in combination, is produced by reduction of its ore by carbon in the electric furnace, or by the Thermit process mentioned on p. 318. It is a brilliant white metal, having a specific gravity of about 6·5, and melting at about 1,800°C. It possesses the property of imparting to iron and steel a high degree of hardness and tenacity, and for that reason has become in recent years of great industrial importance. For this purpose an alloy of iron and chromium (ferro-chrome, produced in the electric furnace) is commonly used; it is cheaper to make, melts at a lower temperature, and is consequently better under control than the pure metal. Stainless steel contains as much as 18 per cent of chromium. The compounds of chromium are also of considerable industrial importance. Chromite, an oxide of iron and chromium, is used very extensively as a refractory material for furnace linings. Other salts, artificially prepared, are used as pigments, and in various industries, such as chromium-plating, dyeing, tanning, photography, etc.

The only source of chromium is *chromite, chrome iron ore*, $FeCr_2O_4$. The world production of chromite is between four and five million tons annually; the chief producing countries are U.S.S.R., Rhodesia, South Africa, Philippines, Albania and Turkey. Chromium also occurs in various rock-forming minerals, such as the chrome-spinel, *picotite* (see p. 321) and the chrome garnet, *uvarovite* (see p. 363). *Crocoisite*, lead chromate, $PbCrO_4$, is described with the Lead minerals on p. 465.

TESTS. The best tests for chromium are provided by the beads; chromium compounds produce a fine green colour in both borax and microcosmic salt beads.

CHROMITE, Chromic Iron, Chrome Iron Ore

COMP. Oxide of iron and chromium, $FeCr_2O_4$.

CRYST. SYST. Cubic. COM. FORM. Occurs in octahedra, but is commonly found massive, having a granular or compact structure. COLOUR. Iron-black and brownish-black. STREAK. Brown. LUSTRE. Submetallic, often faint. FRACT. Uneven, sometimes rather flat; brittle. H. 5·5. SP. GR. 4·5–4·8.

TESTS. Infusible in the oxidizing flame, while in the reducing flame chromite becomes slightly rounded on the edges of splinters, which on cooling are magnetic; heated with sodium carbonate on charcoal, chromite is reduced to magnetic oxide; the borax and microcosmic salt beads are a beautiful chrome-green, this colour being rendered more intense if the mineral is first fused on charcoal with metallic tin.

OCCURRENCE. Chromite occurs as a primary mineral of ultra-basic igneous rocks, peridotites, and their modifications, serpentines; it also occurs in basic gabbros. Usually the chromite occurs as small grains, but by the segregation of these grains, ore-bodies may be formed as those of Norway, Rhodesia, Smyrna, and New Caledonia—the ore-bodies consisting of a peridotite extremely rich in chromite. There has been considerable discussion as to the time of crystallization of the chromite, whether early, late or even during a hydrothermal stage. Being very obdurate, chromite occurs in detrital deposits.

MOLYBDENUM MINERALS

Molybdenum (Mo) does not occur in a free state in nature but may be prepared from its sulphide, the mineral *molybdenite*, directly in the electric furnace, or by the reduction of its oxide by means of carbon, or by the Thermit process mentioned under aluminium on p. 318. The metal has a specific gravity of about 10·2, melts at about 2,620°C, and is white or greyish, and brittle. Molybdic acid forms salts known as molybdates, examples of which occur as minerals; thus, *wulfenite* is lead molybdate, and *molybdite* is possibly a hydrated iron molybdate. Molybdic acid is employed only in the laboratory. Ammonium molybdate is a special reagent for the detection of phosphoric acid, a small quantity added to an acid solution containing phosphates producing after some time a yellow precipitate. The principal use of molybdenum is in the manufacture of special steels, and for this purpose ferro-molybdenum alloy is frequently used in place of the metal. Molybdenum alloys are receiving attention, and the metal is also employed in certain electrical apparatus.

The ores of molybdenum are two, the most important being the sulphide, *molybdenite*, MoS_2, the less important being lead molybdate, *wulfenite*, $PbMoO_4$—this latter being described with the Lead minerals on p. 466. Molybdenite occurs in deposits associated with acid igneous rocks, whilst wulfenite is found in the oxidized portions of lead- and molybdenum-bearing deposits. Molybdenite is a widely distributed ore, but frequently occurs in small veins or scattered in tiny flakes through the rocks, so that a concentration by table-dressing, oil-flotation, etc., is necessary. The chief source is the United States (Colorado and New Mexico mostly), less important being the U.S.S.R. Chile and China. Significant amounts come also from Canada, Japan and Norway (Knaben Mine). Sporadic production in small quantity has come from Korea, Mexico, Philippines and elsewhere.

TESTS. The chief tests for molybdenum are the following. When heated on charcoal, molybdenum compounds give in the oxidizing flame a white encrustation, which becomes blue where touched by the reducing flame. Molybdenum compounds colour the microcosmic salt bead yellow when hot, colourless when cold in the oxidizing flame; the bead is green in the reducing flame.

The two chief minerals of molybdenum have been mentioned, namely the sulphide, molybdenite, and the lead molybdate, wulfenite. It was considered till recently that the oxide occurred as the mineral molybdite, but this mineral has been shown to be most likely a hydrated iron molybdate, possibly $Fe_2O_3.3MoO_3.8H_2O$. The molybdenum minerals considered in this book are therefore:

Sulphide   Molybdenite, MoS_2.
Molybdates   Wulfenite, $PbMoO_4$, described with the *Lead* minerals on p. 466.
                     Molybdite, $FeO_2O_3.3MoO_3.8H_2O.(?)$

MOLYBDENITE

COMP. Molybdenum sulphide, MoS_2.

CRYST. SYST. Hexagonal. COM. FORM. Usually in scales; also massive, foliaceous, and sometimes granular. CLEAV. Perfect basal; the laminae are flexible but not elastic. COLOUR. Lead-grey. STREAK. Greenish lead-grey, the greenish tint distinguishing it from that given by graphite. LUSTRE. Metallic; opaque. TENACITY. Sectile, and almost malleable. H. 1–1·5, easily scratched by the fingernail. SP. GR. 4·7–4·8.

TESTS. Flame-test, yellow-green; heated in the open tube, gives sulphurous fumes, and a yellow sublimate of molybdenum oxide; heated on charcoal, yields a strong sulphurous odour, coats the charcoal with an encrustation of molybdenum oxide, which is yellow while hot, and white cold—this encrustation is copper-red near the assay, and, if touched with the reducing flame, becomes blue; the microcosmic salt bead is green.

OCCURRENCE. Molybdenite occurs in deposits that can in most cases be traced to an acid igneous body; thus it is found in granites and pegmatites as an original constituent, though often of late formation, and it occurs in quartz veins connected with such rocks; it is also found as a mineral of contact-metamorphic or pyrometasomatic origin.

USES. Molybdenite is the chief ore of molybdenum.

MOLYBDITE, Ferrimolybdite, Molybdic Ochre, Molybdena

COMP. Formerly considered to be molybdic oxide, MoO_3, but recently shown to be most likely a hydrated iron molybdate, possibly $Fe_2O_3.3MoO_3.8H_2O$.

CRYST. SYST. Orthorhombic. COM. FORM. As an earthy powder or encrustation, or as silky, fibrous and radiating crystallizations. COLOUR. Straw-yellow, or yellow-white. LUSTRE. Silky to adamantine; of amorphous forms, earthy and dull. H. 1–2. SP. GR. 4·5.

TESTS. Heated alone on charcoal, yields a white encrustation, which is yellow while hot, and copper-red round the assay; this encrustation becomes blue when touched with the reducing flame.

OCCURRENCE. As an oxidation product of molybdenite, and often associated with limonite, etc.

TUNGSTEN MINERALS

Tungsten (W) is not found native, but is produced in the form of a greyish black powder, with a specific gravity of about 19. The metal is obtained by reduction of its ores by carbon, or by the Thermit Process mentioned on p. 318.

The chief tungsten minerals are tungstates of iron, manganese and calcium, the tungsten ores being *wolfram*, $(Fe,Mn)WO_4$, and *scheelite*, $CaWO_4$. Wolfram is usually associated with tinstone, and for a long time the separation of these minerals was a matter of some difficulty. It is effected by electromagnetic separation. Scheelite is less often mixed with tinstone, and the separation in this case is performed by roasting the crushed ore with sodium carbonate, by which operation sodium tungstate is formed, and the tin ores can be removed. In the case of several tin mining companies, the presence of wolfram enables profits to be made, which but for the 'mixed minerals' would be impossible. Tungsten ores, chiefly wolfram, appear on the market in the form of concentrates, varying between 60 per cent and 70 per cent tungstic acid, WO_3, and are purchased on the basis of their tungstic acid content.

The world production of tungsten ores, estimated for the purposes of calculation at an average of 60 per cent WO_3, reached nearly 84,000 tons in 1956. China, U.S.S.R. and the Koreas are the chief producers, followed by the United States, Bolivia, Portugal and others; less than a hundred tons of concentrates was produced in Cornwall annually up to 1957.

The chief use of tungsten, either as metallic tungsten or in the form of ferro-tungsten alloys produced in the electric furnace, is for tool steels. Other important uses are for the production of special alloys, and for the manufacture of electric filaments. Tungsten salts are used as mordants, and for fireproofing purposes. Tungsten carbide is employed in the manufacture of cutting-tools.

Two oxides of tungsten are known, WO_2 and WO_3; the latter is tungstic acid, and from it are formed various salts called tungstates, several of which occur as minerals, as in wolfram and scheelite already mentioned. Tungstic acid, WO_3, occurs as the mineral *tungstite* or *tungstic ochre*, and is formed as an alteration product of the mineral tungstates. The mineral tungstates occur for the most part as primary minerals associated with tinstone and in close connection with acid igneous rocks; the calcium tungstate, scheelite, is found in pyrometasomatic deposits at the contact of granitic rocks and limestones.

TESTS. The tests for tungsten are as follow:—Tungsten salts colour the microcosmic salt bead blue in the reducing flame. Tungsten minerals, fused with sodium carbonate, and then dissolved in hydrochloric acid, give, on the addition of a small piece of tin or zinc, a blue solution when heated.

The tungsten minerals considered here are:

Oxide	Tungstite, Tungstic Ochre, WO_3.
Tungstates	Wolfram, $(Fe,Mn)WO_4$.
	Hubnerite, $MnWO_4$.
	Ferberite, $FeWO_4$.
	Scheelite, $CaWO_4$.

TUNGSTITE, Tungstic Ochre

COMP. Tungstic oxide, WO_3; some material shows the composition $WO_3.H_2O$.

CHARACTERS AND OCCURRENCE. A pulverulent and earthy mineral, of a bright yellow or yellowish-green colour, resulting from the alteration of ores of tungsten; heated before the blowpipe alone, it is infusible, but becomes black in the inner flame; the microcosmic salt bead is in the oxidizing flame colourless or yellow, and becomes in the reducing flame violet when cold; tungstic ochre is soluble in alkaline solutions, but not in acids.

WOLFRAM, Wolframite

COMP. Tungstate of iron and manganese, $(Fe,Mn)WO_4$, the ratio of iron to manganese varying—one end-member being *Ferberite*, $FeWO_4$ and the other *Hubnerite*. $MnWO_4$.

CRYST. SYST. Monoclinic. COM. FORM. Prismatic crystals, commonly tabular; also massive and bladed, the latter being very characteristic. CLEAV. Perfect parallel to the clinopinacoid. COLOUR. Chocolate-brown, dark greyish-black, reddish-brown. STREAK. Chocolate-brown. LUSTRE. Submetallic, brilliant shining on cleavage-surfaces, dull on other surfaces and on fractures; subtransparent to opaque. FRACT. Uneven. H. 5–5·5. SP. GR. 7·1–7·9.

TESTS. Fusible; microcosmic bead reddish, borax bead green, sodium carbonate bead in oxidizing flame green; heated with sodium carbonate and carbon, the fused mass treated with hydrochloric acid and powdered tin, gives a blue solution; heated with sodium carbonate yields a magnetic mass; wolfram is characterized by its high specific gravity, streak, cleavage, and the two lustres.

ALLIED SPECIES. *Hubnerite*, $MnWO_4$, usually contains some iron, like wolfram in appearance; *Ferberite* is the iron tungstate, $FeWO_4$, with up to 20 per cent of manganese tungstate; *Reinite*, iron tungstate, is probably a pseudomorph after scheelite.

OCCURRENCE. Wolfram occurs in pneumatolytic veins surrounding granite masses, associated with tinstone and quartz, as in Cornwall (St. Austell), Zinnwald, Malaysia, Bolivia, etc.; also in veins formed at lower temperatures, and in certain gold-bearing quartz-veins; the disintegration of these tin-wolfram veins results in the formation of alluvial or placer tin and wolfram deposits, which are worked at many localities, as in Burma.

SCHEELITE

COMP. Calcium tungstate, $CaWO_4$.

CRYST. SYST. Tetragonal. COM. FORM. Crystals are combinations of tetragonal bipyramids; usually occurs reniform, with a columnar structure; also massive and granular. CLEAV. Good parallel to the bipyramid, (111). COLOUR. Yellowish-white or brownish, sometimes almost orange-yellow. STREAK. White. LUSTRE. Vitreous, inclined to adamantine; transparent to translucent. FRACT. Uneven, brittle. H. 4·5–5. SP. GR. 5·9–6·1.

TESTS. Microcosmic salt bead, in reducing flame, green when hot, blue when cold; the fused mass obtained by heating on charcoal with sodium carbonate and charcoal, when dissolved in hydrochloric acid and tin added, gives a blue solution; high specific gravity distinctive.

OCCURRENCE. Occurs under the same conditions as wolfram, that is, in pneumatolytic and other veins associated with tinstone, quartz, topaz, and other minerals of pneumatolytic origin, as at Caldbeck Fell (Cumberland), Cornwall, Zinnwald, Harz, Dragon Mountains (Arizona), and Trumbull (Connecticut); also occurs in pyrometasomatic deposits, at the contacts of acid igneous rocks with limestones, as in California and elsewhere.

URANIUM (and RADIUM) MINERALS

Uranium (U) which is not found native, is a hard, white metal, having a specific gravity of 18·7, and melting at a white heat. It is a constituent of a number of rare minerals, the most important of which are *pitchblende* and *carnotite*, together with *torbernite*, *autunite*, and various hydrous derivatives. Metallic uranium is prepared by reduction of its oxide with carbon in the electric furnace. Pitchblende and carnotite are worked also for their content of *Radium* (Ra), which is contained in all uranium minerals (see p. 291).

Uranium forms two oxides, and from these a complex group of uranium salts is derived. The uses of the oxides as such is limited; they are employed in glass-staining, for glazes, in dyeing and in photography. The metal has been employed in the production of certain special steels, but is not used for this purpose to any great extent, since steels of similar properties can be produced with less expensive components. There was little market for uranium until the exploitation of atomic fission in 1945 and the development of nuclear reactors; the demand for radium is on the increase, and the minerals are valued according to their percentage of U_3O_8, which always carries a certain proportion of radium.

Radium (Ra) results from the disintegration of uranium, and uranium minerals contain 320 milligrammes of radium per ton of uranium. The chief sources of radium are pitchblende, carnotite, and various decomposition-products of pitchblende. Radium is used in the treatment of cancer, in certain X-ray apparatus, and for luminous paint. The chief producers are most likely the United States, Canada, U.S.S.R., South Africa, Congo, France and Australia, with minor sporadic production in Czechoslovakia, Finland, Madagascar, Argentina, Poland, Portugal and elsewhere.

TESTS. In the microcosmic bead, uranium compounds give a light moss-green colour, both hot and cold, in the oxidizing flame.

The chief uranium minerals are the following:

Oxide	Pitchblende, Uraninite, UO_2.
Hydrous Phosphates	Torbernite, Copper Uranite $Cu(UO_2)_2$ $(PO_4)_2.12H_2O$. Autunite, Lime Uranite, $Ca(UO_2)_2(PO_4)_2$ $10\text{–}12H_2O$.
Hydrous Vanadate	Carnotite, $K_2(UO_2)_2(VO_4)_2.3H_2O$.
Hydrated Sulphates	Zippeite, Uraconite and others of uncertain composition.

PITCHBLENDE, Uraninite

COMP. Uranium dioxide, UO_2, but very variable, being partly oxidized to U_3O_8 and containing also thorium, zirconium, lead, etc., and various amounts of helium, argon, nitrogen, etc.; specific names, *cleveite*, *bröggerite*, etc., have been given to pitchblendes of various compositions.

CRYST. SYST. Cubic. COM. FORM. Commonly occurs massive, botryoidal or in grains; many examples of pitchblende are coloform gels. COLOUR. Velvet-black, greyish or brownish. STREAK. Black, often with a brownish or greenish tinge. LUSTRE. Submetallic, greasy, pitch-like and dull; opaque. H. 5·5. SP. GR. 6·4, massive, to 9·7, crystallized.

TESTS. Heated before the blowpipe alone, infusible; heated with sodium carbonate, it is not reduced, but usually gives an encrustation of lead oxide, and an arsenical odour may be produced; with borax or microcosmic salt, it gives in the oxidizing flame a yellow, and in the reducing flame a green bead of uranium; dissolves slowly in nitric acid when powdered.

OCCURRENCE. Pitchblende occurs as a primary constituent of igneous rocks, granites and pegmatites, or in high temperature veins associated with tin, copper and lead minerals, as at Joachimsthal (Jachymov) in Czechoslovakia, Johanngeorgenstadt in Saxony, in Cornwall, at Great Bear Lake in Canada, and in Katanga in the Congo.

USES. Pitchblende, and hydrous materials derived from it, are the chief sources of radium.

TORBERNITE, Copper Uranite

COMP. Hydrated phosphate of copper and uranium, $Cu(UO_2)_2$ $(PO_4)_2.12H_2O$; arsenic sometimes replaces a proportion of the phosphorus.

CRYST. SYST. Orthorhombic, pseudotetragonal. COM. FORM. Square thin tabular crystals, often modified on the edges; also scaly and foliaceous. CLEAV. Perfect parallel to the basal pinacoid, giving extremely thin lamellae, and in this respect resembling mica, whence the name *uran-mica*, by which the mineral is sometimes known; the laminae are brittle and not flexible, so differing from those of mica; sectile. COLOUR. Emerald or grass-green. STREAK. Rather paler than the colour. LUSTRE. Subadamantine, pearly on cleavage-planes; transparent to subtranslucent. H. 2–2·5. SP. GR. 3·5.

TESTS. Heated in closed tube, yields water; soluble in nitric acid; heated alone, fuses to a blackish mass, and colours the flame green; microcosmic salt bead green, due to copper.

OCCURRENCE. Occurs with other uranium minerals as a secondary product, as at Jachymov (Czechoslovakia), Schneeberg (Saxony), Cornwall, South Australia, etc.

AUTUNITE, Lime Uranite

COMP. Hydrated phosphate of calcium and uranium, $Ca(UO_2)_2$ $(PO_4)_2.10–12H_2O$.

CRYST. SYST. Orthorhombic. COM. FORM. Crystals resemble those of torbernite, described above, the crystals being nearly square; also foliaceous and micaceous. CLEAV. Perfect parallel to the basal pinacoid—thin brittle laminae resulting. COLOUR. Citron to sulphur-yellow. STREAK. Yellowish. LUSTRE. Subadamantine, on cleavage-faces pearly. H. 2–2·5. SP. GR. 3·1.

TESTS. Gives no reactions for copper, otherwise its behaviour with reagents resembles that of torbernite as described above.

OCCURRENCE. With torbernite as a secondary product from other uranium minerals.

CARNOTITE

COMP. Vanadate of uranium and potassium, approximately $K_2(UO_2)_2(VO_4)_2.3H_2O$.

CRYST. SYST. Orthorhombic. COM. FORM. Powdery, or minute crystal plates. CLEAV. Parallel to the basal pinacoid. COLOUR. Canary-yellow.

OCCURRENCE. As an impregnation or as lenses in Jurassic sandstone in Colorado and Utah, where it was mined for radium; also reported from pitchblende deposits in South Australia, Congo, and elsewhere.

ZIPPEITE, URACONITE

COMP. Hydrated sulphates of uranium, of uncertain composition.

CHARACTERS AND OCCURRENCE. Mostly occur as earthy or powdery crusts to other uranium minerals, as at Joachimsthal, in Cornwall, etc.; *uraconite* is of a lemon-yellow or orange colour; *zippeite* contains sometimes over 5 per cent of copper oxide—it occurs in delicate acicular crystals, rosettes and warty crusts, and accompanies uraconite; heated before the blowpipe, zippeite gives with microcosmic salt in the oxidizing flame a yellowish-green, and in the reducing flame an emerald-green bead.

GROUP 6B
SULPHUR, SELENIUM, TELLURIUM

SULPHUR MINERALS

Sulphur (S) occurs native in orthorhombic and monoclinic forms, the orthorhombic variety being the low-temperature common type, the high-temperature monoclinic varieties being rare.

Native sulphur and metallic sulphides, mainly iron pyrites, FeS_2, form practically the sole source of the sulphuric acid of commerce, and may be regarded as the most important minerals in connection with chemical industry.

Native sulphur is purified from the associated gangue by melting in ovens, etc., or by distilling in closed vessels, with the production of cast-stick sulphur, or of flowers of sulphur condensed in flues. The world production of sulphur from all sources is over 24 million tons, the dominant producers being the U.S.A. and Mexico, followed by U.S.S.R., France, Canada, Japan and Italy. The mode of occurrence and origin of sulphur deposits and the uses of sulphur are discussed under the description of that mineral on p. 499. In America, sulphur, over 99 per cent pure, is obtained in great quantity (Texas alone producing over 3 million tons annually) from deep-seated deposits in the Gulf States; the sulphur is extracted by the Frasch process, in which a double tube is driven down to the sulphur beds, and superheated steam or hot air is forced down with the result that the sulphur melts and is forced up the inner tube and run into vats.

For the manufacture of sulphuric acid, *iron pyrites*, FeS_2, which theoretically contains 53·46 per cent of sulphur and which is commonly sold on a guarantee of 45 to 50 per cent, is more used than any other mineral with the exception of sulphur itself. The oxide of iron which is formed during the roasting of pyrites is saleable for its iron content; and if the mineral contains copper, gold or silver, even in small quantities, it is paid for at a higher rate. The production of pyrites possibly amounts to over 12 million tons annually; for details of this production see p. 523.

The value of sulphides when smelting oxidized ores is well known, and the calorific value of burning sulphur is utilized in pyritic smelting. Enormous quantities of sulphur dioxide are present in furnace gases and many large smelters now recover it as sulphuric acid. The common association of arsenic with sulphur in mineral sulphides, and especially in pyrites, necessitates special care in the manufacture of sulphuric acid; but native sulphur, although seldom containing arsenic, is more liable to be contaminated with selenium, which is also objectionable.

Sulphur forms with hydrogen an acid, sulphuretted hydrogen, H_2S. This gas, being readily absorbed in water, is found in certain mineral springs, as at Harrogate, England. By the replacement of the hydrogen of sulphuretted hydrogen by a metal a *sulphide* is formed; mineral sulphides are abundant and exceedingly important minerals. Examples of common mineral sulphides are—galena, PbS, blende, ZnS, cinnabar, HgS, pyrite or iron pyrites, FeS_2. These and other sulphides are described at the appropriate pages under the metallic element contained in them.

With oxygen and hydrogen, sulphur forms many compounds, only one of which is important as an acid in mineralogy; this is sulphuric acid, H_2SO_4. *Sulphates* are formed by the substitution of a metal for the hydrogen of this acid, and are a very important group of minerals; examples of mineral sulphates are—gypsum, $CaSO_4.2H_2O$, anglesite, $PbSO_4$, barytes, $BaSO_4$. The mineral sulphates are described under the headings of their metallic constituents.

TESTS. Sulphur may be recognized when present in a mineral by the silver coin test; the powdered mineral is first fused with sodium carbonate on charcoal—the fused mass placed on a silver coin and moistened produces a black stain. Sulphides give, on roasting in the open tube or on charcoal, a sharp pungent odour of sulphur dioxide, SO_2. Sulphates give a dense white precipitate of barium sulphate on the addition of barium chloride to the solution; also after reduction by heating to the sulphide they give the silver coin test as above.

NATIVE SULPHUR

COMP. Pure sulphur, S, but often contaminated with other substances such as clay, bitumen, etc.

CRYST. SYST. Orthorhombic. COM. FORM. Crystals bounded by acute pyramids; also massive and in encrustations. CLEAV. Imperfect parallel to the prism (110), and to the bipyramid (111). COLOUR. Sulphur-yellow, often with a reddish or greenish tinge. STREAK. Sulphur-yellow. LUSTRE. Resinous; transparent to subtranslucent. H. 1·5–2·5. SP. GR. 2·07.

TESTS. Sulphur may be recognized by its burning with a blue flame, during which suffocating fumes of sulphur dioxide are formed; sulphur is insoluble in water, not acted upon by acids but is dissolved by carbon disulphide; its colour and low specific gravity are distinctive.

VARIETIES. Crude and impure forms of sulphur are purified by heating, the pure material being collected on cool surfaces. This fine sulphur is usually melted and cast into sticks, and is then known as brimstone or roll sulphur.

OCCURRENCE. Sulphur is found in the craters and crevices of extinct volcanoes, and has been deposited by gases of volcanic origin, as in Japan, where it is mined from an old crater lake. Another mode of origin is by the action of hot springs, by which the sulphur is deposited with tufa, etc., as in Wyoming, California and Utah. The most important occurrences of sulphur, from the commercial standpoint, are those in which the element is bedded or layered with gypsum, and in this case a common association of minerals is sulphur, gypsum, aragonite, celestine, and often petroleum. Such occurrences are located on salt domes in the Gulf States (see p. 231). Examples of such sulphur deposits are provided by the famous Sicily and Texas and Louisiana deposits. Opinion is divided as to whether these bedded deposits have been formed by true sedimentation, or by the alteration by various processes of gypsum.

USES. Sulphur is used for the manufacture of sulphuric acid, in making matches, gunpowder, fireworks, and insecticides, for vulcanizing india-rubber, and in bleaching processes involving the use of sulphur dioxide.

SELENIUM MINERALS

Selenium (Se) belongs to the same structural group as sulphur and tellurium. It occurs in native sulphur and in all pyritic ores, though often in negligible traces. Its mineral compounds are salts of the acid H_2Se, which is analogous with sulphuretted hydrogen H_2S; thus, *clausthalite* is lead selenide, PbSe, *berzelianite* copper selenide, Cu_2Se, *tiemannite*, mercury selenide, HgSe, and *naumannite* silver selenide, Ag_2Se. The principal sources of selenium are the deposits in sulphuric acid chambers and the anode mud or slime obtained in the electrolytic refining of copper and matte. Its industrial uses are expanding, and there is a steady demand for what is produced. There are prospects of a considerable quantity being required with the development of television, and of photoelectric apparatus. It is also employed in the production of red glass, enamels and glazes, in rubber-manufacture, and in dye-making; its employment in the production of special steels is increasing.

TESTS. The chief test for selenium compounds depends on the production when they are heated on charcoal before the blowpipe of a curious smell described as that of decaying horse-radish—a smell that can be readily recognized when once it has been encountered.

Selenium-bearing varieties of sulphur and tellurium are called selensulphur and selentellurium respectively. The description of the selenides mentioned above is outside the scope of this book.

TELLURIUM MINERALS

Tellurium (Te) occurs in small quantities free in nature in sulphur and pyrites. It is mostly found, however, combined with metals as *tellurides*, such metals being bismuth, lead and, most important, gold and silver.

Tellurium is obtained with selenium in the anode slime from electrolytic copper refineries. When pure, tellurium has a greyish-white colour and a metallic lustre. It has a specific gravity of 6·3, and melts at 450°C, and boils at 1,400°C. It is used in the rubber, chemical, metallurgical and ceramic industries, in machine-steels and in thermoelectric processes. The production is increasing, being over 350,000 pounds, mainly from the United States, Canada and Peru.

The chief tellurium minerals are the *native metal*, the oxide *tellurite*, TeO_2, and the *tellurides*. The gold tellurides are important ores of gold, and are described with that metal on pp. 267–269; a silver telluride, hessite, is described with the silver minerals on p. 261; here is considered the bismuth telluride, *tetradymite*, $Bi_2(Te,S)_3$. The tellurium minerals considered in this book are therefore:

Element	Native Tellurium, Te.
Oxide	Tellurite, TeO_2.
Tellurides	Tetradymite, bismuth telluride.
	Hessite, silver telluride (see p. 261).
	Sylvanite, gold and silver telluride (see p. 267).
	Calaverite, gold and silver telluride (see p. 268).
	Petzite, gold and silver telluride (see p. 268).
Sulpho-telluride	Nagyagite, gold and lead sulphotelluride (see p. 269).

The gold-bearing tellurides are extremely important gold ores; they occur in veins and replacement deposits. In the upper parts of the veins, the tellurides are decomposed, some tellurium oxide is formed but most is removed in solution.

TESTS. The best test for tellurium is to heat the compound in the closed tube with strong sulphuric acid, which then assumes a brilliant reddish-violet colour.

NATIVE TELLURIUM

COMP. Nearly pure tellurium, Te, with a little gold and iron.

CRYST. SYST. Hexagonal-trigonal. COM. FORM. Crystallizes in hexagonal prisms, with the basal edges modified; mostly found massive and granular. COLOUR. Tin-white. STREAK. Tin-white. LUSTRE. Metallic; brittle. H. 2–2·5. SP. GR. 6·1–6·3.

TESTS. Heated in the open tube, it forms a white sublimate of tellurous acid, which fuses to small transparent colourless drops before the blowpipe; heated with strong sulphuric acid, gives a reddish-violet solution.

OCCURRENCE. In the Maria Loretto Mine, near Zalanthna, Transylvania, where it was formerly worked for the small quantity of gold (less than 3 per cent) that it contained; also reported from Western Australia and Colorado.

TELLURITE

COMP. Tellurium dioxide, TeO_2.

CHARACTERS AND OCCURRENCE. Occurs in small whitish and yellowish orthorhombic prisms, spherical masses and encrustations, resulting from the oxidation of tellurium or tellurides in the upper parts of veins; recorded from Rumania, Colorado, etc.

TETRADYMITE, Telluric Bismuth

COMP. Bismuth telluride; sulphur and selenium sometimes present, and the general composition may be $Bi_2(Te,S)_3$.

CRYST. SYST. Hexagonal-trigonal. COM. FORM. Crystals often tabular; also found massive, granular and foliaceous. CLEAV. Perfect basal; laminae flexible. COLOUR. Pale steel-grey. LUSTRE. Metallic, splendent. H .1·5–2; marks paper. SP. GR. 7·2–7·6.

TESTS. Heated in the open tube, gives a white sublimate of tellurous acid; heated on charcoal, gives off white fumes, and a coating of tellurous acid and orange bismuth oxide, eventually fusing and volatilizing completely.

OCCURRENCE. Associated with gold tellurides (see p. 267).

GROUP 7A
MANGANESE

MANGANESE MINERALS

Manganese (Mn) does not occur in an uncombined state in nature but may, like chromium, be produced in the electric furnace and by the Thermit process (see under Aluminium on p. 318). It is a light, pinky-grey metal, melting at about 1,260°C, and having a specific gravity of about 7·4.'

The chief application of manganese is in the manufacture of alloys, such as spiegel-eisen and ferro-manganese, containing from 15 to 80 per cent of manganese, and silico-manganese, containing from 10 to 20 per cent silicon, all of which are of great importance in the steel industry.

The chief sources of manganese and its salts are the oxide minerals. Pyrolusite, MnO_2, is also used, as such, for a number of purposes such as the decolorization of glass, as a dryer in the manufacture of paint and varnish, and in dry batteries, and very largely for the manufacture of chlorine, bromine and oxygen, and of permanganates and other manganese compounds. The permanganates of sodium and potassium are used as disinfectants.

For chemical uses, a high percentage of manganese in the form of peroxide is demanded, 80 per cent. MnO_2 being taken as the basis price; lime should be present only in quantities of less than 2 per cent. For metallurgical purposes 50 per cent MnO_2 is a common basis. Ores of iron containing manganese are smelted direct for the production of manganese pig-iron, but such ores should be regarded as iron ores, and the manganese would not be paid for except at the same rate as iron.

The world production of manganese ore exceeds 17 million tons annually. The chief producers are U.S.S.R., Congo, China, South Africa, Brazil, Gabon and India, followed by Ghana, Morocco, Japan, Australia and Egypt.

Manganese is very widely distributed, and replaces to a greater or lesser degree two sets of elements; first, the alkaline earths, calcium, barium and magnesium, and second, aluminium and iron. Purely manganese minerals of the greatest importance are the oxides, others of less importance are the carbonate, silicate and sulphide. The manganese minerals dealt with here are:

Oxides	Hausmannite, Mn_3O_4.
	Braunite, Mn_2O_3.
	Manganite, $MnO(OH)$.
	Pyrolusite, MnO_2.
	Polianite, MnO_2.
	Psilomelane, a hydrated oxide with Ba and K.
	Wad, like psilomelane.
	Asbolan, cobaltiferous wad. (see p. 530).
	Franklinite, $(Fe,Zn,Mn)(Fe,Mn)_2O_4$, described with the *Zinc* minerals on p. 302.
Carbonate	Dialogite, Rhodochrosite, $MnCO_3$.
Silicate	Rhodonite, $MnSiO_3$.
Sulphide	Alabandite, MnS.

In addition to these specifically manganese minerals, the element enters into many silicates, of which the manganese garnet, *spessartite*, is described on p. 362.

Manganese minerals occur in varied ways. Dialogite and rhodonite occur as veinstones in some silver lodes, the gossans of which carry the oxide minerals such as pyrolusite and psilomelane. Dialogite also occurs as a metasomatic replacement of limestone. The most important and interesting deposits of manganese, however, are those of the oxides of sedimentary or residual origin now described. These oxides—pyrolusite, psilomelane, polianite, wad, braunite, and manganite—occur in two main types of deposit. In the sedimentary deposits, the manganese has been precipitated in beds or layers of nodules together with iron compounds with which it is invariably associated—a process at the present time being carried on in moderately deep water. By the upraising of these deposits are formed many of the bodies of workable manganese. Another of this type of manganese deposits is formed by precipitation of manganese oxides in lakes, etc., by the action of minute plants, giving rise to the bog-manganese deposits, as in Sweden, Spain and the United States.

The second type of manganese deposit is formed by the alteration of rocks containing manganese-bearing minerals, chiefly silicates. By the weathering of such rocks, the manganese oxides aggregate together as nodules and layers in the residual clay which forms on the outcrop of the weathered rock. Thus is formed a residual or lateritic deposit of the type worked in India, Brazil, Gabon, Ghana and Arkansas. It will be seen that all deposits of manganese oxides have been formed by the breaking-up of the manganese-bearing minerals of igneous and metamorphic rocks.

By the metamorphism of sedimentary or residual manganese deposits there are formed workable deposits of hausmannite, braunite and franklinite as in Sweden, Piedmont and at Franklin Furnace, U.S.A. (see p. 300).

TESTS. Manganese minerals give distinctive bead reactions. The borax and microcosmic-salt beads are reddish-violet in the oxidizing flame, and colourless in the reducing flame. The sodium carbonate bead is bluish-green in the oxidizing flame.

HAUSMANNITE

COMP. Manganese oxide, Mn_3O_4; manganese, 72 per cent.

CRYST. SYST. Tetragonal. COM. FORM. Commonly occurs in pyramidal forms, frequently twinned; also massive and granular. COLOUR. Brownish-black. STREAK. Chestnut-brown. LUSTRE. Submetallic; opaque. FRACT. Uneven. H. 5–5.5. SP. GR. 4–.86.

TESTS. Gives usual manganese reactions with the fluxes; gives chlorine when dissolved in hydrochloric acid; does not give oxygen on being heated.

OCCURRENCE. A primary manganese mineral occurring in veins connected with acid igneous rocks.

BRAUNITE

COMP. Manganese oxide, Mn_2O_3; manganese, 64.3 per cent; braunite usually contains about 10 per cent silica, so that the mineral is sometimes considered to be a manganese manganate and silicate, such as $3MnMnO_3.MnSiO_3$.

CRYST. SYST. Tetragonal. COM. FORM. Crystals of octahedral habit; also occurs massive. COLOUR. Brownish-black. STREAK. Brownish-black. LUSTRE. Submetallic; opaque. FRACT. Uneven; brittle. H. 6–6·5. SP. GR. 4·75–4·82.

TESTS. Gives the usual manganese reactions with the fluxes (see p. 505); gelatinizes when boiled with hydrochloric acid; does not yield oxygen when heated in the closed tube.

OCCURRENCE. Usually of residual or secondary origin, but may occur as a primary mineral in veins.

MANGANITE

COMP. Hydrous manganese oxide, $MnO(OH)$; manganese, 62·5 per cent.

CRYST. SYST. Monoclinic, pseudo-orthorhombic. COM. FORM. Prismatic crystals often grouped in bundles and striated longitudinally; also columnar. COLOUR. Iron-black or dark steel-grey. STREAK. Reddish-brown or nearly black. LUSTRE. Submetallic; opaque. FRACT. Uneven. H. 4. SP. GR. 4·2–4·4.

TESTS. Heated in the closed tube, gives water; gives the usual manganese reactions with the fluxes (see p. 505).

OCCURRENCE. In veins, and associated with other manganese oxides (see p. 504).

PYROLUSITE

COMP. Manganese dioxide, MnO_2; manganese, 63 per cent.

CRYST. SYST. May be orthorhombic, but usually pseudomorphous. COM. FORM. Occurs in pseudomorphs after manganite, polianite, etc.; usually occurs massive or reniform, and sometimes with a fibrous and radiate structure. COLOUR. Iron-grey or dark steel-grey. STREAK. Black or bluish-black, sometimes sub-metallic. LUSTRE. Metallic; opaque. FRACT. Rather brittle. H. 2–2·5; often soils the fingers. SP. GR. 4·8.

TESTS. Borax bead—amethyst in oxidizing flame, colourless in reducing flame; microcosmic-salt bead—red-violet in oxidizing flame; sodium carbonate bead—opaque blue-green; yields oxygen when heated in the closed tube; soluble in hydrochloric acid with evolution of chlorine.

OCCURRENCE. In secondary manganese deposits as described on p. 505.

USE. An abundant manganese ore.

POLIANITE

COMP. Manganese dioxide, MnO_2.

CRYST. SYST. Tetragonal.

CHARACTERS AND OCCURRENCE. A mineral like pyrolusite in appearance and colour, sometimes showing minute tetragonal crystals, isomorphous with tinstone; hardness, 6–6·5, a distinction from pyrolusite, and specific gravity, 5; occurrence as for other oxides of manganese.

PSILOMELANE

COMP. Hydrated oxide of manganese, with or without varying amounts of barium and potassium oxides.

FIG. 149
Psilomelane

CRYST. SYST. Monoclinic. COM. FORM. Massive, botryoidal, reniform and stalactitic (see Fig. 149). COLOUR. Iron-black, passing into dark steel-grey. STREAK. Brownish-black and shining. LUSTRE. Submetallic; looks as if an attempt had been made to polish the mineral; opaque. H. 5–6. SP. GR. 3·7–4·7.

TESTS. Heated in the closed tube gives water; with borax and microcosmic salt, gives the usual amethyst-coloured bead; sodium carbonate bead—opaque bluish-green; soluble in hydrochloric acid, with evolution of chlorine, the solution often giving a precipitate of barium sulphate on the addition of barium chloride.

OCCURRENCE. In secondary manganese deposits, as described on p. 505; it was probably a colloidal precipitate.

USE. With pyrolusite, the important ores of manganese.

WAD

COMP. Very variable, but resembling that of psilomelane.

COM. FORM. Amorphous, earthy, reniform, arborescent, encrusting, or as stains and dendrites; often loosely aggregated. COLOUR. Dull black, bluish, lead-grey, brownish-black. LUSTRE. Dull; opaque. H. 5–6; usually quite soft. SP. GR. 3–4·28.

TESTS. Gives the usual manganese reactions with the fluxes; yields water in the closed tube.

VARIETIES. *Earthy Cobalt* or *Asbolan* (p. 530) is a variety of wad containing sometimes nearly 40 per cent of cobalt oxide—and giving a blue borax bead, due to cobalt; *Lampadite* is another variety which yields sometimes as much as 18 per cent of copper oxide.

OCCURRENCE. Wad results from the decomposition of other manganese minerals, and generally occurs in damp, low-lying places.

USES. Not as valuable as pyrolusite and psilomelane, but is sometimes used in the manufacture of chlorine, and also serves for umber paint.

DIALOGITE, Rhodochrosite

COMP. Manganese carbonate, $MnCO_3$; manganese, 47·8 per cent; often with varying quantities of the carbonates of iron, calcium and magnesium.

CRYST. SYST. Hexagonal-trigonal, Calcite Type. COM. FORM. Crystals rare, rhombohedral; usually massive, globular, botryoidal or encrusting. CLEAV. Perfect parallel to the faces of the rhombohedron, i.e. in three directions. COLOUR. Various shades of rose-red, yellowish-grey, and brownish. STREAK. White. LUSTRE. Vitreous, inclining to pearly; translucent to subtranslucent. FRACT. Uneven; brittle. H. 3·5–4·5. SP. GR. 3·45–3·6.

TESTS. Heated before the blowpipe, infusible, but the mineral changes to grey-brown and black, and decrepitates strongly; gives the usual manganese amethystine colour to the borax and micro-cosmic-salt beads; dissolves with effervescence in warm hydro-chloric acid; on exposure to the air, the red varieties lose colour.

OCCURRENCE. Occurs as a veinstone in lead and silver-lead ore veins; also as a metasomatic replacement of limestones.

RHODONITE, Manganese Spar

COMP. Manganese silicate, $MnSiO_3$; manganese, 41·86 per cent; some iron, calcium, magnesium, or zinc usually present; some carbonate usually present, causing the mineral to effervesce with acid. Rhodonite belongs to the Pyroxenoid family (see p. 389).

CRYST. SYST. Triclinic. COM. FORM. Tabular crystals; commonly massive and cleavable. CLEAV. Perfect parallel to the prisms. COLOUR. Flesh-red, light brownish-red; greenish or yellowish when impure; and often black on exposed surfaces; colour darkens on exposure, often becoming nearly black from additional oxidation of the manganese. STREAK. White. LUSTRE. Vitreous; transparent to opaque; usually translucent or opaque. FRACT. Uneven; occasionally conchoidal; very tough when massive. H. 5·5–6·5. SP. GR. 3·4–3·6.

TESTS. Heated before the blowpipe, blackens and fuses to a black glass with slight intumescence; in the oxidizing flame, gives the manganese reactions with borax and microcosmic salt.

VARIETIES. *Fowlerite* is a variety containing zinc; *Bustamite* contains calcium, and its formula is perhaps $CaMn(SiO_3)_2$.

OCCURRENCE. As a veinstone in lead and silver-lead veins, as at Broken Hill, N.S.W., associated with dialogite and quartz.

USES. Rhodonite, when cut and polished, is used for ornamental work. It may also be used for imparting a violet colour to glass. On stoneware, when mixed with the common salt-glazing, it forms a black or, if sparingly used, a deep violet-blue glaze.

ALABANDITE

COMP. Manganese sulphide, MnS.

CRYST. SYST. Cubic. COM. FORM. Usually granular and massive. CLEAV. Perfect cubic. COLOUR. Iron-black. STREAK. Greenish-black. LUSTRE. Submetallic; opaque. H. 3·5–4. SP. GR. 3·95.

TESTS. Heated before blowpipe, fusible; amethystine bead given with borax in the oxidizing flame, especially with the roasted mineral.

OCCURRENCE. As a primary mineral in veins associated with sulphides and manganese spars.

GROUP 7B
FLUORINE, CHLORINE, BROMINE, IODINE

HALOGEN (Fluorine, Chlorine, Bromine, Iodine) MINERALS

The four elements, fluorine (F), chlorine (Cl), bromine (Br) and iodine (I), constitute a well-marked group in the Periodic Classification (see p. 26), and show a number of similarities and progressive variations in their properties. All unite readily with the metals to form salts such as fluorides, chlorides, etc., many of which are of great importance in economic mineralogy.

The most important occurrence of fluorine is in salts of hydrofluoric acid, HF; the most important of these *fluorides* are fluorspar, CaF_2 (see p. 284), and cryolite, Na_3AlF_6 (see p. 324), both of which are of considerable industrial importance. Fluorine also enters into the composition of some silicates, such as topaz and amblygonite, and the important mineral apatite is calcium fluoride and phosphate. Fluorine is detected in fluorides by heating with strong sulphuric acid, when hydrofluoric acid is liberated which etches glass.

Certain *chlorides*, salts of hydrochloric acid, HCl, are abundant and important minerals, notable examples described in this book being the following: rock-salt, NaCl; sylvine, KCl; carnallite, $KMgCl_3$; cerargyrite, AgCl; calomel, HgCl; oxychlorides occur as minerals, the most important example being atacamite, $Cu_2(OH)_3Cl$. Chloro-phosphates, chloro-arsenates and chloro-vanadates are represented by the minerals of the Pyromorphite set, e.g. pyromorphite, $(PbCl)Pb_4(PO_4)_3$. Finally, the sodium chloride molecule enters into the composition of certain silicates as, for example, sodalite and marialite-scapolite. Chlorine is detected by precipitation of silver chloride on the addition of silver nitrate solution to a solution of the mineral in nitric acid, and by the copper oxide-microcosmic salt test (see p. 41). The uses of the chlorides are dealt with in their descriptions under the various metallic elements contained in them.

Bromine is not an abundant mineral constituent. It occurs as *bromides*, salts of hydrobromic acid, HBr, the most important of which are silver bromides, such as bromyrite, AgBr, and embolite, Ag(Cl,Br), mentioned on p. 262. When heated with potassium bisulphate and pyrolusite, red vapours of bromine are liberated from bromine compounds. The commercial production of bromine compounds comes from sea-water, salt brines, or from the saline residues described on p. 230. Bromine is employed in medicine, photography and for treating petrol; the production in the United States amounts to about 90,000 tons yearly.

Iodine is rare in nature. The chief source is the Chile nitrate deposits (see p. 232), where certain calcium iodates occur in small amount. The possible production exceeds the demand. Iodine is also produced from sea-weed and kelp, and the brines from oil-wells. Iodine is used in medicine and in a number of industrial processes.

GROUP 8A
IRON, COBALT, NICKEL

IRON MINERALS

Iron (Fe) is, next to aluminium, the most widely distributed and abundant metal, constituting about 4·6 per cent of the earth's crust. It is found native in meteoritic masses and in eruptive rocks, mostly associated with allied metals such as nickel and cobalt. In addition to the essentially iron minerals, iron enters into the composition of a great number of rock-forming silicates, the most important of which are mentioned below. Iron and steel form the foundation of modern industry, and are used in enormous and increasing quantities. Thus, the production of iron ore rose from 430 million tons in 1959 to nearly 600 million tons in 1966 and of steel over the same period from 350 million tons to 465 million tons.

Metallic iron is unaffected by dry air, but oxidizes to 'rust' under the influence of moist air. Cast iron, wrought iron and steel are the chief forms in which iron appears in commerce. Their different properties are primarily due to the presence of varying amounts of carbon. Steel is again divided into several classes or grades, each named after its particular properties (mild, hard, etc.), the use to which it is put (tool-steel, etc.), or the metal with which it is alloyed (manganese, chrome, nickel, or tungsten steel). These special steels have been found in practice to be especially valuable for different purposes, such as armour plate, guns, high-speed cutting-tools, rails, springs, etc. Nickel steel is of particular importance, its tensile strength and elasticity being enormously greater than those of ordinary steel.

Pig-iron, from which all the various grades of iron and steel are obtained, is produced in the blast furnace by the reduction of iron ore by coke. Increasing percentages of the raw ores are now improved before dispatch to the smelters by sintering or agglomervation into pellets.

The chief ores of iron are the oxides and the carbonate; *magnetite*, Fe_3O_4, containing 72·4 per cent. Fe, *hematite*, Fe_2O_3, containing 70 per cent, *goethite*, $Fe_2O_3.H_2O$, containing 62·9 per cent and the main constituent of *limonite*, and the carbonate, *siderite*, *chalybite* or *spathic iron ore*, containing 48·3 per cent. Less important ores are the sulphides, *pyrite*, FeS_2, and *pyrrhotite*, $Fe_{1-n}S$, and the complex oxide, *franklinite*, (FeZn,Mn)$(Fe,Mn)_2O_4$. Finally, certain important iron ores are composed to some extent of hydrous iron silicates, such as *chamosite*, *thuringite*, *greenalite* and *glauconite*. Magnetite of exceptional purity occurs in large quantities in Sweden, and is the source of the noted 'Swedish Iron'. Undesirable impurities in iron ores are arsenic, sulphur and phosphorus, except in the case of the manufacture of basic or non-bessemer steels. Iron ores containing 30 per cent. Fe and upwards are profitably smelted, but the value of iron ores (in common with most other ores, of course) depends on their situation, and also on the composition of their gangue in addition to their iron content. For example, an iron ore containing 30 per cent Fe, and a gangue of silica, alumina and lime in such proportions as to make it self-fluxing, may be more valuable than a richer ore containing impurities which it would be necessary to remove. Most iron ores in Europe require some mechanical cleaning, and in some cases electromagnetic separation is employed. Roasting or calcining is also frequently resorted to for the purpose of removing water, carbonic acid and sulphur.

Considerable quantities of the natural oxides and silicates are mined and prepared for the market, for use in the manufacture of paints, and as linoleum fillers, etc. For example, ochres are hydrated ferric oxides; sienna and umber are silicates of aluminium containing iron and manganese; red and brown ochres are the natural hydrated or anhydrous oxides, or they may be produced by the calcination of carbonates, whereby a wide range of shades is obtained.

Iron ores occur in a number of types of deposit. Magmatic segregations, often followed by injection into the surrounding rocks, are exemplified by the great magnetite deposits of Kiruna in northern Sweden. Pyrometasomatic deposits are widespread, as in the Urals and the Western States. Hydrothermal disseminations, veins and replacements are especially common. Important iron ores occur as syngenetic sedimentary beds as, for example, in the Jurassic rocks of the English Midlands. Ancient sedimentary deposits are present in enormous bulk in the Pre-Cambrian rocks of Canada, the Lake Superior district, Brazil, India, South Africa, West Africa and Australia; very rich iron ores are formed when these chemical precipitates undergo leaching and removal of their primary silica. It has been estimated that the reserves of such enriched ores exceed those of all other types combined. Residual deposits of various kinds, especially the limonitic ores found in the gossans of sulphide deposits, are another important type of occurrence.

As already mentioned, the output of iron ore amounted in 1966 to nearly 600 million tons; the chief producers being U.S.S.R. (156 million tons), U.S.A. (90) and France (54) followed by Australia, Canada, China, Sweden, Venezuela, Liberia, India, Brazil, U.K. and Chile—all these producing over 12 million tons each.

TESTS. The following tests are useful in the detection of iron in compounds. With borax, iron compounds give in the oxidizing flame a bead which is yellow hot, and colourless cold, or, if more material is added to the bead, brownish-red hot and yellow cold; in the reducing flame, the usual colour is bottle-green of various shades. The microcosmic salt bead is similar to that of borax in the oxidizing flame, but in the reducing flame it is brownish-red hot and passes on cooling to yellow-green and finally colourless.

The important iron minerals are the following. The *native element* has been recorded in several localities. The chief oxides are *magnetite*, Fe_3O_4, and *hematite*, Fe_2O_3; ilmenite, $FeTiO_3$, chromite, $FeCr_2O_4$, and franklinite $(Fe,Zn,Mn)(Fe,Mn)_2O_4$, are described with titanium, chromium and zinc respectively; hercynite, $FeAl_2O_4$, is a spinel (see p. 321). The chief hydrated oxides are *goethite*, $Fe_2O_3.H_2O$, and *turgite*, $Fe_2O_3.nH_2O$. The carbonate is *siderite* or *chalybite*, $FeCO_3$. Sulphides are represented by *pyrite*, FeS_2, and *pyrrhotite*, $Fe_{1-n}S$, and complex sulphides by chalcopyrite, $CuFeS_2$ (see p. 245) and arsenopyrite, $FeAsS$ (see p. 477). *Copperas* is a hydrated sulphate of iron, $FeSO_4.7H_2O$;

vivianite is a hydrated phosphate, $Fe_3P_2O_8.8H_2O$. Hydrated silicates of various types are *ilvaite*, $CaFe^{2+}Fe^{3+}O(Si_2O_7)(OH)$, and *chamosite, thuringite, greenalite*, and *glauconite* (see p. 404). In addition to its occurrence in such silicates, iron also enters in considerable amount into many of the rock-forming silicates, such as the pyroxenes, hypersthene and acmite, the amphiboles, anthophyllite, glaucophane and riebeckite, the iron garnets, the biotitic micas, and other silicates such as staurolite and chloritoid. Finally, columbite is an iron tantalate and wolfram an iron tungstate.

In this section, the following iron minerals are dealt with:

Element	Native Iron, Fe.
Oxides	Magnetite, Fe_3O_4.
	Hematite, Fe_2O_3.
Hydrated Oxides	Goethite, $Fe_2O_3.H_2O$ (the main component of Limonite).
	Turgite, $Fe_2O_3.nH_2O$.
Carbonate	Siderite, Chalybite, $FeCO_3$.
Sulphides	Pyrite, FeS_2.
	Marcasite, FeS_2.
	Pyrrhotite, $Fe_{1-n}S$.
Hydrated Sulphate	Copperas, $FeSO_4.7H_2O$.
Hydrated Phosphate	Vivianite, $Fe_3P_2O_8.8H_2O$.
Hydrated Silicate	Ilvaite, $CaFe^{2+}Fe^{3+}O(Si_2O_7)(OH)$.

NATIVE IRON

COMP. Iron, Fe; usually alloyed with nickel or some other metal, or mixed with other iron compounds.

CRYST. SYST. Cubic. COM. FORM. Crystals octahedra; generally massive, and with a somewhat granular structure. CLEAV. Perfect cubic. COLOUR. Iron-grey. STREAK. Iron-grey. FRACT. Hackly; it is also malleable and ductile. H. 4–5. SP. GR. 7·3–7·8.

TEST. Strongly magnetic.

OCCURRENCE. Native iron of terrestrial origin has been reported as occurring in Brazil, Auvergne, Bohemia, and in grains disseminated in basalt from Giant's Causeway (Ireland) and elsewhere. The largest masses known to be of terrestrial origin occur in Disko Island and elsewhere on the coast of West Greenland, where masses ranging up to 25 tons in weight weather out from a basalt. Native iron also occurs as grains in some placer deposits, as at Gorge River, New Zealand, and in Piedmont. Native iron may be formed by the alteration of iron minerals as in Eastern Canada.

Meteoritic iron is found in meteorites—masses which have fallen from the outer atmosphere on to the surface of the earth. Meteorites may consist either entirely of iron, or partly of olivine and other silicates. Meteoritic iron is usually alloyed with nickel and small quantities of cobalt, manganese, tin, chromium, sulphur, carbon, chlorine, copper and phosphorus. It rusts much less readily than terrestrial iron, on account of the nickel which it contains. The minerals olivine, enstatite, augite and anorthite have been found in meteorites. In them, too, has been detected a phosphide of nickel and iron, called *schriebersite*, and also a sulphide of iron, known as *troilite*, having the formula FeS.

MAGNETITE, Magnetic Iron Ore

Comp. Iron oxide, Fe_3O_4; iron, 72·4 per cent; the iron is sometimes replaced by a small amount of magnesium or titanium.

Cryst. Syst. Cubic. Com. Form. Octahedra common, also combinations of octahedron and rhombdodecahedron; also occurs granular and massive. Cleav. Poor, octahedral. Colour. Iron-black. Streak. Black. Lustre. Metallic or submetallic; opaque; but in thin dendrites occurring in mica it is sometimes transparent or nearly so. Fract. Subconchoidal. H. 5·5–6·5. Sp. Gr. 5·18, crystals.

Opt. Props. In thin sections of rocks, occurs as small square sections or shapeless grains, black by transmitted and reflected light.

Tests. Heated before the blowpipe, very difficult to fuse; with borax in the oxidizing flame, gives a bead which is yellow when hot and colourless when cold; if much of the material is added to the bead, the bead is red when hot and yellow when cold; in the reducing flame, the borax bead is bottle-green; magnetite is soluble in hydrochloric acid; it is strongly magnetic and often exhibits polarity.

OCCURRENCE. Magnetite occurs as a primary constituent of most igneous rocks. Large deposits are considered to be the result of magmatic segregation, as in the Urals, and Northern Sweden, Kiruna, Gellivaare—in this case, however, it is suggested that the magnetite has moved after its segregation. Workable magnetite deposits occur also as lenses in crystalline schists, as in the Adirondack belt in the eastern United States, the pre-metamorphic nature of these deposits being a matter for discussion. Pyrometasomatic and contact-metamorphic magnetite deposits are widespread but not of great importance in production—examples are seen in the 'skarn' ores of Scandinavia, and in certain deposits in the Western States (e.g. Iron Springs, Utah), and the Urals. Magnetite is also a constituent of many veins, is found in residual clays, and in placer deposits—the 'black sands'—formed by the degradation of earlier deposits.

USES. Magnetite is one of the most valuable ores of iron.

HEMATITE, Specular Iron, Kidney Ore

COMP. Iron oxide, Fe_2O_3; iron, 70 per cent; clay and sandy impurities are sometimes present.

CRYST. SYST. Hexagonal-trigonal, Calcite Type. COM. FORM. Rhombohedron, often modified, and frequently in thin tabular forms; also in micaceous or foliaceous aggregates, reniform, granular, or amorphous; commonly fibrous internally. CLEAV. Poor, parallel to the rhombohedron and to the basal pinacoid. COLOUR. Steel-grey to iron-black; in particles sufficiently thin to transmit light, it appears blood-red; earthy forms are red. STREAK. Cherry-red. LUSTRE. When crystallized metallic and highly splendent—*specular iron*; in fibrous varieties, silky; in amorphous varieties, dull and earthy; opaque except in very thin plates. FRACT. Subconchoidal or uneven. H. 5·5–6·5. SP. GR. 4·9–5·3.

TESTS. Blowpipe reactions as for magnetite; becomes magnetic on heating; soluble in acids.

VARIETIES. *Specular Iron* is a variety occurring in rhombohedral crystals, black in colour, and with a metallic splendent lustre; *Micaceous Hematite* includes the foliaceous and micaceous forms; *Kidney Ore* (see Fig. 150) is a reniform variety with a metallic lustre, especially on the mammillated surfaces; beneath this surface, kidney ore usually displays a radiating or divergently columnar structure; *Reddle* is the most earthy variety of hematite, red in colour, and used in the manufacture of crayons, for polishing glass, and as a red paint; *Martite* is probably a pseudomorph of hematite after magnetite; it mostly occurs in small black octahedra, which give a reddish-brown streak.

OCCURRENCE. Hematite occurs in pockets and hollows, replacing limestone; as an example of this mode of occurrence,

FIG. 150.—Hematite, Kidney Ore.

the deposits of Ulverston, in North Lancashire, may be given; here the hematite occurs in irregular masses in and on the surface of the Carboniferous Limestone, and probably results from the replacement of that limestone by hematite brought from the ferruginous Triassic sandstones which overlie the limestones; metasomatic deposits of a similar nature are found in the Forest of Dean, and Cumberland, and much more important deposits of the same origin occur at Bilbao, Spain, Utah, and elsewhere. The greatest hematite deposits in the world—those of the Lake Superior district—the Mesabi, Marquette, Menominee, etc., iron-ranges—result from the alteration and concentration of iron silicates and carbonate of sedimentary origin. Another notable iron-ore deposit is the Clinton iron-bed of Alabama and neighbouring Eastern States; the hematitic ore is often colitic in structure, and is a thin sedimentary bed interstratified with Silurian rocks. The great Brazilian deposits are metamorphosed sedimentary ores. The hematite deposit of Elba is probably of contact-metamorphic origin. Finally, residual deposits rich in hematite are known—examples being certain in the Appalachian belt of the Eastern United States, and the far more important deposits in Cuba.

GOETHITE

COMP. Hydrous iron oxide, $Fe_2O_3.H_2O$: iron, 62·9 per cent.

CRYST. SYST. Orthorhombic. COM. FORM. In prisms which are longitudinally striated and frequently flattened, so that the crystals assume a tabular form; also massive, stalactitic and fibrous. CLEAV. One good cleavage, parallel to the brachy-pinacoid. COLOUR. Brownish-black, sometimes yellowish or reddish; often blood-red by transmitted light. STREAK. Brownish-yellow or ochre-yellow. LUSTRE. Adamantine; opaque . H. 5–5·5. SP. GR. 4–4·4.

TESTS. As for limonite, below; distinguished from limonite by being crystalline.

OCCURRENCE. Associated with limonite and hematite—some· localities being Lostwithiel and Botallack in Cornwall, Altenberg in Saxony, Lake Onega in Russia, Jackson Iron mine, Michigan, U.S.A. Goethite is the chief constituent of limonite described below.

LIMONITE, Brown Hematite

COMP. Variable, hydrous ferric oxides, mainly goethite ($Fe_2O_3.H_2O$, see above) with absorbed water, clay minerals and other impurities. Limonite is a rock rather than a definite mineral.

COM. FORM. Occurs in mammillated or stalactitic forms, having a radiating fibrous structure, resembling that of hematite; also in dull earthy condition, and in concretions. COLOUR. Various shades of brown on fracture-surfaces; when earthy, yellow or brownish-yellow; the exteriors of mammillated and stalactitic forms frequently exhibit a blackish glazed coating. STREAK. Yellowish-brown. LUSTRE. Submetallic in some varieties; sometimes silky, sometimes dull or earthy. H. 5–5·5. SP. GR. 3·6–4.

TESTS. Heated in the closed tube, gives water; gives the iron reactions with the fluxes; gives a magnetic residue on heating; soluble in hydrochloric acid; distinguished by its streak and form.

VARIETIES. *Bog Iron Ore* is a loose, porous earthy form of

limonite, found in swampy and low-lying ground, often impregnating and enveloping fragments of wood, leaves, mosses, etc.; *Pea Iron Ore* is a variety of limonite having a pisolitic structure; *Ochres* are brown or yellow earthy forms of limonite used for paint.

OCCURRENCE. Limonite and goethite result from the alteration of other iron minerals; from a highly ferruginous rock there may be formed by its degradation weathered residual deposits, consisting largely of ferric hydroxide, mixed with clay and other impurities; these iron caps are common over the outcrops of pyrites and iron oxide deposits, and the lateritic iron ores are formed in an analogous manner; examples of important deposits of this character are the Bilbao, Spain, and the Cuba deposits. The bog iron ores are formed on the floors of some lakes, as in Sweden, where a layer of 7 inches in thickness accumulated in twenty-six years; the deposition of the iron compounds from the stream-waters flowing into the lakes may be caused by minute organisms, such as bacteria. Some limonite beds are true chemical precipitates, whilst others result from the alteration of chalybite. Limonite enters into the ore of some of the Jurassic iron-ore fields of England, and is the dominant iron mineral in the great 'minette' iron ores of Alsace-Lorraine.

TURGITE

COMP. Hydrous iron oxide, $Fe_2O_3.nH_2O$.

CHARACTERS AND OCCURRENCE. A mineral of doubtful status, resembling limonite but having a red streak, and distinguished from hematite by containing water; found in small quantity with limonite.

CHALYBITE, SIDERITE, Spathose Iron

COMP. Iron carbonate, $FeCO_3$; iron, 48·3 per cent; often a little manganese, magnesium and calcium are present.

CRYST. SYST. Hexagonal-trigonal, Calcite Type. COM. FORM. Rhombohedra, the faces often curved; also massive, cleavable and granular. CLEAV. Perfect rhombohedral. COLOUR. Pale yellowish or buff-brownish, and brownish-black or brownish-red.

STREAK. White. LUSTRE. Pearly or vitreous; opaque, rarely translucent. FRACT. Uneven; brittle. H. 3·5–4·5. SP. GR. 3·7–3·9.

TESTS. Heated before the blowpipe, it blackens and becomes magnetic; gives the iron reactions with the fluxes; acted upon very slowly by cold acids, but in hot hydrochloric acid, it effervesces very briskly.

VARIETIES. *Clay Ironstone* is an impure iron carbonate occurring as beds and nodules especially in the Coal Measures of many countries; it is common in most of the British coalfields and in those of Pennsylvania, Ohio, etc., in the United States; occasionally it exhibits a curious radiately disposed, rude, sub-columnar structure, causing it, when struck, to fall to pieces in conical masses which envelop or cap one another, and to which the name of *cone-in-cone* structure has been given. Formerly these clay ironstones constituted valuable ores of iron; besides occurring in the Coal Measures, clay ironstones are also found in layers and nodules in other formations; *Blackband* is a dark often carbonaceous type of clay ironstone; *Oolitic Ironstone* is an iron carbonate which has replaced the calcium carbonate of an oolitic limestone, retaining the structure of the original rock, as in the celebrated Cleveland iron ore; in many examples of oolitic ironstone, however, it has been shown that the iron carbonate did not replace calcium carbonate, but was formed at the same time as the oolitic structures, these resulting from colloidal processes.

OCCURRENCE. Some siderite deposits are formed by direct precipitation either in lakes, as in some bog iron ore deposits, or in the sea. These sedimentary deposits of siderite may be altered into limonite, hematite and magnetite, and siderite has contributed to the great Lake Superior hematite deposits. The sedimentary siderites include those of the Coal Measures mentioned above—typical localities being South Wales, South Staffordshire, and the other British coalfields, and the Eastern United States coalfields. The metasomatic siderites are extremely important deposits; in these the iron carbonate has replaced calcium carbonate of limestones, retaining many of the original features of the rock (oolitic structure, fossils, etc.); to this class belong the Cleveland ores. Siderite is an important vein mineral and deposits of this type have been worked in Germany. The minette ores of Alsace-Lorraine (see p. 521) contain siderite, as do the Mesozoic ores of the English Midlands.

PYRITE, PYRITES, IRON PYRITES, Mundic

COMP. Iron sulphide, FeS_2; iron, 46·6 per cent.

CRYST. SYST. Cubic; Pyrite Type (see p. 104). COM. FORM. Cube, and pyritohedron (see Figs. 41, 44); the faces of the cube are often striated, the striae of one face being at right angles to those of the adjacent faces (see Fig. 43); it also occurs massive, in nodules which have generally a radiating structure, in finely disseminated crystals or particles, and it occasionally replaces the calcite in fossil shells. COLOUR. Bronze-yellow to pale brass-yellow. STREAK. Greenish or brownish black. LUSTRE. Metallic, splendent; opaque. FRACT. Conchoidal, uneven; brittle; strikes fire with steel, for which reason it was formerly used instead of flint in the old wheel-lock fire-arm, a steel wheel being made to revolve rapidly by means of clockwork against a piece of pyrites, from which sparks were thrown off into the pan of the weapon. H. 6–6·5; compare copper pyrites, p. 245. SP. GR. 4·8–5·1.

TESTS. Heated in closed tube, gives sublimate of sulphur, leaving a magnetic residue; borax bead, yellow in oxidizing flame, bottle-green in reducing flame; heated on charcoal with sodium carbonate, gives magnetic residue; black stain in silver coin test; soluble in nitric acid, insoluble in hydrochloric acid.

OCCURRENCE. Pyrite is a mineral of very common occurrence in many parts of the world. It occurs not commonly as an accessory mineral in igneous rocks; more often it is secondary. It is a common constituent of many ore-veins, and large deposits of varying types are worked, chiefly for sulphur or for the small percentage of copper or other metal contained in the pyrite. Many of the large pyrite masses are difficult to interpret. Some are due to magmatic segregation, and in these the pyrite is accompanied by pyrrhotite. Others are of contact-metamorphic or pyrometasomatic origin, whilst some are considered to be injected bodies. The great deposits of Rio Tinto in Spain, Mt. Lyell in Tasmania,

Rammeslberg in the Harz, and elsewhere, are low-temperature deposits associated with calcite, barytes, quartz, etc. Pyritic deposits considered to be of sedimentary origin may be illustrated by the oolitic pyrite of the Cleveland Hills, England. In some cases, pyrite contains enough gold to pay working for that metal. As already stated, pyrite is not worked directly as an ore of iron, the sulphur which it contains rendering it comparatively worthless for that purpose; but a good deal of the sulphuric acid and sulphate of iron, copperas, is derived from its decomposition; when present in shales or clays, its decomposition and oxidation give rise, upon the roasting of the pyritous clay or shale, to sulphuric acid, which combines with the alumina present; on the addition of potassium compounds, alum, a hydrous sulphate of aluminium and potassium is formed; in this way some of the alum of commerce is made. Sulphur is also procured from pyrites, but now the market for both sulphur and sulphuric acid is largely supplied from native sulphur. The world production of pyrite amounts to between 17 and 18 million tons annually, the chief producers being Japan, U.S.S.R., Poland, Italy, Spain, Norway, Finland and Sweden.

MARCASITE, White Iron Pyrites

COMP. Iron sulphide, FeS_2, identical with pyrite.

CRYST. SYST. Orthorhombic. COM. FORM. Tabular crystals, often repeatedly twinned, producing pseudo-hexagonal forms

FIG. 151.   Marcasite, Twinned

(see Fig. 151); the names *cockscomb pyrites* and *spear pyrites* are given to some of the aggregates of these twinned crystals, which occur in the Chalk and other sedimentary deposits; also occurs in radiating forms, externally nodular. COLOUR. Bronze-yellow, paler than that of pyrites. STREAK. Greyish. LUSTRE. Metallic. FRACT. Uneven; brittle. H. 6–6·5. SP. GR. 4·9.

TESTS. As for pyrite; it is paler than pyrite and decomposes more readily.

OCCURRENCE. Formed at lower temperatures than pyrite, usually in concretions in sedimentary rocks, such as the English Chalk, or accompanying galena, blende, etc., in replacement deposits in limestones.

USES. Used for the same purposes as pyrite, and was formerly cut and polished for ornaments.

PYRRHOTITE, PYRRHOTINE, Magnetic Pyrites

COMP. Iron sulphide, $Fe_{1-n}S$, with n varying from 0 to 0·2; it often contains nickel, sometimes up to 5 per cent, and is then valuable as a source of that metal.

CRYST. SYST. Hexagonal. COM. FORM. Sometimes occurs in hexagonal prisms, which are frequently tabular; generally massive. CLEAV. Sometimes distinct parallel to the basal pinacoid. COLOUR. Reddish or brownish, bronze or copper-colour; readily tarnishes on exposure. STREAK. Dark greyish-black. LUSTRE. Metallic; opaque. FRACT. Uneven or imperfectly conchoidal; brittle. MAGNETISM. Magnetic. H. 3·5–4·5. SP. GR. 4·4–4·65.

TESTS. Soluble in hydrochloric acid, with evolution of sulphuretted hydrogen; heated before the blowpipe fuses in the reducing flame to a black magnetic globule—in the oxidizing flame it is converted into a globule of red iron oxide; pyrrhotite is distinguished from pyrites by its inferior hardness and by its colour, from copper pyrites by its colour, and from kupfernickel by its specific gravity and blowpipe reactions.

OCCURRENCE. The most important occurrence of pyrrhotite is that of Sudbury, Canada; here the mineral is accompanied by the nickel-bearing sulphide pentlandite (p. 535), and the deposits constitute the world's largest known source of nickel. The pyrrhotite ore-bodies are genetically connected with a gabbro mass, the so-called 'norite', and occur as marginal deposits fringing the norite, as impregnations and as vein-like masses; opinions differ concerning their origin—suggestions being that they are due to magmatic segregations, to the injection of sulphides as a magma, or to hydrothermal agencies. Other less important localities of pyrrhotite are Morocco, Finland and Norway and it occurs at Botallack (Cornwall), Beer Alston (Devonshire), Dolgelly (Wales), Kongsberg (Norway), Andreasberg in the Harz and elsewhere.

USES. The most valuable ore of nickel.

COPPERAS, MELANTERITE, Green Vitriol

COMP. Hydrated iron sulphate, $FeSO_4.7H_2O$.

CRYST. SYST. Monoclinic. COM. FORM. When crystallized, it occurs in acutely prismatic crystals, but it is more commonly found massive or pulverulent, and also botryoidal, reniform, or stalactitic. COLOUR. Various shades of green to white; exposed surfaces generally of a yellowish or yellowish-brown colour, and with a vitrified or glazed appearance, at times resembling a furnace slag. STREAK. Colourless. LUSTRE. Vitreous; subtransparent or translucent. FRACT. Conchoidal; brittle. TASTE. Sweetish, astringent and metallic, nauseous. H. 2. SP. GR. 1·9.

TESTS. Soluble in water—on the addition of barium chloride to the solution a white precipitate is thrown down; heated before the blowpipe, becomes magnetic; gives a green glass with borax.

OCCURRENCE. Copperas results from the decomposition of pyrites in the zone of oxidation, and is found in small quantities wherever pyrite occurs—notable localities being Copperas Mount, Ohio, U.S.A., and Goslar in the Harz.

USES. Copperas is used by tanners, dyers and ink manufacturers, as it yields a black colour with tannic acid. When treated with potassium ferrocyanide (yellow prussiate of potash) it forms the pigment known as Prussian Blue.

VIVIANITE, Blue-iron Earth

COMP. Hydrated iron phosphate, $Fe_3(PO_4)_2.8H_2O$; iron peroxide is sometimes present.

CRYST. SYST. Monoclinic. COM. FORM. Crystals modified prisms, generally very small, and often forming divergent aggregations; also occurs radiating, reniform, and as encrustations.

CLEAV. Perfect parallel to the clinopinacoid. COLOUR. White, or nearly colourless when unaltered, but usually deep blue or green; or often a dirty blue. STREAK. Bluish-white, sometimes colourless, soon changing to indigo blue; colour of dry powder, liver-brown. LUSTRE. Pearly to vitreous; transparent to translucent, turning opaque on exposure. FRACT. Not observable; sectile; thin laminae flexible. H. 1·5–2. SP. GR. 2·66.

TESTS. Heated before the blowpipe, fuses, loses its colour, and becomes converted to a greyish-black magnetic globule; with the fluxes gives reactions for iron; heated in the closed tube, it whitens, exfoliates and yields water; soluble in hydrochloric acid; fused with sodium carbonate, ignited with magnesium, and moistened, gives phosphoretted hydrogen.

OCCURRENCE. Vivianite is found associated with iron, copper, and tin ores; it also occurs in clay, mud and peat, and especially in bog iron-ore; sometimes it is found in or upon fossil bones or shells; some localities are—several mines in Cornwall and Devon, in peat swamps in Shetland, in the Isle of Man (occurring with the horns of elk and deer), Bodenmais in Bavaria, Orodna in Transylvania, etc.

ILVAITE, Lievrite, Yenite

COMP. Silicate of iron and calcium, $CaFe^{2+}Fe^{3+}O(Si_2O_7)(OH)$; a little oxide of manganese is frequently present.

CRYST. SYST. Orthorhombic. COM. FORM. Prismatic crystals, the sides often deeply striated longitudinally; also compact and massive. COLOUR. Black or brownish-black. STREAK. Black, brownish or greenish. LUSTRE. Submetallic; opaque. FRACT. Uneven; brittle. H. 5·5–6. SP. GR. 3·8–4·1.

TESTS. Heated before the blowpipe, fuses to a black magnetic globule; with borax, yields a dark green and nearly opaque bead; soluble in hydrochloric acid, forming a jelly.

OCCURRENCE. Associated with magnetite, and with zinc and copper ores; some localities are Elba, the Harz, Tyrol, Saxony, Norway and Rhode Island.

COBALT MINERALS

Cobalt (Co) is a malleable metal closely resembling nickel in appearance. It has a high melting point, approaching 1,500°C. It can be produced by the reduction of its oxides by carbon. The uses of cobalt and its compounds are of three chief classes; it is employed in the production of a valuable series of alloys used in rustless and high speed steels and certain non-ferrous alloys; it is used to a certain extent in electro-plating, and finally, its compounds are extensively used in the manufacture of pigments, especially blues, employed in the glass, enamel and pottery industries.

The chief producers of cobalt in the West are the Congo and Zambia as a by-product in copper smelting (see p. 252), Canada where it is associated with arsenic, nickel and silver and its profitable exploitation often depends on the price of the last-named metal, Morocco and West Germany. Cobalt deposits of economic value occur in three main ways: as cobalt veins carrying smaltite and cobaltite (Saxony, Cobalt, Ontario), as cobaltiferous pyrrhotite (Sudbury, Ontario), and as asbolite which results from the weathering of cobaltiferous basic and ultrabasic rocks—this last origin being analogous to the garnierite deposits of nickel (New Caledonia). The primary cobalt sulphides, sulpharsenides and arsenides weather in the oxidation-zone into oxides and hydrated oxy-salts. The chief cobalt minerals are therefore:

Arsenide	Smaltite, $(Co, Ni)As_{3-n}$.
Sulphide	Linnaeite, Co_3S_4.
Sulpharsenide	Cobaltite, $CoAsS$.
'Bloom', Hydrated	
Arsenate	Erythrite, $Co_3(AsO_4)_2.8H_2O$.
'Oxide'	Asbolite: oxide of manganese, sometimes containing up to 40 per cent cobalt oxide.

TESTS. Cobalt minerals colour both borax and microcosmic salt beads a rich blue. The residue obtained by heating cobalt minerals with sodium carbonate and charcoal is feebly magnetic. Cobalt minerals weather on their exterior to pinkish cobalt 'blooms'—the *cobalt indicators*.

SMALTITE, Tin White Cobalt

COMP. Cobalt nickel arsenide, $(Co,Ni)As_{3-n}$ forming passages to chloanthite, $(Ni,Co)As_{3-n}$; some iron usually present also. CRYST. SYST. Cubic, Pyrite Type. COM. FORM. Crystals show octahedron, cube and rhombdodecahedron, variously modified; usually occurs massive or reticulated. CLEAV. Octahedral, and tolerably distinct; cubic, less distinct. COLOUR. Tin-white, approaching steel-grey when massive; tarnishes on exposure, sometimes iridescently. STREAK. Greyish-black. LUSTRE. Metallic. FRACT. Granular and uneven; brittle. H. 5·5–6. SP. GR. 6·4.

TESTS. Heated in the closed tube, gives a sublimate of metallic arsenic; heated in the open tube, gives a sublimate of arsenious oxide; heated on charcoal, gives off an arsenical odour, and fuses to a globule, which yields reactions for cobalt with borax; the presence of cobalt is often indicated by the occurrence of a pinkish coating (cobalt bloom) on the surface of the mineral.

OCCURRENCE. Smaltite occurs in hydrothermal veins, associated with calcite, barytes, quartz and silver, nickel and copper minerals; the main source of supply is Cobalt, Ontario; other localities are Schneeberg, Freiberg and Annaberg, Saxony, and less important are several mines in Cornwall, Jachymov in Czechoslovakia, etc.

LINNAEITE

COMP. Cobalt sulphide, Co_3S_4.

CRYST. SYST. Cubic.

CHARACTERS AND OCCURRENCE. A steely grey mineral, tarnishing coppery red, occurring in octahedral crystals or massive forms, in sulphide veins, associated with chalcopyrite, pyrite, etc.; heated in the closed tube, gives sulphur after a time, and the roasted material gives the cobalt blue in the borax bead.

COBALTITE

COMP. Sulpharsenide of cobalt, CoAsS; some iron often present.

CRYST. SYST. Cubic, Pyrite Type. COM. FORM. Crystals, cubic or pyritohedral; usually found massive, granular and compact. COLOUR. Silver-white, with a reddish tinge. STREAK. Greyish-black. LUSTRE. Metallic. H. 5·5. SP. GR. 6–6·3.

TESTS. Heated in the closed tube, remains unaltered; heated in the open tube, it yields a sublimate of arsenious oxide and gives off sulphurous fumes; gives the blue cobalt bead with borax; decomposed by nitric acid.

OCCURRENCE. With smaltite (see above); also in metasomatic contact-deposits.

USE. An ore of cobalt.

ERYTHRITE, Cobalt Bloom

COMP. Hydrated cobalt arsenate, $Co_3As_2O_8.8H_2O$; calcium, nickel and iron oxides are frequently present.

CRYST. SYST. Monoclinic. COM. FORM. Crystals prismatic, uncommon; occurs mostly earthy, pulverulent, or encrusting, sometimes globular or reniform. CLEAV. Perfect parallel to the clinopinacoid, giving rise to a foliaceous structure, the laminae being flexible. COLOUR. Peach-red, or crimson-red, occasionally greyish or greenish. STREAK. Same as colour, but rather paler, and the powder when dry is lavender-blue. LUSTRE. Of cleavage-planes, pearly; of other faces, adamantine or vitreous; in massive specimens, dull and lustreless. H. 1·5–2·5. SP. GR. 2·95.

TESTS. When heated slightly in the closed tube, it yields water, and on additional heating, it gives a sublimate of arsenious oxide; heated before the blowpipe, alone, fuses, and colours the flame light blue; with borax, gives a deep blue bead; soluble in hydrochloric acid, forming a rose-red solution.

OCCURRENCE. Erythrite is the *cobalt bloom* formed in the upper parts of veins, etc., by the weathering of cobalt ores, and is found associated with smaltite and cobaltite.

ASBOLITE, ASBOLAN, Earthy Cobalt, Black Oxide of Cobalt

COMP. This mineral is essentially wad (see p. 508) or hydrated oxide of manganese, containing a variable percentage of cobalt oxide mechanically mixed with it, sometimes amounting to nearly 40 per cent; sulphide of cobalt and oxides of copper, iron and nickel are at times present.

COM. FORM. Amorphous, earthy. COLOUR. Black or blue-black. STREAK. Black, shining and resinous.

TESTS. Heated in closed tube, yields water; borax bead blue, due to cobalt; sodium carbonate bead, opaque green, due to manganese; soluble in hydrochloric acid.

OCCURRENCE. Occurs with the chief ores of cobalt in the oxidation zone, and with manganese ores, as at Mine La Motte, Missouri, U.S.A.; an important occurrence is in New Caledonia, where the asbolite deposits represent the superficial alteration of a cobaltiferous serpentine, analogous to the garnierite deposits formed from a nickeliferous serpentine (see p. 537).

NICKEL MINERALS

Nickel (Ni) which never occurs native, is a white malleable metal, unaffected by moist or dry air, and capable of taking a high polish. It is obtained by the reduction of its oxide, or by the 'Mond' process, which consists of the formation of a volatile nickel carbonyl produced by passing carbon monoxide over heated nickel oxide, and the dissociation of this compound at a higher temperature into nickel and carbon monoxide, which can be used again. The ore is usually first treated for the production of matte and, besides, the copper-nickel ores of Canada are smelted for the direct production of 'monel metal', an alloy of nickel and copper, whose applications are of great industrial importance.

Nickel is used in the coinage of a large number of countries. It is used, though not so extensively as formerly, in electroplating (nickel plating); it is employed in the construction of certain storage batteries, and several of its salts are used in chemical industry. The main use of nickel, however, is in its applications in the form of alloys with other metals—for example, German silver (an alloy of copper, nickel and zinc in varying proportions), white metal, nickel bronzes, etc. The manufacture of nickel-steel alloys containing from 2·5 to as much as 79 per cent of nickel absorbs the largest proportion of the nickel produced. The properties of the alloys vary remarkably with the amount of nickel, but, in general, nickel steel has a greater hardness and tensile strength than carbon steel, and is used for a great number of purposes—for armour-plate, aircraft construction, motor-cars, etc.

Nickel deposits may be divided into three types similar to the cobalt deposits—that is, veins, nickeliferous pyrrhotite, and nickeliferous serpentines (garnierite). The most important source of nickel is the mineral pentlandite, which is commonly associated with pyrrhotite, giving the so-called 'nickeliferous pyrrhotite' ores as at Sudbury, Ontario, Canada, which supplies 60 per cent or more of the world total; with the pyrrhotite are associated arsenic, copper and cobalt, and often a considerable amount of silver, and a minute proportion of platinum—certain of these metals now constituting an important part of the metal output of the Sudbury field. In 1966 the Canadian production of nickel amounted to over 234,000 tons. New Caledonia has an annual production of contained nickel metal of over 34,000 tons. The U.S.S.R. is an important producer, followed by Cuba and the United States. Relatively small amounts come from Finland, Rhodesia and elsewhere, and developments are expected in Guatamala and Australia in particular.

The primary nickel minerals are sulphides and arsenides, and these are oxidized in the upper parts of the deposits into the *nickel blooms*, of a green colour. The minerals of nickel dealt with here are:

Arsenides	Kupfernickel, Niccolite, NiAs.
	Chloanthite, $(Ni,Co)As_{3-n}$.
Antimonide	Breithauptite, NiSb.
Sulphides	Millerite, NiS.
	Pentlandite, $(Fe,Ni)S$.
'Blooms'	Emerald Nickel, Zaratite, $NiCO_3.2Ni(OH)_2.4H_2O$.
	Nickel Vitriol. Morenosite, $NiSO_4.7H_2O$.
	Nickel Bloom, Annabergite, $Ni_3(AsO_4)_2.8H_2O$.
Silicates	Garnierite, and Genthite—hydrated nickeliferous magnesium silicates.

Tests. Blowpipe reactions for nickel are poor. In the borax bead, nickel compounds give a reddish-brown colour in the oxidizing flame, which changes to an opaque grey in the reducing flame; in the microcosmic salt bead, the colours are reddish-browns. Nickel compounds give a feebly magnetic residue when heated on charcoal with sodium carbonate. Nickel ores oxidize on the surface to a green colour, the *nickel blooms*, due to the formation of oxysalts of the metal.

KUPFERNICKEL, NICCOLITE, Copper Nickel, Arsenical Nickel

Comp. Nickel arsenide, $NiAs$; antimony, and traces of cobalt, iron and sulphur are sometimes present.

Cryst. Syst. Hexagonal. Com. Form. Usually found massive. Colour. Pale copper-red, like that of a new penny; sometimes with a tarnish. Streak. Pale brownish-black. Lustre. Metallic; opaque. Fract. Uneven; brittle. H. 5–5·5. Sp. Gr. 7·3–7·6.

Tests. Heated before the blowpipe, gives arsenic fumes, and fuses to a globule, and this fused substance, if subsequently heated in the borax bead, affords reactions for nickel—oxidizing flame, reddish-brown, reducing flame opaque grey, and also for cobalt and iron; soluble in nitric acid, giving an apple-green solution.

Occurrence. Kupfernickel is usually associated in hydrothermal veins with cobalt, silver and copper ores, and is a source of metallic nickel. Prominent localities are Cobalt (Ontario), and many mines in Cornwall, Saxony, the Harz, etc. (see also the occurrences of smaltite on p. 529).

CHLOANTHITE, White Nickel

Comp. Nickel cobalt arsenide, $(Ni,Co)As_{3-n}$; at times much cobalt replaces the nickel, causing the mineral to grade into smaltite, as stated on p. 529; sometimes iron is also present.

Cryst. Syst. Cubic, Pyrite Type. Com. Form. Crystals, cube; usually massive. Colour. Tin-white. Streak. Greyish-black. H. 5·5–6. Sp. Gr. 6·4–6·7.

Tests. As for kupfernickel.

OCCURRENCE. Usually occurs with smaltite at localities cited for that mineral (see p. 529), especially at Cobalt, Ontario, and is a valuable nickel ore.

BREITHAUPTITE, Antimonial Nickel

COMP. Nickel antimonide, NiSb; often with a considerable amount of lead sulphide.

CRYST. SYST. Hexagonal. COM. FORM. Crystals rare, usually massive. COLOUR. Light copper-red when freshly broken. LUSTRE. Highly metallic. H. 5·5. SP. GR. 7·5.

TESTS. Heated on charcoal gives a white coating of antimony oxide; after roasting, yields nickel reactions in the borax bead.

OCCURRENCE. Occurs with kupfernickel, as at Andreasberg in the Harz; found also at Cobalt, Ontario.

MILLERITE, Nickel Pyrites, Capillary Pyrites

COMP. Nickel sulphide, NiS; traces of cobalt, copper and iron often present.

CRYST. SYST. Hexagonal-trigonal. COM. FORM. Usually occurs in capillary crystals of extreme delicacy, whence the name *capillary* or *hair pyrites*; sometimes but rarely occurs in columnar tufted coatings; also in rhombohedra, rarely. CLEAV. Perfect rhombohedral. COLOUR. Brass-yellow to bronze-yellow; often tarnished. STREAK. Greenish-black. LUSTRE. Metallic. H. 3–3·5. SP. GR. 5·3–5·6.

TESTS. Heated in the open tube, gives sulphurous fumes; heated before the blowpipe, after roasting, it gives with borax and with microcosmic salt, a violet bead in the oxidizing flame, and a grey bead, owing to the reduction to metallic nickel, in the reducing flame—the impurities also often give reactions in the beads; heated on charcoal with sodium carbonate and charcoal, gives a metallic magnetic mass of nickel.

OCCURRENCE. Occurs as nodules in clay ironstone, as in South Wales; also in veins associated with other nickel and cobalt minerals as at Cobalt (Ontario), Cornwall, Saxony, and especially at the Gap Mine, Lancaster Co., Pennsylvania, U.S.A.

PENTLANDITE

COMP. Nickel iron sulphide, (Fe,Ni)S, often approximating to 2FeS.NiS.

CRYST. SYST. Cubic. COM. FORM. Usually massive or granular. COLOUR. Bronze-yellow. STREAK. Black. LUSTRE. Metallic. FRACT. Uneven; brittle. H. 3·5–4. SP. GR. 5·0.

TESTS. Heated in the open tube gives sulphurous fumes; after roasting, gives nickel reactions in the borax bead; before the blowpipe, yields a magnetic mass; soluble in nitric acid, the solution giving a reddish-brown precipitate of ferric hydroxide on the addition of ammonia.

OCCURRENCE. Occurs intergrown with pyrrhotite in the Sudbury, Ontario, nickel deposit (see p. 525), and elsewhere, associated with nickeliferous pyrrhotite, kupfernickel, millerite, etc.

USE. The chief ore of nickel.

Nickel Blooms

The hydrated and oxidized nickel minerals form on the exterior of the primary nickel minerals, and are known as *nickel blooms* or *nickel indicators*. They are all green in colour, and the main species are:

Emerald Nickel, Zaratite, $NiCO_3.2Ni(OH)_2.4H_2O$.
Nickel Vitriol, Morenosite, $NiSO_4.7H_2O$.
Nickel Bloom, Annabergite, $Ni_3As_2O_8.8H_2O$.

EMERALD NICKEL, Zaratite

COMP. Hydrated carbonate of nickel, $NiCO_3.2Ni(OH)_2.4H_2O$; in some of the paler varieties, a little of the nickel is replaced by magnesium.

COM. FORM. Amorphous; occurs as an encrustation, sometimes minutely mammillated and stalactitic; also massive and compact. COLOUR. Emerald-green. STREAK. Paler than the colour. LUSTRE. Vitreous; transparent to translucent. H. 3. SP. GR. 2·5–2·6.

TESTS. Heated in the closed tube, gives off water and carbon dioxide, leaving a dark, magnetic residue; gives the usual nickel reactions in the borax bead; dissolves with effervescence when heated in dilute hydrochloric acid.

OCCURRENCE. Occurs as a coating to other nickel minerals, and is associated with chromite-bearing serpentines, as in Unst, Shetland Islands.

NICKEL VITRIOL, Morenosite

COMP. Hydrated sulphate, $NiSO_4.7H_2O$.

CRYST. SYST. Orthorhombic. COM. FORM. Acicular crystals and fibrous crusts. COLOUR. Apple-green and pale-green. H. 2–2·5. SP. GR. 2.

TESTS. Gives nickel reactions in borax bead; soluble in hydrochloric acid, the solution giving a dense white precipitate on the addition of barium chloride solution.

OCCURRENCE. Associated with other nickel blooms as a weathering product of primary nickel minerals.

ANNABERGITE, Nickel Bloom

COMP. Hydrous nickel arsenate, $Ni_3(AsO_4)_2.8H_2O$.

CRYST. SYST. Monoclinic.

CHARACTERS AND OCCURRENCE. Occurs as a coating of apple-green capillary crystals, and results from the decomposition of nickel minerals; in the borax bead gives the nickel reactions and when heated in the closed tube with charcoal gives water and an arsenic mirror.

GARNIERITE, Noumeite

COMP. Essentially a hydrated nickel magnesium silicate, but very variable.

COM. FORM. Amorphous; soft and friable. COLOUR. Apple-green to nearly white. LUSTRE. Dull. H. 3–4. SP. GR. 2·2–2·8.

TESTS. Adheres to the tongue; heated in the closed tube, yields water and blackens; in borax bead, gives nickel reactions; in the microcosmic bead, gives a nickel reaction, and leaves an insoluble skeleton of silica, indicating a silicate.

VARIETY *Genthite* is a hydrated nickel magnesium silicate, related to garnierite.

OCCURRENCE. Occurs in serpentine near Noumea, New Caledonia, in veins associated with chromite and talc; a residual deposit, rich in nickel, is formed by the lateritic decay of the nickeliferous serpentine; also found at Riddle (Oregon), Webster (North Carolina) and Revda (Urals).

USES. Garnierite is an important source of nickel, and the New Caledonia deposits, before the development of the Sudbury nickeliferous pyrrhotite deposit, were the chief source of this metal.

GROUP 8B
RUTHENIUM, RHODIUM, PALLADIUM, OSMIUM, IRIDIUM, PLATINUM

PLATINUM GROUP MINERALS

The members of the platinum group of metals—platinum, palladium, osmium, iridium, rhodium and ruthenium—occur together in nature as the native metals or alloys. The most abundant of these metals is platinum, the others occurring in small quantities with this. The metals are used in jewellery, in the electrical trades, in electro-plating, in chemical industries, in dentistry, for certain photographic purposes and especially as catalysts in the chemical and petroleum industries.

Platinum

Platinum (Pt) occurs *native*, and in that form constitutes the most important source of the metal. It is a greyish-white lustrous metal, having a specific gravity of 21.46 and melting at 1,760°C. It is malleable and ductile, and may be welded at a bright red heat. Its resistance to acids, and to chemical influence generally, renders it of particular use in the laboratory, and in the electrical and other industries. It is also largely used as a catalytic agent in the manufacture of chemicals by the contact process, and in dentistry and jewellery. Platinum is refined and separated from associated metals by a somewhat complicated series of operations.

In addition to the native metal, an important source of platinum is *sperrylite*, $PtAs_2$, as this occurs in the important platinum deposits of Sudbury, Ontario. Platinum occurs in a number of ways. It is found disseminated as small original grains in basic and ultra-basic igneous rocks such as olivine-gabbros and peridotites, as in the Urals; it occurs also in similar rocks in the Bushveld norite complex in South Africa, and in chromite-rich layers in the same complex; in the pyrrhotite deposits of Sudbury, Canada, which are possibly magmatic segregations (see p. 525), sperrylite and the platinum group metals afford a large share of the world output; platinum occurs in quartz veins in the Transvaal, and a small amount is present in many copper deposits. From all these types of occurrence, placer or alluvial deposits are formed, and this type of deposit up to recently supplied the greater part of the output. The chief producers of platinum group metals are U.S.S.R., South Africa and Canada. In 1966, it is estimated that U.S.S.R. produced 1,700,000 ounces troy, South Africa 600,000 ounces and Canada nearly 400,000 ounces.

The platinum minerals considered here are two:

Element   Native Platinum, Pt.
Arsenide   Sperrylite, $PtAs_2$.

NATIVE PLATINUM

COMP. Platinum, alloyed with iron, iridium, osmium, gold, rhodium, palladium, and copper; in 21 analyses cited by Dana, the amount of platinum ranges from 45 to 86 per cent.

CRYST. SYST. Cubic. COM. FORM. Crystals rare; usually found in grains and irregularly shaped lumps; a nugget found in the Urals weighed 21 pounds troy. COLOUR. White, steel-grey. STREAK. White steel-grey. LUSTRE. Metallic; opaque. FRACT. Hackly; ductile. H. 4–4·5. SP. GR. 21·46, chemically pure.

TESTS. Sometimes exhibits magnetic polarity, some Uralian specimens being said to attract iron filings more powerfully than an ordinary magnet; the high specific gravity, the infusibility and insolubility of platinum serve to distinguish it from other minerals.

OCCURRENCE. The occurrence of platinum has been sum-marized in the introduction to the platinum minerals above.

SPERRYLITE

COMP. Platinum arsenide, PtAs$_2$.

CRYST. SYST. Cubic, pyritohedral. COM. FORM. In tiny cubes or large combinations of cube and octahedron. COLOUR. Tin-white. STREAK. Black. LUSTRE. Metallic. H. 6–7. SP. GR. 10·6.

OCCURRENCE. In the pyrrhotite deposits of Sudbury, Ontario, and in the Bushveld norite in large crystals, and in many detrital platinum deposits.

Palladium

Palladium (Pd) occurs native in crude platinum, and in small quantities in cupriferous pyrites, especially those containing nickel and pyrrhotite. It is a silver-white metal as hard, but not so ductile, as platinum; it oxidizes more readily than that metal. Its hardness is 4·5–5, specific gravity, 11·3–12, and melting point, 1,546°C.

Palladium is much used in dental alloys, as a catalyser, for coating the surfaces of silver reflectors used in search-lights, etc., and in the construction of delicate graduated scales.

The chief source of supply is the copper-nickel ore of Sudbury (see p. 525), from whose matte it is recovered. The metal is also found with platinum in Brazil, the Urals and elsewhere, and usually contains iridium as well as platinum.

Osmium and Iridium

Osmium. Osmium (Os) occurs native in crude platinum and in osmiridium, an alloy with iridium described below. Osmium is a bluish-grey metal, and has a specific gravity of 22·48. It is the heaviest of metals, and fuses at 2,200°C. No reliable statistics are available as to production. It is of little commercial importance, and the supply is in excess of the demand.

Iridium. Iridium (Ir) occurs native in crude platinum, and alloyed with osmium as *osmiridium* or *iridosmine*. It is a steel-white metal, having a specific gravity of 22·4, and melting at about 2,290°C. Its chief source is from crude platinum, but about 5,000 ounces troy of osmiridium are produced annually from gold ores in South Africa. Its consumption is increasing, its chief application being in the dental, electrical and jewellery trades.

IRIDOSMINE, Osmiridium

Comp. An alloy of iridium and osmium in variable proportions.

Cryst. Syst. Hexagonal, rhombohedral. Com. Form. Chiéfly in small, flattened grains. Colour. Tin-white to steel-grey. Lustre. Metallic. H. 6–7. Sp. Gr. 19·3–21·12.

Occurrence. Both this mineral and iridium are found in the gold washings of the Urals, Bingera (New South Wales), Brazil, and in Canada; a considerable but varying production comes from placer deposits from Tasmania, the Sudbury, Ontario, recoveries from platinum refineries, and the South African gold-ores.

Rhodium

Rhodium (Rh) occurs in native platinum and in the pyrrhotitic ores of Sudbury, Ontario. It is a white metal, ductile and malleable at a red heat. It has a specific gravity of 12·1 and melts at about 2,000°C. Its chief source is crude platinum, in which it is said to exist to the extent of 2 per cent on the average. It has few applications, but is used in the manufacture of thermal couples, for crucibles, and there are possibilities in the development of rhodium plating.

Ruthenium

Ruthenium (Ru) occurs native in crude platinum, in the refining of which it is recovered. It is a white hard and brittle metal, of a specific gravity of 12·2. It has few or no industrial applications.

Page numbers in **heavy type** refer to the description of the mineral.

Z